Geographies of Peace and Armed Conflict

This collection addresses the impact of armed conflict and explores pathways to peace across the world. Topics range from geopolitics to the effects of armed conflict on the environment, resources, health, children, and transnational migration. Others explore the social processes involved in post-conflict situations, and others still the lessons for achieving effective peace. The geographical concepts addressed include the notion of "conflict space", landscapes of terror, the relationship between violence and justice, the conditions for peace, and the dynamics of post-conflict. Methods include landscape analysis, interviews with a range of citizens, mapping and geographic information science, and policy analysis. Several papers address the situation of children in conflict zones, the impact of conflict on patterns of migration, the role of gender in achieving peace, the concept of territory as a basis for conflict and for negotiation of peace, as well as the economic impact of conflict. The studies cover several world regions, including Africa, the Middle East, South and Southeast Asia, and eastern Europe.

This book was originally published as a special issue of *Annals of the Association of American Geographers.*

Audrey Kobayashi is a Professor and Research Chair of Geography at Queen's University, Kingston, Ontario. She served as Editor of the People, Place, and Region section of thc *Annals of the Association of American Geographers* from 2002–2011, and is currently President of the Association of American Geographers. Her research, writing, and teaching address issues of human rights, especially around questions of migration, racism, and gender.

Geographies of Peace and Armed Conflict

Edited by
Audrey Kobayashi

LONDON AND NEW YORK

First published 2012
by Routledge
2 Park Square, Milton Park, Abingdon, Oxfordshire OX14 4RN

Simultaneously published in the USA and Canada
by Routledge
711 Third Avenue, New York, NY 10017

First issued in paperback 2014

Routledge is an imprint of the Taylor and Francis Group, an informa business

This book is a reproduction of *Annals of the Association of American Geographers*, vol. 99, issue 5. The Publisher requests to those authors who may be citing this book to state, also, the bibliographical details of the special issue on which the book was based.

British Library Cataloguing in Publication Data
A catalogue record for this book is available from the British Library

ISBN 978-0-415-69658-6 (hbk)
ISBN 978-1-138-85336-2 (pbk)

Typeset in Baskerville
by Taylor & Francis Books

Publisher's Note
The publisher would like to make readers aware that the chapters in this book are referred to as articles as they had been in the special issue. The publisher accepts responsibility for any inconsistencies that may have arisen in the course of preparing this volume for print.

Contents

Geographies of Peace and Armed Conflict: Introduction

Audrey Kobayashi

A decade into the 21st Century there seems no utopian end of warfare in sight. Although we have not seen the nuclear holocaust that in the late 20th Century many believed would be the outcome of decades of Cold War, armed conflicts are current within and on the part of more countries than were involved in either of the great wars of the 20th Century. The total armed forces of the world number over 70 million, of whom at least 18 million are in active duty,[1] supported by annual military budgets that total something between two and three trillion dollars. Many countries have military budgets that exceed those for health or education.[2] Although the number of military casualties has decreased dramatically in recent armed conflicts compared to earlier decades, tens of thousands of civilian casualties annually result from practices such as ethnic cleansing and "collateral" civilian deaths and injuries, as well as from the complex social, economic, political, and environmental perils that armed conflicts instigate. The United Nations estimates that some 42 million people are currently forcibly displaced from their homes, 80 percent of them from developing countries, and that the repatriation of displaced persons is decelerating (UNHCR 2011). Armed conflict results in mass destruction of infrastructure, and the disruption of transportation, communication, health services, and education. Armed conflict is responsible for some of the most serious environmental degradation, and exacerbates other processes such as climate change, drought, or natural resource scarcity. Developing countries suffer disproportionately from these effects, and the poorest citizens, including those displaced from their homes, suffer most of all.

This volume brings together some of the best geographical researchers to address armed conflict and its effects, as well as possibilities and prospects for peace. They cover a wide range of geographical sub-disciplines to address questions of population, resources, political relations, boundaries, environmental change, forging peace, and post-conflict conditions. They address conflicts in all parts of the world, at a variety of scales, including both international and intranational disputes. They speak from a variety of methodological and theoretical perspectives. One thing they have in common, however, is a united commitment to play a role in creating a world based on peace, not war.

Geographers and Peace and Conflict

Geographers have a long history of studying armed conflict, and have by no means always been dedicated to peace. Rachel Woodward (2004, 2009) divides the history of military geographies into two approaches: those that contribute to warfare through the application of geographical concepts and tools to military problems, and those who study the effect of military actions, or armed conflict, upon society and the environment. Although the second approach is much more common nowadays, it is by no means universal, and the role of military geographies varies in different parts of the world. In other words, military geographies, and geographies of peace and armed conflict, have their own geography.

Throughout the 19th and most of the 20th Centuries, as geography became an established part of university curricula in Europe and North America, most geographers who studied the effects of war adopted a perspective on war as a natural phenomenon (e.g., Ratzel 1897, 1903; Mackinder 1904; Brunhes and Vallaux 1921). As the discipline developed, few spoke out to support a role for the discipline in achieving peace (e.g., Kropotkin 1885; Atwood 1935) [3] and most of the discipline evinced at best a morally neutral, functionalist perspective. In a detailed review of the role of geographers and war over the past two centuries, Mamadouh (2005, 34) suggests that "academic geographers were divided into a war-minded camp that saw war as the legitimate expression of competition between states and a peace-minded camp that promoted international cooperation. … Still, geography was mainly an aid to war-waging states."

As the modern world has emerged, geographers have often played a direct role in affecting the course of armed conflict. At the beginning of the 19th Century, Napoleon led one of the most effective campaigns using geographers in the pursuit of military aims (Godlewska 1994). In the 20th Century, perhaps the most infamous case is that of the German school of *Geopolitik* (Mamadouh 2005, 31), some of whose members were directly involved in strategy for the expansion of Nazism from the 1930s by using geographical principles (Minca and Barnes forthcoming). In the U.S. during World War II, geographers played a significant role in the Office of Strategic Services (OSS), the intelligence gathering precursor to the Central Intelligence Agency (CIA). In some countries, such as Chile under military rule, geographers were directly involved in strategy and planning even as it became difficult for scholars who did not support the military regime to practice, and many ended up in exile (Caviedes 1991). Indeed, Augusto Pinochet taught geopolitics at the National Military Academy, and published a book on the topic (1968).

Not only have academic geographers played both a direct and an indirect role in armed conflict, but also their work affects fundamentally the larger scholarly context, our understanding of science and its purposes, and the place of our ideas in human society. Barnes and Farish (2006) advance sociologist Andrew Pickering's claim that World War II represented an "epochal change" (a Foucauldian term) in the nature of science:

> The intersection of science and the military in World War II can thus be understood as a macromangling that encompassed both an inner transformation of these two macro-actors and an outer transformation in their relationship to one another. The way of doing science changed from small to big science; the military shifted its tactics and basic disciplines; both institutions were topologically transformed in a reciprocal transformation of shape. … and all of these transformations were interactively stabilized in relation to transformations in machinic culture (symbolized by developments in radar technology). (Pickering 2005, 239, quoted in Barnes and Farish 2006, 809)

Developing the notion of institutional mangling, Barnes and Farish show that the concept of the "region," surely one of the most intellectually challenging and contested concepts upon which the discipline is based, was much more than academic. Stemming from "belief in the instrumental power of science in achieving national security interests" (p. 810) they chart the relationship between military enterprise and R&D efforts in which regional geography plays a strongly politicized role. Not only did academic geographers play a direct role in the events of World War II (in the U.S. and elsewhere), they also played a major role in fashioning the discipline during the subsequent Cold War, when the science of regions was "mangled" with military strategy, national ideology, and the development of key ideas and methods that transformed the discipline:

> In demonstrating that the discipline was very much caught up in the military-industrial-academic complex, the cyborg regime of technoscience, or whatever provocative descriptors are appropriate to the era in question, we point to the external, to the world outside where geography and life itself reside. (Barnes and Farish 2006, 821)

If geographers have been mangled within the study and promulgation of armed (or, in the case of Cold War, potentially armed, conflict), so too have they been mangled in movements for peace. The Cold War period also awakened a strong tradition of peace scholarship (Roder 1973; Hudson 1977; Lacoste 1976, 1977; Hewitt 1983; Kofman 1984; Openshaw and Steadman 1983; Pepper and Jenkins 1983). Pepper and Jenkins (1985) brought much of this peace work into a single volume, a thoughtful and pathbreaking set of analyses by geographers,to draw attention to the spatial implications and human consequences of potential nuclear hostilities, and to point towards the possibility of geographers working directly to counter that possibility.

The end of the Cold War brought an end to scholarship on the specter of nuclear war, but subsequent events have been of equally troubling concern for geographers seeking world peace. Concern has shifted to a new specter, that of terrorism. Especially since 9/11, geographers have refocused their concerns not only on the effects of new technologies of violence facilitated by sophisticated locational tracking systems and geospatial databases, but also upon the potential for geographical technology in preventing and mitigating violence. The line between the two, which represents the dialectic of peace and war, is certainly a troubled one. A volume of papers that walk this line, edited by Cutter, Richardson, and Wilbanks (2003), illustrates the paradoxical nature of our discipline. The editors issue a *cri de coeur* for research to investigate the root causes of terrorism in all its forms, including social practices of exclusion and inclusion and the geographical dimensions of power and geopolitics. Many of the authors provide compelling evidence of the ways that geographers can contribute in times of emergency using the very technologies that have been instrumental in waging war for waging peace. The range of opinions among geographers internationally over this precarious relationship surely represents one of the most urgent discourses in the history of our discipline. We can expect continued controversy as a result over the role of geographical technologies, the place of geographers in formulating public policy, the relationship between geographers and the state, and the potential for geographers to contribute to peace.

These questions occupy much of the contemporary agenda in the sub-field of political geography. The *Geopolitics Reader* (Ó Tuathail, Dalby, and Routledge 2006) brings together a historical collection of political statements—including some by geographers, other scholars, and politicians—that represent key moments in the history of geopolitics and conflict. This volume is widely read by students, the academic geographers of the future, for whom gaining a perspective on the relationship between dominant ideologies, forceful personalities, and sociopolitical context is vital (see Sparke 2000).

Increasing numbers of political geographers (several of them represent the current volume) are reading between the lines of such statements. In another recent volume, *Violent Geographies,* Derek Gregory and Allan Pred (2006) bring together a group of geographers interested in just such between-the-lines analysis that shows that violence breeds violence in ways that are historically rooted, spatially expressed, and exceedingly complex and interwoven with social and political circumstances. The complexity of the dialectics of violence prompts Michael Watts to ask:

> Why has the tyranny, oppression, and ruthless austerity of a global neoliberal order in the service of secular American empire generated such a powerful form of resistance (a variant of modern terror), in equal measure ruthless, tyrannical and fanatically single-minded that draws from the deep well of modern Islam? (Watts 2006, 178)

Others have mapped out the complexity of violence and peace in detail, challenging the Manichean compulsion of contemporary geopolitics. Derek Gregory (2005) charts the complicated journey of international relations that mark the "colonial present" in the War on Terror. Feminist geographers also resist the compulsion to good and evil by pointing out that violence is always a choice, often taken in a context of masculine pride and control (Hyndman 2003; Woodward 2006).

Political geographers have made their most compelling contribution to understanding peace and armed conflict by showing that conditions of violence and non-violence need to be *placed*, that context matters, and that human relations of violence are spatialized. As Matt Coleman (2007) has eloquently argued, conflict involves geopolitics enacting on the ground in specific ways, with reverberations from actual theaters of war to home bodies thousands of miles away, as expressions of engagement of human bodies through the deployment of power, whether of the state, of civil society, or of stateless groups that choose violence as a means of achieving their ends. The more we understand about the ways in which human relations are spatialized, the more we will understand about the complexity of warfare and its effects.

The Geographical Agenda

One of our greatest challenges is to identify and assess the many complex and overlapping factors that influence the development of peace or the escalation of conflicts where geography and life reside. Climate change represents one of the most significant but least understood risks for escalating armed conflict. A recent in-depth study of the Levant, a region with a high level of armed conflict as well as high susceptibility to the deleterious effects of climate change, suggests that risks for increased militarization and conflict come from competition for water resources that complicates peace processes, food insecurity, worsening economic and social stability, increases in forced migration, increased militarization of strategic natural resources, and increased resentment and distrust of the West (Brown and Crawford 2005). All these issues are of immense concern to geographers working in different parts of the world.

Warfare is certainly spatialized. Tables 1 and 2 depict those countries most involved in international and intrastate conflicts since World War II. Table 1 indicates that the colonial powers remain by and large most significantly engaged in international conflict, reproducing the uneven political and economic power that has regulated the world for several centuries. Table 2 indicates the extent of intrastate conflicts, which are often isolated from the rest of the world. What these tables do not show, however, is the scale at which conflict occurs.

Countries such as Burma/Myanmar have large numbers of conflicts but the scale is much smaller than in countries such as Iraq or Afghanistan, where the "War on Terror" has resulted in large numbers of international players involved in what were once internecine conflicts.

Nor do these figures provide a sense of the differential impacts of conflict upon the countries involved, including the differential impact upon civilians, as well as differential concern from the rest of the world. As Colin Flint has pointed out (2005, 3), the "hot war" instigated after the 11 September 2001 attacks on the United States has so dominated international geopolitics in this decade that other conflicts, such as that in the Congo where deaths number not in the thousands but in the millions, are sidelined. This point also applies for many of the civil conflicts represented by most of the scenes of violence in the world today.

number of war	countries
21	United Kingdom
19	France
16	United States of America
9	Russia (Soviet Union)
7	Australia
7	Netherlands
6	Israel
6	Egypt
6	China
6	Thailand

Table 1. Countries engaged in the largest number of international armed conflicts, 1943–2003. Source: The *Human Security Report 2005,* 26.

number of conflict-years	country
232	Burma (Myanmar)
156	India
88	Ethiopia
86	Philippines
79	Israel
77	United Kingdom
66	France
60	Iraq
60	Vietnam, Democratic Republic of,
51	Russia (Soviet Union)
49	United States of America
48	Iran
44	Angola
42	Colombia

Table 2. Countries experiencing the most conflict years, 1946–2003 Source: The *Human Security Report 2005,* 27.

Those countries most affected are also the poorest, either because they are located in part of the developing world where historical effects of colonialism have left them poor throughout modern history, or because—and usually in addition—conflict has ravaged resources and infrastructure, so that the map of economic development and the map of conflict largely coincide. For example, the three countries hosting the largest number of refugees, the majority of whom are fleeing conflict or its complications, are Pakistan, The Islamic Republic of Iran, and the Syrian Arab Republic. Those hosting the largest number relative to their permanent populations, however, are Pakistan again, the Democratic Republic of Congo, and Kenya (UNHCR 2011).

Scale, a fundamental geographical concept, is important in assessing the effects of warfare. From the global to the local, the significance of scale is manifest technologically, in the differences between carpet bombing and suicide bombs, or in the difference between the use of conventional firearms and the body itself as a weapon, as occurred most notoriously in the former Yugoslavia. Scale is also important in understanding the complex sets of discourses through which conflict is initiated, sustained, and resolved, from local conflicts to global international geopolitics. International coalitions such as NATO work at a different scale and using very different forms of political exchange than do international covert organizations such as Al-Qaeda. International and civil wars dictate very different scalar approaches to both conflict and conflict resolution, as well as to the scale, or extent, of political hegemony and to the extent of death and destruction. As O'Loughlin shows (2005), civil wars often occur in a "hegemonic shadow" that covers a larger portion of the world than just a few decades ago.

Buhaug et al. (2009, 3) make a significant contribution to our understanding of the geography of civil war, showing that although civil wars "tend to be less intense and claim fewer direct casualties than do interstate wars, such conflicts are often very persistent and have proven more difficult to settle. Beyond the direct casualties from conflict, efforts to evaluate the indirect consequences of civil war furthermore find that the negative human and social impacts of civil war are so severe as to constitute 'development in reverse' [Collier et al. 2003]." Emphasizing the importance of scale of analysis, their study of intrastate local variations in wealth distribution shows that risk of violence is increased in areas with low income, especially in those states that have large relative discrepancies between high- and low-income regions, typical of many developing countries.

Notwithstanding the 9/11 attacks, for most of the colonial powers, conflict still happens "over there," and the impact on national populations is therefore disproportionate. The theme of sending brave soldiers "over there" (Figure 1) has been a dominant one from the time of the Crusades and birth of chivalry, through the major colonial wars of the 16th to 19th Centuries, and the two World Wars, to add to the mythology of war and gallantry, national identities, and to the disproportionate effects, in terms of both lives lost and resources depleted, between the colonial powers and Others at a distance. The poster is based on the popular American recruiting song, "Over There," which begins "Johnny get your gun, get your gun," and invokes masculinity and family, along with concepts of blood, sacrifice, and nation, to romanticize the act of war. Similar sentiments are depicted in a range of media from music to painting to advertising (see Dijkink 2005). The theme of war as a defining context for masculinity and feminity is taken up by geographers. Although parts of Europe have certainly been battlegrounds throughout modern history, especially since the advent of air combat during World War I, European powers, like the U.S., have engaged much more often in distant battles. Just as the politics and the technology of engagement are hugely variable, so are the means used to evoke nationalism, bring citizens on board in sufficient numbers to support armed conflict and, ultimately, re-define human relations—or, rather, *dehumanizing relations*—to the extent that the death or wounding of the other is seen as triumph rather than tragedy.

For those on the ground, however, conflict is often a source of daily terror, waged at the scale of individuals, households, and communities. That terror takes many forms, from bombs that rain impersonally from the sky, to rocket attacks, house-to-house conflict in urban warfare, to ethnic cleansing in brutal attacks against people who may once have been neighbours, but the vast majority of countries whose citizens experience daily terror also rank among the least developed on the United Nations Human Development Index (UNDP 2007). Scale and place are also gendered, as men and women experience the violence of war in different ways: most obviously, men are killed and wounded in direct battle, whereas women are often victims of rape, mutilation, and forced displacement.

Bodies matter. Bodies are *placed* in conflict, sometimes in strategic and sometimes in accidental ways. Bodies bleed. Bodies present themselves to others as friend or enemy, as human beings with whom a common destiny can be forged, or as inhumanized objects to be terminated without compunction. Encounters between bodies mean that somebody always makes a decision over life and death. "Making the body count," in the words of James Tyner (2009), is as important in planned death as it is in strategies for lasting peace, albeit to radically different ends. And, as Cynthia Enloe (2004, Part II) points out in compelling geographical terms, "wars are never 'over there.'"

Figure 1. "Lick Them Over There" Canadian World War II Poster.
Source: *Library and Archives Canada, Acc. No. 1983-30-236.*

The Contributions of this Book

Regional and spatial variations in warfare and its effects, scale, the relative impact of conflict, environmental effects, social interactions within landscapes of war and peace, and the spatial relations of conflict and post-conflict situations form the major themes of this book. In order to ensure broad representation within the confines of a single issue, authors were asked to limit their contributions to a total of 5,000 words, a limitation that may leave some readers searching for more detail, or more theoretical development, than these relatively short pieces allow. The trade-off, however, is that we cover a range of topics across a large part of the world. Nonetheless, some parts of the world, such as the Balkans and the Levant, have received more attention from geographers than have others. South America is not represented, and there are only three chapters on Africa. The chapters on Southeast and South Asia, however, represent a growing geographical concern with that region. The methodologies are very diverse, spanning research that is mainly theoretical, archival and policy-directed research, natural resource analysis, GIS and statistical analysis, and qualitative methodologies such as interviews.

The chapters do not fit well into clear topical or subdisciplinary categories, but the order of chapters follows a general pattern. It begins with chapters focussed on questions of territory, nationality, and armed conflict, including geopolitics, the role of natural resources in conflict, and the impact of armed conflict on particular groups such as children. This section is followed by a series of chapters that address conflict resolution and peace processes, with a range of degrees of optimism, depending on the situation. The final chapters deal with post-conflict situations, which, if they no longer involve conditions of "hot" armed conflict, generally remain conflictual. One of the very inspiring aspects of putting this collection together was to discover how much dedicated work is being done by geographers in post-conflict situations, much of it on the ground working directly with survivors.

Most of these authors write from what has become known as a "critical" geography perspective. Critical is a contested term that is not tied to one theoretical approach, but loosely encompasses those perspectives critical of the political status quo—including everything from totalitarian to neoliberal regimes—that allows armed conflict to take place. It also implies a commitment to social change based on fundamental re-thinking of political economy. In that sense, all of the authors here express a strong commitment to change, and they show the extent to which dominant attitudes have changed among geographical scholars over the past several decades. In the post-World War II era that ushered in the Cold War, there was a genuine belief that the betterment of humankind lay in efforts to use neutral and detached science, in particular the science of spatial location, to discern both the root causes of spatial pattern and the theoretical patterns that would optimize, especially economic development. Two fundamental changes have occurred since that time. The first is that with the advent of poststructuralism as the dominant theoretical paradigm, few social scientists now believe that there are structural causes of human conditions independent of human action. Human conditions are almost universally believed to be social constructions, although there remain widespread opinions about the efficacy of individual agency in determining outcomes. The most important conclusion to be drawn from this point is that just as the practice of warfare is profoundly changeable in response to changing conditions, from technology to strategy to environmental interaction, so too the human response to warfare, and to the creation of the conditions of peace, is highly contingent, the product not of necessity but of collective and often concerted will.

The emphasis on social construction has led to a second recognition, which is that science is not value-free. It occurs within a historically produced normative framework, and that framework is subject to change according to the actions of scholars, who work within a broader social context. All of the authors in this volume have chosen to place their values on the line to contribute to overcoming armed conflict and its effects, just as I and the members of the editorial board who were instrumental in making decisions over what would be included in the collection chose to use the potential for contributing to peace as one of the criteria for publication. This commitment does not in any way alter the basic criteria for publication—excellent research, contribution to the advancement of the discipline of geography, and effective writing—but it does signal a recognition that the discipline does not act in a social vacuum. The work we do as geographers has the potential to make a real difference in the world.

Acknowledgements

I wish to acknowledge all the individuals who have helped to bring this collection to publication. The chapters here were originally published in the first special issue of the *Annals of the Association of American Geographers* (95, 5, 2009). The idea of a fifth annual issue had been strongly encouraged by the Council and Executive of the Association of American Geographers and supported by our publishers, Taylor & Francis. We initially put out a call for abstracts and received nearly 200 responses, of which forty-four were invited to submit manuscripts. Although the final tally is twenty-three chapters, the enthusiastic response from so many geographers concerned with the impact of armed conflict on the earth was very gratifying. Nearly one hundred reviewers provided constructive and in-depth comments that aided immensely in improving the quality of the issue. All the members of the *Annals* editorial board pitched in to help vet abstracts, seek reviewers, or review manuscripts. All of the production was professionally and efficiently managed by Robin Maier, Managing Editor of the *Annals*. And all of the authors worked very hard to conform not only to the theme of the issue, but also the 5000-word limit that was imposed in order to accommodate so many excellent articles in one volume. Finally, I wish to acknowledge all the millions of people affected by armed conflict across the world, as well as the much smaller number of people, geographers included, who are working for sustainable peace. For all those people the contents of this book are so much more than academic.

Notes

1. The number of independent states has tripled since World War II, and the number of international conflict zones has not changed much, but the number of sites and the frequency of civil conflict have increased. The result, however, is that the likelihood of armed conflict *per state* has gone down quite dramatically. This point is often used to make the argument that armed conflict has dropped dramatically, but such claims need to be qualified. Similarly, claims that the number of military deaths have declined, largely due to new technologies of warfare, often fail to take into account the civilian toll. Notwithstanding these cautions on the use of data, it is generally agreed that there was a significant drop in armed conflicts with the end of the so-called Cold War (from 1989) and there has been an increase since 11 September 2001.
2. Data on the size of armed forces and budgets are not reliable, but these figures are probably close. They are based on estimates from the Center for International Strategic Studies, the Jaffa Institute for Strategic Studies, and web sites for several hundred countries, compiled by Wikipedia at http://en.wikipedia.org/wiki/List_of_countries_by_size_of_armed_forces (last accessed 22 July 2009). The United Nations *Human Development Report* uses the same data. Of course many caveats apply to their use including caution about the general reliability of Wikipedia data (even when sources are cited), and recognition that the role of armed forces may in many countries extend beyond their combat capacity to include a range of bureaucratic functions, policing, and in some cases political roles.
3. Atwood's presidential address to the Association of American Geogaphers is remarkable for its time, presented during the escalation of international animosity that preceded World War II. Although the dominant tone espouses a form of environmental determinism that sees the discipline of geography as the study of how people adapt to environmental conditions, as well as how they can better adapt to environmental conditions, he makes a strong plea for understanding the negative effects of imperialism and unequal power relations, as well as a plea for the United States to become less isolationist. His main point is that "Our supreme responsibility today is to overcome the ignorance among our own people of the other peoples of the world and to remove all vestiges of hatred." (Atwood 1935, 15).

References

Atwood, W. W. 1935. The increasing significance of geographic conditions in the growth of nation-states. *Annals of the Association of American Geographers* 25 (1): 1–16.

Barnes, T. J. and M. Farish. 2006. Between regions: science, militarism, and American geography from World War to Cold War. *Annals of the Association of American Geographers* 96(4): 807–826.

Brown, O. and A. Crawford. 2005. *Rising temperatures, rising tensions: Climate change and the risk of violent conflict in the Middle East.* Winnipeg: International Institute for Sustainable Development. Available on-line at: http://www.humansecuritygateway.info/documents/IISD_ClimateChange_RiskViolentConflict_MiddleEast.pdf (last accessed 26 July 2008).

Brunhes, J. and C. Vallaux. 1921. *La géographie de la paix et de la guerre sur terre et sur mer* [The geography of peace and war on land and sea]. Paris: Alcan.

Buhaug, H., K. S. Gleditsch, H. Holtermann, G. Østby, and A. F. Tollefsen. 2009. Revolt of the paupers or the aspiring? Geographic wealth dispersion and conflict. Oslo: International Peace Research Institute. Available on-line at:http://www.humansecuritygateway.info/documents/PRIO_GeographicWealthDispersionConflict.pdf (last accessed 21 July 2009).

Caviedes, C. N. 1991. Contemporary geography in Chile: a story of development and contradictions. *The Professional Geography* 43 (3): 359–362.

Coleman, M. 2007. A geopolitics of engagement: Neoliberalism, the War on Terrorism, and the reconfiguration of US immigration enforcement. *Geopolitics* 12: 607–634.

Collier, P., V. L. Elliott, H. Hegre, A. Hoeffler, M. Reynal-Querol and N. Sambanis. 2003. *Breaking the conflict trap: Civil war and development policy.* Washington DC: World Bank.

Cutter, S. L., D. Richardson, and T. Wilbanks, eds. 2003. *The geographical dimensions of terrorism: action items and research priorities.* New York: Routledge.

Dijkink, G. 2005. Soldiers and nationalism: The glory and transcience of a hard-won territorial identity. In *The geography of war and peace: from death camps to diplomats,* ed. C. Flint, 113–132. New York: Oxford University Press.

Enloe, C. 2004. *The surprised feminist.* Berkeley, Los Angeles, and London: University of California Press.

Flint, C., ed. 2005. *The geography of war and peace: from death camps to diplomats.* New York: Oxford University Press.

Gilbert, E. and D. Cowan, D. eds 2008. War, citizenship, territory. *London:* Routledge.

Godlewska, A.1994. Napoleon's geographers: imperialists and soldiers of modernity. In Geography and empire: Critical studies in the history of geography. ed. A. Godlewska and N. Smith, 31–53. Oxford: Blackwell.

Gregory, D. 2005. *The colonial present: Afghanistan, Palestine, Iraq.* Oxford and Malden, Mass.: Oxford University Press.

Gregory, D. and A. Pred, eds. 2006. *Violent geographies: fear, terror and political violence.* Oxford and New York: Routledge.

Hewitt, K. 1983. Place annihilation: area bombing and the fate of urban places. *Annals of the Association of American Geographers* 73: 257–84.

Hudson, B. 1977. The new geography and the new imperialism. *Antipode* 9 (1): 12–19.

Human Security Centre. n.d. *The Human Security Report 2005.* Vancouver: The University of British Columbia Liu Institute for Global Issues. Available at: http://www.humansecurityreport.info/ (last accessed 1 October 2009)

Hyndman, J. 2003. Beyond either/or: a feminist analysis of September 11th. *Acme* 2 (1): 1–13.

Kofman, E. 1984. Information and nuclear issues: The role of the academic. *Area* 16 (2): 166.

Kropotkin, P. 1885. What geography ought to be. *Nineteenth Century* 18: 940–956. Reprinted in *Antipode* 10 (1979/1980): 6–15.

Lacoste, Y. 1976. La geographie, ca sert, d'abord, a faire la guerre. Paris: Maspero.

______. 1977. The geography of warfare. An illustration of geographical warfare: Bombing of the dikes on the Red River, North Vietnam. In *Radical Geography,* ed. R. Peet, 244–262. London: Methuen.

Mackinder, H. J. 1904. The geographical pivot of history. *Geographical Journal* 23: 421–437.

Mamadouh, V. 2005. Geography and war, geographers and peace. In *The geography of war and peace: from death camps to diplomats,* ed. C. Flint, 26–60. New York: Oxford University Press.

Minca, C. and T. Barnes. Forthcoming. Nazi spatial theory: The dark geographies of Carl Schmitt and Walter Christaller. *Annals of the Association of American Geographers.*

O'Loughlin, J. 2005. The political geography of conflict: Civil wars in the hegemonic shadow. In *The geography of war and peace: from death camps to diplomats,* ed. C. Flint, 85–112. New York: Oxford University Press.

Ó Tuathail, G., S. Dalby, and P. Routledge, eds. 2006 [1998]. *The geopolitics reader* 2nd ed. London and New York: Routledge.

Openshaw, S. and P. Steadman. 1983. The geography of two hypothetical nuclear attacks on Britain. *Area* 15 (3): 193–201.

Pepper, D. and A. Jenkins**.** 1983. A Call to Arms: Geography and Peace Studies. *Area* 15 (3): 202–8.

Pepper, D. and A. Jenkins. 1985. *The geography of peace and war.* New York: Basil Blackwell.

Pinochet Ugarte, A. 1968. *Geopolitica.*

Ratzel, F. 1897. *Politische geographie* [Political geography]. Munich and Leipzig: Oldenbourg.

Ratzel, F. 1903. *Politische geographie oder die geographie der staaten, des verkehres, und des krieges* [Political geography, or, the geography of the state, traffic, and war]. Munich: Oldenbourg.

Roder, W. 1973. Effects of guerilla war in Angola and Mozambique. *Antipode* 5 (2): 14–21.

Sparke, M. 2000. Graphing the geo in geo-political: *Critical Geopolitics* and the re-visioning of responsibility, *Political Geography* 19 (3): 373–380.

Tyner, J. A. 2009. *War, violence, and population: Making the body count.* New York: Guilford.

United Nations Human Development Index (UNDP). 2007a. Fighting climate change: human solidarity in a divided world. *Human Development Report-2007/2008.* UNDP. Available at: http://hdr.undp.org/en/. Last accessed 26 July 2009.

———. 2007b. Armed forces, index. *Human Development Report-2007/2008.* UNDP. Available at: http://hdrstats.undp.org/indicators/262.html. Last accessed 26 July 2009.

United Nations High Commission on Refugees (UNHCR). 2011. *Global trends: Refugees, asylum seekers, returnees, internally displaced and stateless persons* 2010. Geneva: UNHCR.

Watts, M. 2006. Revolutionary Islam. In *Violent geographies: fear, terror and political violence,* eds. D. Gregory and A. Pred, 175–204. Oxford and New York: Routledge.

Woodward, R. 2004. Military geographies. Oxford: Blackwell Publishers, RGS–IBG book series.

———.2006. Warrior heroes and little green men: soldiers, military training and the construction of rural masculinities. In Country boys: masculinity and rural life, eds. M.M. Bell and H. Campbell. Philadelphia: Pennsylvania State University Press.

———. 2009. Military geographies. In *The international encyclopedia of human geography,* eds. R. Kitchin and N. Thrift, 122–127.

Conceptualizing ConflictSpace: Toward a Geography of Relational Power and Embeddedness in the Analysis of Interstate Conflict

Colin Flint, Paul Diehl, Juergen Scheffran, John Vasquez, and Sang-hyun Chi

The concept of ConflictSpace facilitates the systematic analysis of interstate conflict data. Building on relational theories of power, we identify the spatiality of conflict as a combination of territorial and network embeddedness. The former is modeled through spatial analysis and the latter by social network analysis. A brief empirical example of the spread of World War I illustrates how the position of states within physical and network spaces explains their roles within a broader geography of territorial settings and network relations. *Key Words: interstate war, social network analysis, spatial analysis, World War I.*

冲突空间的概念促进了对国家间冲突数据的系统分析。在权力关系的理论基础上，我们确定了冲突的空间性是领土和网络内嵌性的一种组合。通过空间分析对前者加以建模，通过社会网络对后者加以分析。本文将第一次世界大战的蔓延作为一个简短的经验例子加以分析，揭示了国家的空间物理位置和网络空间位置是如何在一个更广泛的地理领土背景和网络关系体系里定义其角色的。*关键字：国家间的战争，社会网络分析，空间分析，第一次世界大战。*

El concepto de Conflicto-Espacio facilita el análisis sistemático de datos de los conflictos entre estados. Construyendo desde las teorías relacionales del poder, identificamos la espacialidad del conflicto como una combinación de la compenetración territorial y la de redes. Lo primero se modela mediante análisis espacial, lo segundo por medio de análisis de redes sociales. Mediante un breve análisis empírico de la propagación de la Primera Guerra Mundial, se ilustra cómo la posición de los estados dentro de espacios físicos y sistemas de redes explica sus papeles en una geografía más amplia de escenarios territoriales y relaciones de redes. *Palabras clave: guerra entre estados, análisis de redes sociales, análisis espacial, Primera Guerra Mundial.*

The geography of conflict is an expression and a medium for the myriad power relations that cause and facilitate violence. With a focus on interstate violent conflict, we explore the interaction of geographic and network spaces to understand how war spreads through the exercise of a variety of power relations. Traditional conflict analysis, especially in the field of political science, takes a singular view of power and geography, considering them only as fixed attributes in the context of dyadic relations (between a pair of states). Our approach provides a panoramic conception of relational space that uses dyads as a building block to describe the surface across which conflict spreads, as well as to help us understand how states become involved in an ongoing conflict. Furthermore, as an illustrative example, we are able to analyze how the embeddedness or situation of particular power relations shapes the construction of broader social networks to define the multidimensional and multiactor contexts that facilitated the diffusion of World War I.

The purposes of this conceptualization and application are founded on developments within the discipline of geography and also the potential and difficulties of enhancing interdisciplinary conversations between geography and the international relations (IR) discipline. As new theoretical understandings of power have taken root in geography and other social sciences, the study of interstate conflict has lagged behind (Allen 2003; Bosco 2006a); however, little of this language or ontology has pollinated the work of geographers analyzing conflict in a systematic manner. Hence, the application of relational power to conflict has been restricted to case studies in geography that are not framed within the normal science projects typical of IR.

Simultaneously, the field of IR has renewed its interest in "geography" but has not developed a sophisticated understanding of the term. Instead, physical distance, contiguity, absolute location, and simplified incorporations of terrain predominate (O'Loughlin 2000). The limitations on the understanding of geography are a function of predominant ontologies and the associated understanding of power. The state is the sole actor in the hegemonic data sets used by IR scholars, and conflict is viewed in dyadic terms; various static relationships between two states in a given year (known as the dyad-year) are the focus of analyses.

The opportunity for a number of theoretical and methodological developments is ripe. Specifically, we focus on the dyad as an initial building block in a more complex theorization and analysis of relational power that is more attuned to social networks and actor network theory. The tenet of relational power that is found in actor network theory drives the conceptualization of this research, but the nature of the data and systematic analysis is not conducive to actor network theory methodology. ConflictSpace uses the discussion of embeddedness in human geography, which was stimulated by actor network theory, to derive a way to model the spatial and social position of states with specific reference to their decision to join an ongoing war. Placing dyads within a social network analysis facilitates the simultaneous analysis of geographic and network spaces and in so doing moves the analysis of interstate conflict away from relying on simple contiguity by integrating multiple dimensions of power across a broad network of actors while still incorporating the importance of relative geographic location.

Spatial Analysis and Conflict

The scientific study of interstate conflict has produced a body of knowledge that has advanced our understanding of the causes of war and processes of escalation and diffusion. On the whole, these questions have been addressed within an established ontology and a related notion of power. States have been identified as the primary, or sole, geopolitical actors, with attributes and bilateral interstate relations the key explanatory variables. The related methodology has been systematic (large-*N*) quantitative analysis spreading across decades and historical and spatial contexts (Vasquez 2000). The goal has been to identify general patterns of behavior that are treated as universal tendencies. The dominant unit of analysis is the dyad-year, the presence or absence of interactions (alliances, conflict, trade, etc.) between two countries in a particular year.

This dyad-year approach has produced a vibrant body of normal science that structures theory construction and empirical analysis in ways that political geography lacks. Scholars who adopt such an approach to the study of war and conflict are able to build on established and empirically supported foundational hypotheses and engage contested or unresolved hypotheses in an attempt to advance knowledge. The benefits of this approach lie in facilitating statistical analyses that can evaluate hypotheses as a step toward laws of conflict behavior. There are also some constraints and concerns in the approach, however, that need to be addressed. The essence of these problems is the dominance of a simple understanding of geography and an outdated underlying conceptualization of power.

The quantitative study of interstate conflict has been slow to recognize the role of geography, and even when considered, the specification has been crude: absolute or relative location. The former is beginning to be addressed through the construction of a new data set that records the geographic coordinates of conflicts, rather than merely the state participants (Braithwaite 2005). The latter is the most commonly used path that explores the role of contiguity between states in processes of war (O'Loughlin and Anselin 1992; Buhaug and Gleditsch 2008).

Even after the initial work of O'Loughlin and Anselin (1992), IR research has remained tied to a crude notion of contiguity. The empirical reality of being a neighboring state has not been matched with an understanding of the processes of interaction that promote conflict. As a result, emphasis on physical contiguity leads to a large number of "false positive" predictions, as thankfully most neighbors do not go to war against each other. Therefore, simple contiguity as an empirical fact must be integrated with a more sophisticated understanding of the processes of conflict, especially the operation of power.

A change in the understanding of the geography of interstate conflict is a necessary and parallel step to engaging the new understanding of power. Bilateral relations captured by dyads can be expanded into a complete network of relations among all of the states involved in a conflict and, equally important, those not involved. The geography, or space, of a conflict is not an aggregation of what are treated as separate bilateral relations measured by a single variable. Instead, the space of a conflict is best conceptualized as a network that identifies the complexity of connections among all the states involved in a conflict and identifies and incorporates into the analysis multiple forms of relations (Maoz et al. 2006). Using dyads as a building block within a multiactor social network, we are able to integrate processes of power (discussed in detail in the next section) with the related patterns of contiguity and network relations that span the globe. The result is an analysis of interstate conflict that builds on Tobler's (1970) First Law of Geography: The relation of everything to everything else is visualized within a social network approach, but the "nearness" of multiple actors is a matter of physical and network spaces.

ConflictSpace, Relational Power, and Networks

Defining ConflictSpace

The purpose of introducing the term *ConflictSpace* is to incorporate the analysis of multiple spatialities (Leitner, Sheppard, and Sziarto 2008) into the analysis of war. The context within which wars are initiated and enacted is a complex mixture of relationships and structures that can be addressed only through the consideration of multiple geographies: relative physical location and positions with networks of relationships. ConflictSpace defines war as a function of political decisions made within the context of bordering states and the constraints and opportunities of a state's economic relations, political alliances, and cultural ties that are simultaneously local and global. The concept is intended, primarily, to operationalize variables and aid in the integration of multiple methodologies in the quantitative analysis of a large number of cases. Relative physical location has been addressed through the consideration of contiguity in the diffusion of war (Buhaug and Gleditsch 2008), and networks of alliances and rivalries have also been considered (Maoz et al. 2006). ConflictSpace aims to integrate these multiple spatialities so that the situation of actors in wars is one of network position and relative physical location. The empirical illustration in this short article integrates the spatialities of relative location and network position, although there is the potential to introduce other spatialities, such as scale or environmental features (watersheds and mountain ranges, for example).

Moreover, the geography of neighbor-to-neighbor interactions and cross-border flows and alliances, and other spatialities, is seen as an active feature of conflict. The changing geography of ConflictSpace is a dynamic surface over which the multiple logics of war play out, and hence the geography of war is conceptualized as a multivariate and recursive process in which conflict behavior shapes the multiple spatialities and vice versa. The surface is also part of the modeling, as different logics of conflict can operate in different ConflictSpaces and change over time. For example, economic relations might produce different logics of war diffusion than do cultural ties at different stages of a conflict, or the role of contiguity might be relevant in one region but not in another.

Recent scientific research on war has advanced understanding of both the role of social position in networks, on the one hand, and spatial context, on the other. Nevertheless, these two paths of research are separate and reflect a persistent inability to conceptualize fully and model the complexity of geographic context (O'Loughlin 2000). ConflictSpace integrates physical and network spaces to understand the possibilities and constraints a state faces in its decision to enter or avoid war. The ability of a state to avoid war or its spread is only partially a function of its shared border. In addition, political calculations based on military alliances, persistent rivalries, or a history of political and cultural empathy will play a role in a state's calculations to declare entry into an ongoing war or not. The advantages of integrating the spatial and political contexts are that the separate but related imperatives could pull states into decisions that are suboptimal or unexpected when considering only one set of factors. For example, Switzerland's neutrality in World War I is contrary to the expectations gained when considering only the spatial context.

ConflictSpace qualifies and adds to the contextualization of conflict in quantitative studies through the analysis of contiguity by grounding the network ties of political actors within geographies of relative physical location. Simple notions of shared borders are integrated within mosaics of political connections between actors. Contiguity becomes just one expression of the geography of conflict and is seen to interact with network relationships. In combination, ConflictSpace encompasses the array of different relations among actors, not merely physical ones, including trade, alliances, and other components. Thus, the analytical geography of interstate conflict required to capture relational power combines networks of social relations with the context of geographic location. We call this analytical geography ConflictSpace.

Power and Networks

Two theoretical imperatives underlie ConflictSpace. The first is an attempt to integrate a relational approach to power within the study of war. The second is to incorporate the theoretical concepts of positionality and embeddedness into the quantitative analysis of the spread of war through the simultaneous analyses of physical and network spaces.

To explain and expand on the basic definition of ConflictSpace, we utilize Allen's (2003) identification of power as relational rather than a set of attributes that are possessed by a social actor (whether the actor is defined as a state, an institution, or an organization). The ability to affect behavior is seen not only as the product of an actor having resources or attributes to enforce the behavior of another but the product of a host of relations and understandings that facilitate compliance, with varying degrees of willingness (Allen 2003). To make this approach applicable to studies of interstate war, we interpret the traditional variables analyzed by IR scholars, such as alliances and rivalries, as the resources underlying arrangements of power "which are mobilized to produce a succession of mediating *effects* in space and time" (Allen 2003, 97). In terms of the diffusion of conflict, for example, the alliances and rivalries are resources, or techniques of power, that are mobilized through networks and become effects on other actors that produce outcomes regarding the decision to join a war or not. The additional explanatory power we offer is situating separate dyadic relationships within the full network of actors' interactions and hence seeing bilateral relations as a component of a wider network of relations.

The traditional view of war and power is a zero-sum game of two competing actors gaining and losing power, with the result that one can get the other to do something that they would not willingly do (Hinchliffe 2000, 220). This approach has limited value, confined to an idealized premodern era in which two states, isolated from other social actors, compete over the same limited resources until one conquers the other. In a globalized world, rarely are conflictual contexts unidimensional in terms of issues, nor is the dispute confined to two actors, with no impact from other players. By reconceptualizing rivalries and alliances, ConflictSpace takes a traditional IR approach to resources and translates it into politics of networked relationships. Rivalry and alliances are specific resources mobilized as sets of arrangements of power that have specific effects; namely, decisions to go to war or not (Allen 2003, 97).

To apply the relational perspective to an analysis of interstate conflict we utilize a network approach in which "power is the consequence of A's abilities to bring B into its program of action. It is through action, and in this case through B acting in accordance with the aims of A, that power is composed" (Hinchliffe 2000, 222). The ontology of power is not one of the attributes of actors (such as states) but of network relations: "power is not possessed but to exercise power is to distribute power across a network" (Hinchliffe 2000, 224). Thus, ConflictSpace is concerned not only with A's direct interactions with B but also with how they map within the broader web of interactions involving those parties and actors C through Z. For example, Chad's entry into the Congo War cannot be explained without reference to its ties to France and Francophone African states.

There is an additional key consideration in discussing relational power in the form of networks. Allen (2003, 97) stresses that power is *mediated* in different geographic and historical settings. ConflictSpace addresses mediation through the incorporation of contiguity into the quantitative analysis of networks. Such an approach builds on criticisms of relational approaches that do not incorporate structural power relations and geographic context, or embeddedness (Dicken et al. 2001). We follow the goals of Dicken et al. (2001) and the potential they see in the analysis of network relations "which, when realized empirically within distinct time- and space-specific contexts, produce observable patterns" (91).

Dicken et al. (2001, 92) identify three central issues to a network methodology: (1) networks are relational processes through which power is exercised, (2) the inclusion of multiple geographical scales, and (3) the territorial embeddedness of networks. We explore these ideas through the concepts of positionality (Sheppard 2002) and embeddedness (Hess 2004), or the interdependencies between actors. Sheppard critiques social network analysis for its focus on structures of inequality through a static and unidimensional concentration on the structure of a network. ConflictSpace takes this critique seriously, with an application to the study of wars, through a social network analysis that is sensitive to temporal dynamics and social and geographic positionality. We aim to ameliorate the neglect in social network analysis of change, struggle, and agency (Sheppard 2002, 317), while also emphasizing structural power within geographic contexts.

Sheppard's idea of positionality is essential to the conceptualization of ConflictSpace. Specifically it forces consideration of two ideas. First, the manner in which different states are connected is a consequence and cause of inequitable power relations. Second, "the conditions of possibility in a place do not depend primarily on local initiative or on embedded relationships splayed across scales, but just as much on direct interactions with distant places" (Sheppard, 2002, 319). Including positionality in the modeling of interstate conflict through a combination of social network and spatial analysis facilitates jointly analyzing the context of local power relations and global connections.

Positionality in a network has been conceptualized using the metaphor of the rhizome (Hess 2004), in which the dynamic topology of actors and the networks they generate are constituted through three forms of embeddedness: societal, network, and territorial. Societal embeddedness refers to the social history of actors, or perhaps more simply their distinctive and shared attributes. Network embeddedness refers to the ties an actor has within a social network, and territorial embeddedness refers to "the extent to which an actor is 'anchored' in particular territories or places" (Hess 2004, 177). Bosco (2006b) emphasizes the way embeddedness becomes, or is constructed, and hence his contribution is most useful for qualitative studies; however, the statement that "network processes are affected by, and cannot be divorced from, the conditions governing the context in which they are produced and in which they operate" (360) grounds the goals of ConflictSpace: to analyze simultaneously territorial and network embeddedness in conflict processes. Actors in a conflict are situated within historical, network, and territorial circumstances that must be analyzed simultaneously.

In summary, a state's decision to wage war, and which side to support in an ongoing war, is an outcome of relational power that is simultaneously local (near) and global (far) and also a function of territorial, network, and social attributes (Sheppard 2002). Furthermore, such decision making and positionality can be modeled quantitatively. Our modeling approach is to combine spatial and social network analyses; the former incorporates territorial embeddedness and the latter network embeddedness. Societal embeddeness can be incorporated into both, as can temporal dynamics.

In a quantitative analysis of interstate conflict, societal embeddedness is represented through commonly used variables (such as participation in alliances, trading relations, common International Organization memberships, etc.). Network embeddedness can be analyzed using social network analysis techniques to identify topological features such as connectivity and centrality. Territorial embeddedness can be represented simply, as in this article, through various measures of proximity, but more sophisticated spatial econometric techniques (such as spatial autocorrelation measures and spatial regression) may be adopted in the future. Such advanced techniques would allow for the integration of clustering techniques that reflect the pattern of all cases and not just immediate neighbors, mirroring the goal of considering the whole network of relations and not just separate dyads. In combination, the integration of these analytical techniques with the concepts of positionality and embeddedness in network and physical spaces forms our definition of ConflictSpace.

Empirical Example: The Spread of World War I

The following brief empirical example is part of a much larger project analyzing the spread of World War I through a combination of spatial analysis and social network analysis. World War I began simply enough with war initially between only Austria and Serbia, but it eventually spread to include thirteen other states, including all the major powers. Rather than offering a complete analysis, the emphasis in this brief article is on illustrating ways in which the three different forms of embeddedness identified by Hess (2004) can be used to frame the design of empirical studies of interstate conflict and integrate multiple actors across a panoramic network of relations with relative geographic position.

We begin with the simplest and most conventional conception of physical contiguity (two states sharing borders), interpreted as spatial embeddedness, before moving on to show that this form of positionality is best understood as a combination of network and societal embeddedness. In considering all war participants and against whom they fought, the traditional conception of spatiality that we argued was inadequate in the previous section has limited value; only fourteen of thirty-seven pairs of warring states can be accounted for by shared borders.

We expand this conception of physical location through a tripartite conception of spatial embeddness by considering whether a state has a "hostile" or "friendly" neighbor; these refer to when a state has a hostile or friendly relationship with a bordering state already at war with a third state. If state B is already at war with state C then we are interested in the relationship between B and another state, A; whether A and C's relationship is friendly or hostile. Considering all three forms of contiguity, only two entrants to the war, Portugal and the United States, were not, at the moment of their entry, contiguous in some fashion to a state already in the war. For example, the spread of the war westward can be identified in the contiguity of France with Germany, which brought Britain into the conflict in the "friendly contiguity" category.

Still, the more complex forms of contiguity do not appear to be a significant driver of joining the ongoing conflict. Hostile and friendly neighbor contiguity are related to the formation of new conflict dyads between states already in the war but do not necessarily add new states to the conflict. In other words, although having hostile neighbors is related to the establishment of conflict between Turkey and France in October 1914, both states were already participants in World War I. In addition, Denmark, The Netherlands, and Switzerland did not join World War I despite the risk or opportunity based on their physical location. Simple contiguity, so common in extant studies, is inadequate in accounting for conflict processes, and more sophisticated indicators of physical proximity do not redress the problems encountered. Territorial embeddedness alone does not determine the spread of World War I, as societal embeddedness can ameliorate purely geographic conditions.

Network embeddedness requires consideration of the network relations as a whole as well as the position of a state within the network. Also, states are embedded in a number of networks at any one time. In this example we examine two networks: targeted alliances and rivalries between neighboring states (Figure 1). A rivalry between states is defined by repeated militarized disputes over a period of time (Klein, Goertz, and Diehl 2006). In our analysis a rivalry between states at any time between 1890 and 1913 resulted in a linkage in the rivalry network. The targeted alliance network is a directional network with the arrows in the network showing whether a state was being targeted by other states or was part of an alliance targeting a particular state in 1913. Alliance networks are a useful illustration as they not only tie one state to another in a formal relationship, but the strategy of alliance formation is made with respect to other actual and potential alliances within the whole interstate system. Social network analysis visualizes the network as a whole and allows a state to be placed within particular linkages, as well as noting the patterns of exclusion.

The two networks illustrate the benefits of considering physical and network spaces simultaneously. The network of rivalries between neighboring states illustrates the pervasiveness of this form of conflict behavior, which explains its poor performance in predicting the spread of conflict. On the other hand, the network of targeted alliances is restricted to the European continent, plus imperial rivalry in East Asia. The simultaneous politics of spatial and network embeddedness requires identifying the pivotal states in these networks. Also, the disconnect of The Netherlands, Denmark, and Switzerland, contiguous states that did not enter the war, shows that network embeddedness is a better explanation of conflict behavior than contiguity alone.

The counterintuitive conclusion that the network analysis provides is the relative unimportance of Germany and the United Kingdom. The Ottoman Empire, Austria-Hungary, Russia, and Italy were deeply embedded in both spatial and network power relations just before the outbreak of the war. Traditional power dimensions such as military and economic strength were not able to control the dynamics associated with the network power that others had developed. The "eastern question," or the geopolitical issues revolving around the potential collapse of the Ottoman Empire and Austria-Hungary, is evident. Also counterintuitive is that for the Balkan states, spatial embeddedness is relatively unimportant. Bulgaria, Serbia, Greece, and Romania were much more integral to the targeted alliance network than to the contiguous rivalry network. In other words, their network embeddedness played a greater role than spatial embeddedness. The Balkans are commonly seen as a hotbed of national–territorial politics, but the analysis shows that the embeddedness in a network of alliances was integral.

Pivotal states are the ones that are embedded both spatially and through the networks and play a role in connecting belligerents that are more embedded in one network rather than the other. The way some states were embedded in a spatial and a network sense meant that issues of physical space were connected with the politics of alliances that were a matter of network spaces.[1] The Ottoman Empire, Austria-Hungary, Russia, and Italy were the key states linking these forms of embeddedness. The spread of World War I was a product of the integration of two forms of embeddedness: the territorial embeddedness expressed as political rivalries between neighbors and the network embeddedness of alliances driven by competition between the European Great Powers.

A more sophisticated interpretation of the role, or embeddedness, of states within network spaces is obtained by measuring the centrality of particular states in the alliance network (Figure 2); targeting refers to whether a state is a member of an alliance targeting a particular state and targeted refers to whether a state is itself a target of a multistate alliance. By calculating and comparing centrality, it becomes possible to evaluate the relative influence of individual actors in a network system.[2] By looking at territorial and network embeddedness simultaneously we discover some counterintuitive findings. Despite its perception and rhetoric

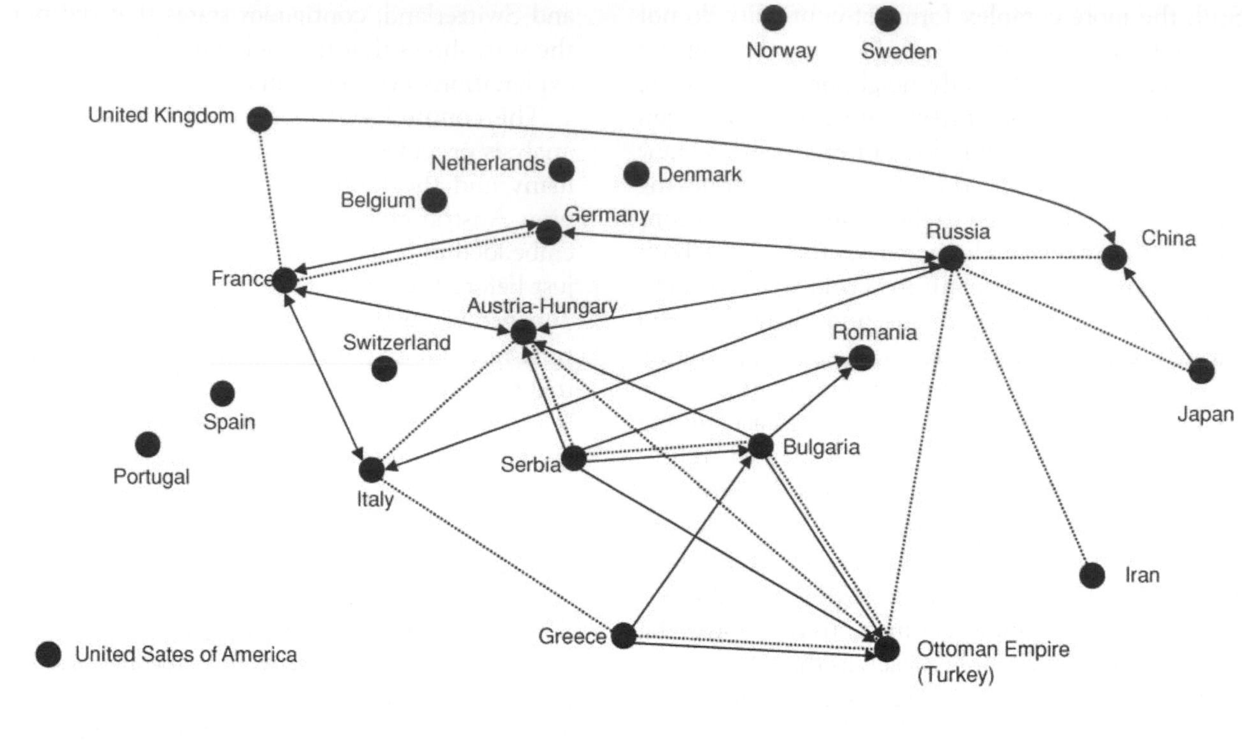

Figure 1. Networks of World War I targeted alliances and rivalries between neighboring states.

Figure 2. Measures of centrality for states in World War I in network of targeted alliances.

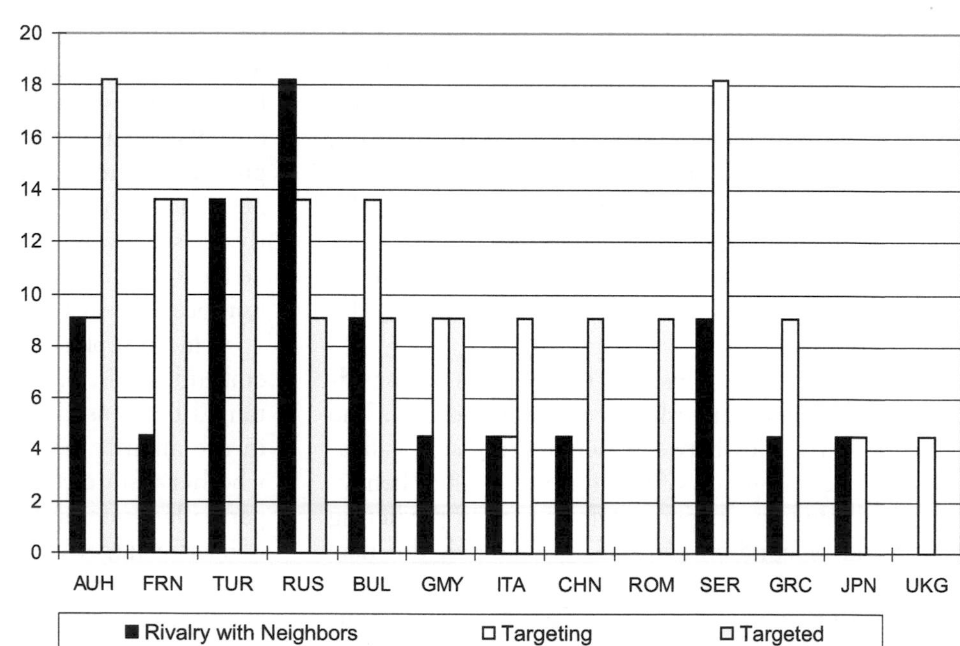

to the contrary, Germany was not particularly targeted; other countries were just as central in the network of alliance targets. The centrality of Serbia and Bulgaria in the network of states forming alliances against other states, compared to their position in the contiguous rivalry network, reinforces the counterintuitive finding of the Balkan conflict as creating networked conflict ties that spread across Europe, rather than a regionally limited territorial issue. The centrality of Russia, for example, in both rivalry and alliance networks indicates that it was faced with territorial opportunities and fears given its relationship with the Ottoman Empire but could not separate these from its competition with Germany, Austria-Hungary, and Italy. Territorial issues were inseparable from Great Power competition and therefore World War I spread as states saw these issues as either causes or opportunities for war with a number of perceived enemies simultaneously.

Alliances and rivalries are only two components of the multidimensional set of networks that embed actors. In a broader project on war diffusion, we operationally define ConflictSpace as not only including physical contiguity, rivalry, and networks of alliances but networks of other power relationships as well, including trade flows, common memberships in international organizations, and diplomatic ties. Such an analysis would, for example, highlight the way that disruption of established trade networks influenced the decision of the United States to join the war. It is the integration of these multiple network spaces, and their dynamism, with geographic contiguity that constitute the surface across which conflict can spread.

Summary

Decisions on whether to go to war are made within complex political geographic settings that can be conceptualized as multiple positionalities or integrated forms of embeddedness within diverse power relations. ConflictSpace conceptualizes such positionality and embeddedness in a manner that facilitates the quantitative analysis of dyadic data sets commonly used by IR scholars. ConflictSpace emphasizes the multiplicity of contextual settings that have, to date, been either ignored or modeled in a simple way. By first analyzing contiguity, the relationship of a state with one or two geographic neighbors is understood, but the subsequent interpretation of the whole network of states increases the number of nodes, or actors, and the multidimensionality of the context. The result is a multiactor, multidimensional surface that simultaneously incorporates physical and network spaces. Our illustration of interstate war has the potential to be expanded to include nonstate actors (such as terrorist groups) and other spatialities (such as scale).

In the case of World War I, territorial embeddedness was best conceived of as a component of political relations, through the concepts of friendly and hostile neighbors rather than simple contiguity. Alliances were interpreted as network relations that had geographic expression in their ability to tie different geographic regions together. By considering these multiple spatialities we made counterintuitive findings regarding the role of key states in World War I. ConflictSpace is primarily an analytical tool to situate actors in network and territorial settings and hence add a spatial component to network analyses of conflict. In addition, the concept and related methodological techniques will provide a new wave of analysis that might raise important questions for policymakers: Why do some wars occur? How do they spread or not? Why do some countries become involved and others do not? By considering both network and physical settings the likelihood of conflicts spreading in unexpected ways and to seemingly disconnected actors, a better understanding of such processes might have made early twentieth-century policymakers more reticent about using military force and might caution political figures who contemplate what are perceived to be quick and easy attacks against a single enemy in regions such as the Middle East.

Acknowledgments

Funding for this project was provided by the Critical Initiatives in Research and Scholarship program at the University of Illinois at Urbana–Champaign.

Notes

1. Discussion of the dynamism of the networks is beyond the scope of this short article.
2. Many different ways of measuring centrality exist. The particular centrality index used depends on features of the network (see Wasserman and Faust 1994, 169–219). We used degree centrality because it enabled us to avoid making the many specific assumptions about actors and relationships required for other centrality indexes.

References

Allen, J. 2003. *Lost geographies of power*. Oxford, UK: Wiley-Blackwell.

Bosco, F. J. 2006a. Actor-network theory, networks, and relational approaches in human geography. In *Approaches to human geography*, ed. S. Aitken and G. Valentine, 136–46. London: Sage.

———. 2006b. The Madres de Plaza de Mayo and three decades of human rights activism: Embeddedness, emotions, and social movements. *Annals of the Association of American Geographers* 96:342–65.

Braithwaite, A. 2005. Location, location, location... Identifying hot spots of international conflict. *International Interactions* 31:251–73.

Buhaug, H., and K. S. Gleditsch. 2008. Contagion or confusion? Why conflicts cluster in space. *International Studies Quarterly* 52:215–33.

Dicken, P., P. F. Kelly, K. Olds, and H. W.Yeung. 2001. Chains and networks, territories and scales: Toward a relational framework for analyzing the global economy. *Global Networks* 1:89–112.

Hess, M. 2004. "Spatial" relationships? Towards a reconceptualization of embeddedness. *Progress in Human Geography* 28:165–86.

Hinchliffe, S. 2000. Entangled humans: Specifying powers and spatialities. In *Entanglements of power*, ed. J. Sharp, P. Routledge, C. Philo, and P. Paddison, 219–37. London and New York: Routledge.

Klein, J. P., G. Goertz, and P. Diehl. 2006. The new rivalry dataset: Procedures and patterns. *Journal of Peace Research* 43:331–48.

Leitner, H., E. Sheppard, and K. M. Sziarto. 2008. The spatialities of contentious politics. *Transactions of the Institute of British Geographers* 33 (2): 157–72.

Maoz, Z., R. D. Kuperman, L. Terris, and I. Talmud. 2006. Structural equivalence and international conflict: A social network analysis. *Journal of Conflict Resolution* 50:664–89.

O'Loughlin, J. 2000. Geography as space and geography as place: The divide between political science and political geography continues. *Geopolitics* 5:126–37.

O'Loughlin, J., and L. Anselin. 1992. Geography of international conflict and cooperation: Theory and methods. In *The new geopolitics*, ed. M. D. Ward, 11–38. Philadelphia: Gordon and Breach.

Sheppard, E. 2002. The spaces and times of globalization: Place, scale, networks and positionality. *Economic Geography* 78:307–30.

Tobler, W. R. 1970. A computer movie simulating urban growth in the Detroit region. *Economic Geography* 46:234–40.

Vasquez, J., ed. 2000. *What do we know about war?* Oxford, UK: Rowman and Littlefield.

Wasserman, S., and K. Faust. 1994. *Social network analysis: Methods and applications*. Cambridge, UK: Cambridge University Press.

Oil Prices, Scarcity, and Geographies of War

Philippe Le Billon and Alejandro Cervantes

Many commentators warn that oil scarcity increases the likelihood of war; we question this simplistic concept of scarcity-driven wars. Questioning the relationship between violence, scarcity, and oil begins from reconsidering the causal relationship between high prices and war: Wars can arise in the context of low prices, and the oil-related dimensions of conflicts that occur in the context of high oil prices cannot be solely reduced to struggles over dwindling resources. Based on a succinct review of recent studies, a discussion of major hypotheses, and a brief case study of Sudan, we suggest that scarcity is in part a narrative constructed for and through prices. Power relations resulting in massive financial windfalls mediate this narrative and its selective geographies of war and peace. We outline several hypotheses, and—drawing on critical geopolitics and political ecology—explore avenues for further studies incorporating spatially disaggregated analyses. *Key Words: oil, peace, scarcity, Sudan, war.*

许多评论家警告说，石油的短缺会增加战争的可能性，我们对这种物质短缺驱动战争的简单概念提出了质疑。我们对暴力，匮乏，和石油关系的置疑是从重新评估高价位和战争之间的因果关系开始的：战争可能会出现在低价位的背景下， 而在油价高价位的背景下所发生的与石油有关的冲突不能单独地简化成争夺日益减少资源的斗争。基于对最近相关研究的摘要综述，对主要假设的讨论，以及对苏丹个例的简要研究，我们认为石油匮乏是为了价格并且是通过价格而被构建的事实论述的一部分。产生了巨额暴利的权力关系传达了这种论述事实和它所选择的战争与和平的地理布局。我们概述了几种假说，并从批判地缘政治学和政治生态学的角度，探索了利用空间分类分析进行进一步研究的途径。*关键词：石油，和平，匮乏，苏丹，战争。*

Para muchos comentaristas, la escasez de petróleo puede incrementar la probabilidad de guerra; ponemos en duda este concepto simplista de guerras causadas por escasez. Para empezar el cuestionamiento de la relación entre violencia, escasez y petróleo se debe reconsiderar las relación causal entre altos precios y guerra: Las guerras pueden empezar en el contexto de precios bajos, y lo que pueda tener que ver con petróleo en conflictos que ocurren en el contexto de altos precios del crudo no pueden circunscribirse meramente a pugnas sobre recursos limitados. A partir de la sucinta revisión de estudios recientes, una discusión de las principales hipótesis y el breve estudio del caso de Sudán, sugerimos que la escasez es en parte una narrativa construida en función de los propios precios. Las relaciones de poder que resultan en ganancias financieras masivas fuera de orden se interponen entre esta narrativa y sus geografías selectivas de guerra y paz. Bosquejamos varias hipótesis y—apoyándonos en geopolítica crítica y ecología política—exploramos las posibilidades de estudios adicionales que incorporen análisis desagregados espacialmente. *Palabras clave: petróleo, paz, escasez, Sudán, guerra.*

Oil reached a historical price peak of US$147.27 per barrel on 11 July 2008, a day after Iran tested medium-range missiles. Initiated in 2002, this latest oil boom abruptly ended amid a fast-spreading recession, with the price bottoming out at US$30.28 on 23 December 2008. As one oil trader stated, "The only thing that can drive oil higher in the short term is geopolitical news because we are loaded with crude" (Agence France Presse 2009). Israel's attack on Gaza on 27 December 2008 reinflated prices, yet with limited effect, as Iran did not respond by openly escalating the conflict.

Many commentators argue that oil scarcity increases the likelihood of armed conflicts, with oil price considered to be an index of scarcity (Klare 2004). The causes and implications of oil booms and busts are, of course, multiple. Occurring in the midst of a U.S. hegemony crisis expressed in part through reckless credit policies and militarism, the recent boom is related to U.S. dollar depreciation, speculation, and precautionary demand in a context of strong growth, prospects of declining reserves, and instability in several production areas. This brief article focuses on relationships among oil scarcity, oil prices, and armed conflicts, pointing to the utility of incorporating geographical approaches—including geographic information system (GIS)-based quantitative analyses, critical geopolitics, and political ecology (Le Billon 2007). The following section briefly reviews studies linking oil and conflicts, focusing on price and the main related hypotheses, emphasizing the value of spatially based analyses.

The Oil Curse, Oil Conflict, and Conflict Oil Arguments

Three main arguments applicable to oil relate resources and conflicts (Le Billon 2008). The *oil curse* argument suggests that oil dependence negatively affects the quality of institutions and results in economic shocks and long-term underperformance. The *oil conflict* hypothesis posits that oil exploration, exploitation, and consumption increase various forms of violence, ranging from disputes over rent allocation as well as social and environmental impacts, to international hostilities over oil access and control. Finally, according to the *conflict oil* argument, oil shapes the tactics, opportunities, and behavior of belligerents by financing their activities and influencing their relations with local populations and external actors.

Recent studies suggest that both oil dependence (oil exports as a percentage of gross domestic product [GDP]) and oil abundance (rent per capita) are positively correlated with the risk of war (Fearon 2005; Ross 2006). Humphreys (2005) finds that oil increases risk for weaker states but reduces it for stronger ones, production is riskier than discovery, and medium levels of dependence and abundance present a higher risk—as high abundance trumps negative effects of high dependence (Basedau and Lay 2009). The location of oil and type of conflict also matter; overlapping conflict and oil areas are associated with longer governmental conflicts (over central government) but not with territorial (secessionist) ones (Figure 1; Lujala, Rød, and Thieme 2007). The presence of oil in conflict areas also increases the number of deaths resulting directly from hostilities, whereas the presence of oil within the country but outside the conflict area tends to decrease it (Lujala 2009).

Debating Price Linkages: Booms, Busts, and Conflicts

In the context of "peak oil" debates, rising energy consumption, and the Iraq war, oil scarcity has recently been a focal point for geopolitical accounts of "resource wars" (Klare 2004; Heinberg 2005). Whereas the number of conflicts declined after the Cold War, it doubled among oil producers between 1989 and 2005 (Ross 2008), and the proportion of conflict zones overlapping oil-producing areas increased from about 20 percent to 40 percent (see Figure 2). The scarcity argument also seems partly vindicated when simplistically

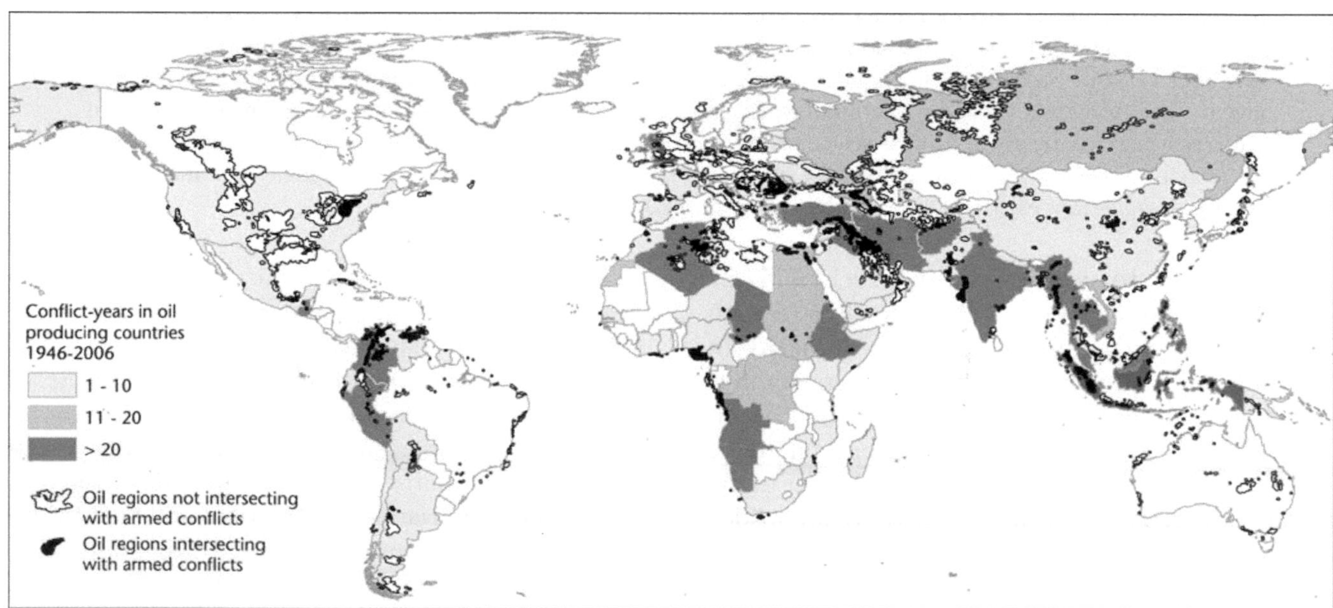

Figure 1. Conflict in oil countries, 1946–2006. Source: Uppsala Conflict Data Program/International Peace Research Institute Armed Conflicts Dataset v. 4–2007, Conflict Site Dataset v.2.0, and Petroleum Dataset v.1.1.

Figure 2. Oil and conflicts, 1946–2005. Source: Fuel rent Ross (2006); Uppsala Conflict Data Program/International Peace Research Institute Armed Conflicts Dataset v. 4–2007, Conflict Site Dataset v.2.0, and Petroleum Dataset v.1.1.

operationalized through price (see Huber 2009). Observation of oil prices, number of conflicts, and annual rate of conflicts between 1960 and 2006 suggests an association between higher oil prices and higher occurrences of conflicts in oil-rich countries compared to oil-poor countries (see Figure 3). Yet we do not observe such a fit between oil prices and the ratio of conflicts overlapping with oil-producing areas. Nor do we observe a clear pattern in the relationship between annual price variation and the number of conflicts, with a zero-, one-, or two-year lag. The direction of any causal relationship is also not clear. Out of seventy-five major events identified by the U.S. Energy Information Administration as affecting oil prices between 1970 and 2006, fifteen were conflicts and all were followed by price increase (Energy Information Agency 2007). The latest rise in conflicts in oil areas started during the 1990s slump, suggesting that oil glut and low oil prices might be conducive to conflicts.

Several quantitative studies have examined price, rent, and conflict relationships. Rising prices in high rent and low employment resource sectors, such as oil, would increase the likelihood and duration of conflicts, contrary to low rent and high employment sectors such as cash crops (Dube and Vargas 2007; Demuynck and Schollaert 2008). Yet oil booms do not negatively affect the durability of oil regimes (Smith 2007), whereas negative economic growth shocks increase the likelihood of conflict, including for oil-producing countries (Miguel, Satyanath, and Sergenti 2004), a decline in nonfuel primary commodity prices increases conflict risk in nondemocratic states in sub-Saharan Africa (Brückner and Ciccone 2007), and primary commodity price decline reduces conflict duration (Collier, Hoeffler, and Soderbom 2004). Ross (2006) finds that higher fuel rents increase the risk of conflict onset, which is offset if fuel rents are translated into higher GDP (about double the value of the rent), whereas negative price shocks are linked with separatist conflicts and positive ones only with governmental conflicts.

We present hypotheses requiring statistical analyses to extend their temporal scale to the recent crisis and its aftermath, disaggregate institutional variables and types of conflict, assess the institutional effects of (distinct) oil boom and bust cycles, and include spatial variables pertaining to the location of oil fields and conflict events. Four hypotheses link higher prices to higher conflict onset or incidence. First, higher prices would increase the number of oil-producing countries, and thus potential conflicts. Figure 4 suggests that oil production does increase with price rise but not much more rapidly than in its absence. Figure 5 shows that high prices also promote oil exploration, although less so in the current boom than in previous ones and with a lag, as in the mid-1980s price collapse. Second, higher prices would promote oil ventures in regions at greater risk of conflict. This risk affordability factor would compound the risk associated with a higher number of oil countries. There is, however, anecdotal evidence that

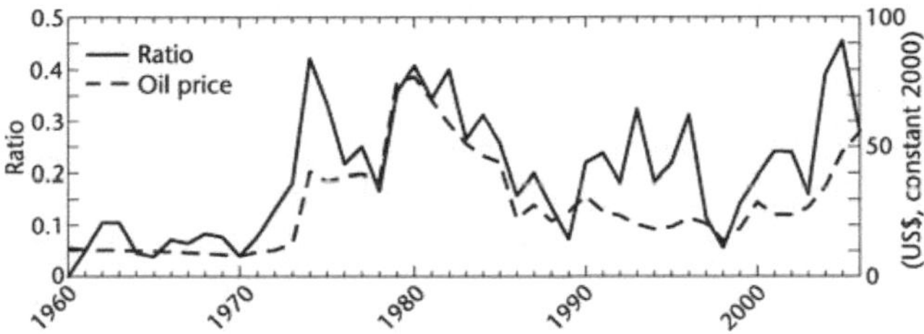

Figure 3. Oil price and ratio of conflicts in oil-rich to oil-poor countries 1960–2006. Source: Fuel rent Ross (2006); Uppsala Conflict Data Program/International Peace Research Institute Armed Conflicts Dataset v. 4–2007, Conflict Site Dataset v.2.0, and Petroleum Dataset v.1.1. Minimum oil wealth criterion is fixed at US$100 of oil rent per capita.

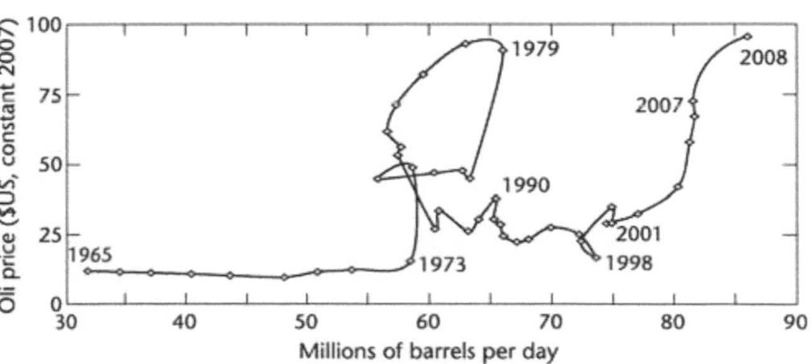

Figure 4. Oil prices and oil production, 1965–2008. Source: BP Energy Outlook 2008.

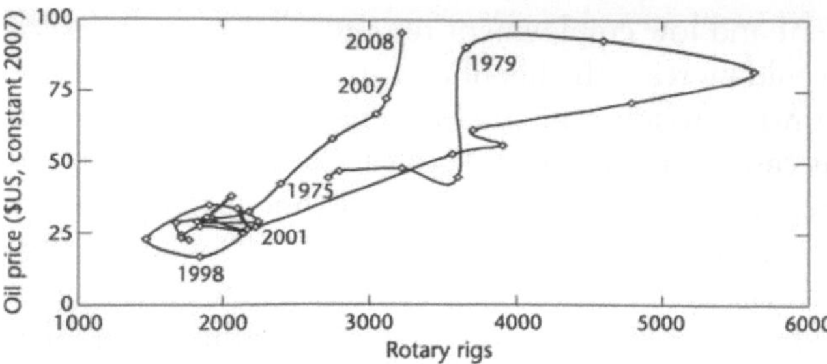

Figure 5. Oil prices and oil exploration, 1975–2008. Source: BP Energy Outlook 2008 and Baker Hughes international rotary rig count, available from http://investor.shareholder.com/bhi/rig_counts/rc_index.cfm.

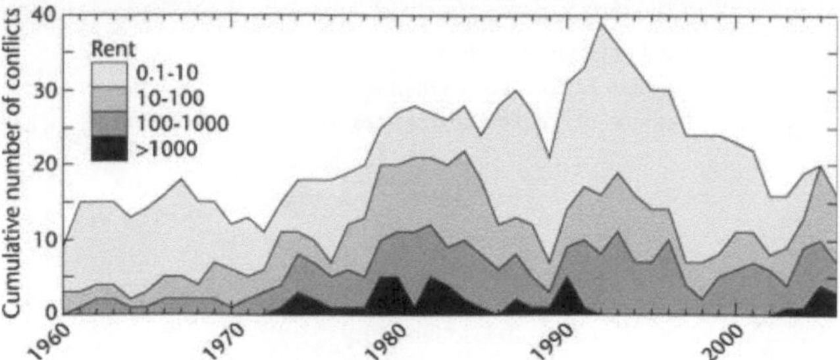

Figure 6. Oil rent level and occurrences of armed conflicts, 1960–2006. Source: Fuel rent Ross (2006); Uppsala Conflict Data Program/International Peace Research Institute Armed Conflicts Dataset v. 4–2007, Conflict Site Dataset v.2.0, and Petroleum Dataset v.1.1.

"political risks" are also taken during bust years, at least in terms of exploration, notably because of the long exploitation time horizons, lower costs of development, and more favorable policy environment (e.g., liberalization). Third, price increases would exacerbate a mix of "greed" and "grievances," increasing the likelihood of civil war, coup, secession, or even foreign intervention through growing nationalism, allegations of corruption, or desire to capture a (greater) share of larger revenues. Fourth, high prices would increase the resilience of countries already at war to external initiatives (e.g., mediation, peacekeeping), thereby possibly prolonging conflicts and adding to the incidence if not the onset of wars (Ross 2008).

Two hypotheses link lower prices to higher conflict onset. First, low oil prices would undermine the capacity of oil regimes. As Figure 6 suggests, countries with lower oil rents experienced a higher incidence of conflicts (with conflict rate of 5 percent for oil rent > US$1,000 compared to about 15 percent for lower rents). Lower revenues for security, patronage, and public welfare should increase conflict risk (Jaffe and Manning 2000), and low prices might decrease incentives for external assistance in favor of the government. Negative revenue shocks might disproportionately affect populations in production areas, lower state capacity, and precipitate the fall of authoritarian governments, thereby fostering separatist conflicts (Ross 2006). Although low oil prices are associated with higher levels of protest, leadership in oil-dependent countries is relatively durable (Smith 2007), with oil sustaining authoritarianism (Ross 2001; but see Dunning 2008). This pattern of regime durability does not necessarily mean an absence of conflicts but ineffectiveness in changing leadership. Building on McGowan (2003), we examine this proposition using the number of successful coup attempts in Africa but observe an approximately equal rate of coup success for oil producing countries (40 percent) versus other countries (41 percent). Second, low oil prices would facilitate a convergence of interests—largely on the part of oil producers, oil companies, and "defense" companies—to see their financial position improve as a result of conflict (Bichler and Nitzan 2004; Boal 2005). If industry earnings can be negatively affected by war at the individual level, conflicts affecting major oil producers are often positive (if politically sensitive) for industry earnings (Adelman 1995; Han and Wang 1998). As Collier (2000, 91) argues—although focusing on rebel movements—one can be "doing well out of war."

Doing Well out of War: Scarcity and Geographies of War and Peace

Doing well out of war depends in part on scarcity and selective geographies of war and peace. A critical geopolitics of oil wars can help deconstruct the narratives and practices linking conflicts, fears of scarcity, oil prices, and profits (for some), thereby going beyond the "facts and figures" on which positivist analyses are based (see Bridge 2001). Stability is arguably a favorable context for a capital-intensive industry with long time horizons. Instability for firms relates primarily to government policy changes (e.g., tax regimes, nationalization), erratic levels of corruption, and conflicts (e.g., communal unrest, civil wars, sanctions, and territorial disputes). Conflicts can directly affect a company's assets and oil flows, as well as the ability to secure financing. Oil and conflicts overlaps are widespread, as visualized in Figure 1. Yet wars do not always affect the contractual and infrastructural stability of oil exploitation (e.g., Angola). Furthermore, war generally disadvantages governments in their negotiations with firms. War can also provide competitive advantages for specific firms (Frynas and Mellahi 2003) and attract risk-taking companies more inclined to use bribery, military force, and minimal corporate social responsibility measures. Peace, in turn, can paradoxically be interpreted negatively by financial markets (Guidolin and La Ferrara 2007).

Thus, the geography of "oil wars" is not simply that of "scarce oil." This geography is expressed at various scales and through different forms of violence, which are differentially characterized as "war" according to distinct criteria employed by different constituencies. Geopolitical constructs of oil wars based on misleading notions of "national interest" match the geography of "abundant oil," places where the size of production and reserves, and thus the risks of scarcity, false promises of "energy security," and financial rewards, are greatest. The politicizing effects of oil dependence amplify and distort imagined geographies of vulnerability, collusive friendship, enmity, and endangerment (Sidaway 1998). Upon these constructs are superimposed or opposed environmental, communitarian, and social perspectives from which emerge, for example, alternative geographies of "progressive" coalitions seeking to improve the human rights record of the oil sector through "good governance" (Watts 2005). Selective reporting and underreporting of particular forms of violence and resistance both construct misleading geographies of peace amenable to capital and oil flows.

Oil companies, governments, consumers, communities, and the media also shape profitable geographies of war and peace. In the case of oil companies, corporate-led destabilization efforts have targeted oil-nationalizing governments (e.g., Mexico in 1938 and Iran in 1951), with stabilization efforts similarly directed toward supportive ones (Kobrin 1985). Failure by governments and consumers to reduce oil demand reinforces scarcity, and an associated politics of tension around "strategic" oil areas leads to the banalization of geopolitics of fear and force, such as in the Middle East. This "banal" violence is also expressed at the individual oil project level. Faced with the spatialities of oil concessions and operations, and constitutive of petro-states' political geography, "communities" and political entrepreneurs shape geographies of contestation and insecurity to improve access to revenues from and conditions of oil exploitation (Watts 2004). In the Niger Delta, such geographies have included street demonstrations and peaceful occupation of oil infrastructure, as well as stoppages, sabotage, kidnappings, illegal but licit oil bunkering, and separatism. The media's fascination with and selective reporting of violence makes this variable potentially more significant than other sources of scarcity and risk, such as Organization of the Petroleum Exporting Countries (OPEC) oil cuts and production incidents. Yet underreporting of violence—ranging from state repression, communal conflicts, and poverty—reflects the symbolic violence of imagined geographies of producing regions, where oil wealth, corruption, poverty, and brutality are expected and banalized. Oil markets are quick to translate these selective geographies into financial opportunities: discourses of war geographies shift from depicting oil as a use-value to an exchange-value when filtered and transformed through discourses of violence propagated among analysts, traders, and investors.

Iraq has been a prime object of representation and subject of (in)stability leading to profitable price fluctuations (Blair 1976). Under foreign control achieved through violence, Anglo-American companies limited Iraqi production to advantage their production elsewhere. These companies also resisted the process of oil nationalization through production cuts in fields still under their control, while coup attempts and Iranian-backed Kurdish insurrection received the support of some of their home governments. The Iran–Iraq war was actively supported, and United Nations (UN) sanctions were maintained on Iraq (as well as U.S. sanctions on Iran, Libya, and Sudan) during a period of oil glut. The U.S. administration (under President George W. Bush) perceived that Saddam Hussein's rule under a dwindling sanction regime had limited advantages compared to his forceful removal (Le Billon and El Khatib 2004). From an international oil company perspective, the war in Iraq could "at best" yield lucrative access to vast oil reserves and possibly grant (U.S.) oil companies the swing supplier status that they lost in the early 1970s, and "at worst" lead to flooding of the oil market by a reinvigorated Iraqi national oil company. Over five years of war, oil prices and corporate earnings sharply rose, and the business prospects of Western oil companies in Iraq slowly consolidated. War in Iraq was matched with stabilization efforts elsewhere, notably through the reopening of U.S. trade relations with Libya, which has the largest oil reserves in Africa, not to mention many Persian Gulf states (Mitchell 2002).

Oil-related geographies of war and scarcity are most visible to Western consumers during major oil crises. Previous crises were portrayed as relating to a single geographical (if multidimensional and global in impact) site, the Middle East. In contrast, the geopolitical landscape of the current oil crisis brought together multiple political processes and sites. To the multiplicity of conflicts (mostly Iraq, Iran, the Niger Delta, and Venezuela) was added multipolar economic growth (discursively centered on China) and the (ab)use of financial instruments (focused on hedge funds). Combining narratives of newfound African oil wealth, Chinese dominance of its oil sector, decades of civil war, genocide, and (past) support for terrorism, Sudan has been a prime—if somewhat overlooked due to its minor proven oil reserves—part of the puzzle of the recent crisis.

The Price of Oil: Scarcity, War, and Oil Exploitation in Sudan

Sudan's current petroleum history stems partly from the Yom Kippur War and the oil crisis of 1973 when a wave of nationalization and calls for increased profit-sharing in oil-producing countries spurred Western oil companies to diversify their reserve portfolios. Supported by a U.S. government opposing neighboring "communist" Ethiopia, Chevron obtained an oil and gas exploration license for Sudan's interior in 1974. Chevron's venture also followed the 1972 Addis Ababa Peace Agreement ending nearly two decades of secessionist struggle in southern Sudan. Chevron's oil discoveries in the south led the central Sudanese authorities to renege on the Peace Agreement, redrawing provincial borders and creating a new "Unity" province around the main oil fields (Rone 2003). This move contributed to the resurgence of rebellion in the south, with a southern secessionist group killing three Chevron employees in 1984. By then, global oil development activities were on the retreat (Figure 5). In a context of low prices, and with the reported incentive of a $550 million U.S. government tax write-off, Chevron withdrew in 1992, selling its concessions to a Sudanese company for about US$25 million (Drohan 2003). Oil discovery, in other words, helped to reignite a conflict that undermined foreign investments estimated at US$1 billion but also allowed Chevron to suspend its operations and recoup some of its investment thanks to a "geography of war" justifying divestment and government support.

Following Chevron's departure, "junior" oil companies (Canadian, Swedish, and Austrian) moved in, often partnering with Asian state oil companies (Chinese, Malaysian, and Indian; Patey 2007). The "peace" required by oil companies in production areas was in part secured through the 1997 Khartoum Peace Agreement (which included oil revenue sharing) between the central authorities and southern secessionist military factions (Young 2003). The resulting "peace" allowed for major new investments, with the first oil export purchased by Shell in August 1999 following completion of a 1,600 kilometer pipeline to Port Sudan. Yet this "peace" was also the result of brutal population displacements, including government-supported raids that intensified with the arrival of Chevron and escalated after 1992, including aerial bombing of civilian targets (Rone 2003). From the perspective of a junior company, which underreported violence involving its operations and whose activities were repeatedly suspended because of hostilities, oil nevertheless "represented an incentive for peace insofar as oil activities could not be pursued in a war context" (Batruch 2004, 160). Through selective reporting and speculating on the future impact of oil development, the company was able to construct a "geography of peace" sharply contrasting with the situation on the ground.

Ultimately, military opposition by the Sudan People's Liberation Army (SPLA), repression by government forces and its allies against local populations, infighting between southern factions, and the reluctance by Sudan's government to respect the agreement led to renewed hostilities.[1] Oil development continued, however, in part because of the prospect of rising prices, lower exposure to reputational risk for Western juniors and Asian parastatal oil companies, and China's push for access to oil reserves. In contrast, major Western companies holding concessions refrained from reinvesting despite offers of military support by the government of Sudan.[2] Hostilities forestalled several oil projects but did not prevent the completion of the main ones, including the Port Sudan pipeline, with much of the violence directed by armed forces at the civilian population or pitting various armed groups against each other for the rewards of oil field protection (Rone 2003).

The period of "peace" that allowed the development of oil fields in southern Sudan demonstrates the pragmatic compromises and extent of human rights abuses associated with oil exploitation. Central authorities in Khartoum clearly saw a strategic victory in oil exports that extended well beyond the south. Fast rising oil revenues did consolidate the government of Sudan's security apparatus, but also intensified wealth incentives for southern groups. New peace agreements in 2002 and 2005 included oil wealth sharing and a future referendum over southern sovereignty, but renewed fighting undermined these. The precariousness of this peace was recently demonstrated by fighting between SPLA and government of Sudan forces in the disputed border area between north and south where much of the oil is located. Overlapping with grazing areas involving their own logic of disputes between agrarian and pastoralist groups (Human Rights Watch 2008), disputed areas like Abyei had already seen predatory practices that contributed to rebellion, including cattle raiding and famine instrumentalization by army personnel and northern traders (Keen 1994).

This peace also fostered hostilities in other parts of Sudan, with both the Sudan Liberation Army (SLA) in Darfur and the Beja Congress in the east denouncing the inequities of an agreement in effect dividing fast rising oil revenues between the ruling party in Khartoum and the SPLA (Young 2005). Internationally, lower oil prices would have done little to prevent China from vetoing stronger pressure on the Sudanese government to end conflicts and genocide in Darfur, given the stakes of Chinese parastatal oil companies in Sudan, but high prices did not help either. In the end, events in Sudan seemed to follow the oil curse pattern presented earlier (Van Dyck 2008). Meanwhile, in neighboring Chad, a different kind of stability was sought, this time through an internationally promoted scheme that allowed an Exxon-led consortium to develop oil fields in the southwest with the reputational protection of the World Bank. As this scheme failed in the context of a growing proxy war between Chad and Sudan, the possible presence of oil in Darfur added a further layer of complexity to the geography of war in the region.

Conclusion

The oil scarcity, price, and conflict argument is partly valid. Scarcity and its effects on prices can promote destabilizing ventures by oil companies into politically sensitive areas. Scarcity can also foster state and corporate coalitions to advance particular interests—ideological, geostrategic, or commercial—through conflict. Yet the scarcity argument misses important points. First, scarcity is in part a narrative constructed for and through prices. Many conflicts take place in a context of oversupply and low prices, rather than scarcity and high prices. Oil glut and associated low prices can be in part addressed through conflict-induced scarcity (or the threat thereof). Second, rising oil prices can help end conflicts through military victory (through better funded armies or financial enticements). Third, the scarcity argument is too narrowly focused on long-term calculations of interstate conflicts, overlooking other forms and scales of violence. Finally, attributing conflict to scarcity rather than the host of factors already discussed reinforces a counterproductive vision of energy security dismissive of other concerns such as human rights.

Our analysis thus implies that a more comprehensive understanding of the recent "oil crisis" requires engagement not only with price but also with international tensions over oil reserves and transportation corridors, as well as the multiscalar geographies of war and peace influenced by scarcity narratives and the spatialities of the oil sector. The selective geographies of war and peace through which oil acquires symbolic value also deserve greater attention, as does the associated symbolic violence through which poverty and brutality in many oil production activities are justified or at least banalized. How might we go about doing so? GIS-based and quantitative spatial analyses bring greater precision and new variables to large-sample quantitative studies. But they must be combined with critical geopolitics deconstructing geographies of "resource wars" and political ecology perspectives expanding the historical scope and forms of violence studied, contextualizing power relations, and drawing attention to commodity flows and transnational spaces. In so doing, simplistic narratives of scarcity-driven oil wars might finally be laid to rest, and greater clarity can be brought to bear on the politically urgent task of addressing the conflicts, violence, and inequities associated with the oil sector.

Acknowledgments

Thanks to Halvard Buhaug, Paivi Lujala, Patrick McGowan, and Michael Ross for their databases; to Eric Leinberger for the graphs; and to the Social Sciences and Humanities Research Council for funding.

Notes

1. Interview with Taban Deng Gai, former governor of Unity Province, 2001.
2. Interview with European oil major company manager, June 2006.

References

Adelman, M. A. 1995. *The genie out of the bottle: World oil since 1970*. Cambridge, MA: MIT Press.

Agence France Presse. 2009. Oil prices fall close to $40. Agence France Presse 9 January. http://www.zimbio.com/AFP+Business/articles/2273/oil+prices+fall+close+to+40 (last accessed 7 September 2009).

Basedau, M., and J. Lay. 2009. Resource curse or rentier peace? The ambiguous effects of oil wealth and oil dependence on violent conflict. *Journal of Peace Research* 46 (6): 1–20.

Batruch, C. 2004. Oil and conflict: Lundin petroleum's experience in Sudan. In *Business and security: Public–private sector relationships in a new security environment*, ed. A. J. K. Bailes and I. Frommelt, 148–60. Oxford, UK: Oxford University Press.

Bichler, S., and J. Nitzan. 2004. Dominant capital and the new wars. *Journal of World-Systems Research* 10 (2): 255–327.

Blair, J. M. 1976. *The control of oil*. New York: Pantheon.

Boal, I. A. 2005. *Afflicted powers: Capital and spectacle in a new age of war*. London: Verso.

Bridge, G. 2001. Resource triumphalism: Post-industrial narratives of primary commodity production. *Environment and Planning* A 33:2149–73.

Brückner, M., and A. Ciccone. 2007. Growth, democracy, and civil war. Discussion Paper 6568. London: Centre for Economic Policy Research.

Collier, P. 2000. Doing well out of war: An economic perspective. In *Greed and grievance: Economic agendas in civil wars*, ed. M. Berdal and D. M. Malone, 91–111. Boulder, CO: Lynne Rienner.

Collier, P., A. Hoeffler, and M. Soderbom. 2004. On the duration of civil war. *Journal of Peace Research* 41 (3): 253–73.

Demuynck, T., and A. Schollaert. 2008. International commodity prices and the persistence of civil conflict. Working Paper 2008/518, Ghent University, Belgium.

Drohan, M. 2003. *Making a killing: How and why corporations use armed force to do business*. Toronto: Random House.

Dube, O., and J. F. Vargas. 2007. Commodity price shocks and civil conflict: Evidence from Colombia. Unpublished working paper, Harvard University and UCLA, Cambridge, MA, and Los Angeles.

Dunning, T. 2008. *Crude democracy: Natural resource wealth and political regimes*. Cambridge, UK: Cambridge University Press.

Energy Information Agency. 2007. Annual oil market chronology. In *Country analysis briefs*. Washington, DC: Energy Information Agency. http://www.eia.doe.gov/emeu/cabs/AOMC/Overview.html (last accessed 7 September 2009).

Fearon, J. D. 2005. Primary commodities exports and civil war. *Journal of Conflict Resolution* 49 (4): 483–507.

Frynas, J. G., and K. Mellahi. 2003. Political risks as firm-specific (dis)advantages: Evidence on transnational oil firms in Nigeria. *Thunderbird International Business Review* 45 (5): 541–65.

Guidolin, M., and E. La Ferrara. 2007. Diamonds are forever, wars are not—Is conflict bad for private firms? *American Economic Review* 97 (5): 1978–93.

Han, J. C. Y., and S. Wang. 1998. Political costs and earnings management of oil companies during the 1990 Persian Gulf crisis. *The Accounting Review* 73 (1): 103–17.

Heinberg, R. 2005. *The party's over: Oil, war and the fate of industrial societies*. Gabriola Island, BC, Canada: New Society Publishers.

Huber, M. T. 2009. The use of gasoline: Value, oil, and the "American way of life." *Antipode* 41 (3): 465–86.

Human Rights Watch. 2008. *Abandoning Abyei: Destruction and displacement*. New York: Human Rights Watch.

Humphreys, M. 2005. Natural resources, conflict, and conflict resolution: Uncovering the mechanisms. *Journal of Conflict Resolution* 49 (4): 508–37.

Jaffe, A. M., and R. A. Manning. 2000. The shocks of a world of cheap oil. *Foreign Affairs* 79 (1): 16–29.

Keen, D. 1994. *The benefits of famine: A political economy of famine and relief in Southwestern Sudan, 1983–1989*. Princeton, NJ: Princeton University Press.

Klare, M. T. 2004. *Blood and oil: The dangers and consequences of America's growing dependency on imported petroleum*. New York: Metropolitan Books.

Kobrin, S. J. 1985. Diffusion as an explanation of oil nationalization (or the Domino Effect rides again). *Journal of Conflict Resolution* 29 (1): 3–32.

Le Billon, P. 2007. Geographies of war: Perspectives on "resource wars." *Compass* 1 (2): 163–82.

———. 2008. Diamond wars? Conflict diamonds and geographies of resource wars. *Annals of the Association of American Geographers* 98 (2): 345–72.

Le Billon, P., and F. El Khatib. 2004. From free oil to "freedom oil": Terrorism, war and US geopolitics in the Persian Gulf. *Geopolitics* 9 (1): 109–37.

Lujala, P. 2009. Deadly combat over natural resources: Gems, petroleum, drugs, and the severity of armed civil conflict. *Journal of Conflict Resolution* 53 (1): 50–71.

Lujala, P., J. K. Rød, and N. Thieme. 2007. Fighting over oil: Introducing a new dataset. *Conflict Management and Peace Science* 24 (3): 239–56.

McGowan, P. J. 2003. African military coups d'état, 1956–2001: Frequency, trends and distribution. *Journal of Modern African Studies* 41 (3): 339–70.

Miguel, E., S. Satyanath, and E. Sergenti. 2004. Economic shocks and civil conflict: An instrumental variables approach. *Journal of Political Economy* 112 (4): 725–53.

Mitchell, T. 2002. McJihad. *Social Text* 20:1–18.

Patey, L. A. 2007. State rules: Oil companies and conflict in Sudan. *Third World Quarterly* 28 (5): 997–1016.

Rone, J. 2003. *Sudan, oil, and human rights*. Washington, DC: Human Rights Watch.

Ross, M. 2001. Does oil hinder democracy? *World Politics* 53 (3): 325–62.

———. 2006. A closer look at oil, diamonds, and civil war. *Annual Review of Political Science* 9:265–300.

———. 2008. Blood barrels: Why oil wealth fuels conflict. *Foreign Affairs* 87 (3): 2–8.

Sidaway, J. D. 1998. What is in a gulf? From the "arc of crisis" to the Gulf War. In *Rethinking geopolitics*, ed. G. Ó. Tuathail and S. Dalby, 224–39. London and New York: Routledge.

Smith, B. 2007. *Hard times in the lands of plenty*. Ithaca, NY: Cornell University.

Uppsala Conflict Data Program/International Peace Research Institute. 2007. *Armed Conflicts Dataset v. 4–2007, Conflict Site Dataset v. 2.0, and Petroleum Dataset v. 1.1*. http://www.prio.no/CSCW/Datasets (last accessed 7 September 2009).

Van Dyck, J. 2008. *International development association interim strategy note for the Republic of Sudan*. Washington, DC: World Bank.

Watts, M. 2004. Antinomies of communities: Some thoughts on geography, resources and empire. *Transactions of the Institute of British Geographers* 29 (2): 195–216.

———. 2005. Righteous oil? Human rights, the oil complex and corporate social responsibility. *Annual Review of Environment and Resources* 30:373–407.

Young, J. 2003. Sudan: Liberation movements, regional armies, ethnic militias and peace. *Review of African Political Economy* 30 (97): 423–34.

———. 2005. Sudan: A flawed peace process leading to a flawed peace. *Review of African Political Economy* 32 (103): 99–113.

Mobilizing Rivers: Hydro-Electricity, the State, and World War II in Canada

Matthew Evenden

World War II drove an unprecedented search for resources at a global scale to supply military activity in Europe, Asia, and beyond. This article contributes to recent debates about the environmental consequences of global warfare by examining how total war imposed political pressures on states, industries, and citizens that conditioned the resource development process and overrode preexisting constraints. Examining Canada's role as a supply warehouse for the Allies, the analysis focuses on a keystone resource, hydro-electricity, which powered a wide range of resource processing and manufacturing activity. To mobilize the economy, the Canadian state ordered the damming of rivers, interconnection of systems, and rationalization of industrial and consumer demands. Drawing on a wide range of archives, I examine the wartime politics of resource development, the centralizing strategy of the federal government, and the role of international governments in shaping the development agenda. Global war redrew the rules of river development and hydro power, created the opportunity for new state powers, and weakened the position of river development critics. The case allows us to consider the direct and indirect effects of global military activity in shaping resource development at a number of geographical and temporal scales. *Key Words: Canada, hydro-electricity, war and resources, World War II.*

第二次世界大战开了一个先例，即在全球范围内寻找资源，以支持在欧洲，亚洲和其他地区的军事活动。通过研究全面战争是如何强加给国家，行业和公民的政治压力，以及这种情况下对资源发展进程的影响和对此前限制条件的推翻，本文对最近关于全球战争的环境后果的争论将有所贡献。本文分析了加拿大作为同盟国供应仓库的历史角色，分析的重点是水电这种基础资源，它为各种广泛的资源加工和制造活动提供了动力。为了调动经济，加拿大国家下令在河中筑坝，对电力系统进行联网，并对工业用电和普通消费需求进行合理规划。经过对大量档案的分析，本文作者研究了资源开发方面的战时政治，联邦政府的集中战略，以及国际政府在制定发展议程的角色和作用。全球战争重新规划了河流开发和水电的规则，给新的国家权力创造了机会，削弱了对河流开发的批评立场。这个例子有助于我们了解全球军事活动对资源开发在不同时空尺度上的直接和间接影响。*关键词：加拿大，水电，战争和资源，第二次世界大战。*

La Segunda Guerra Mundial generó una inusual búsqueda de recursos, a escala global, para solventar las actividades militares en Europa, Asia y otras partes. El presente artículo es una contribución a los debates recientes sobre las consecuencias ambientales de la guerra global, examinando la manera como la guerra total presiona políticamente a estados, industrias y ciudadanos, condicionando así el proceso de desarrollo de los recursos y sorteando obstáculos preexistentes. Al examinar el papel cumplido por Canadá como despensa abastecedora de los aliados, centré el análisis en un recurso clave, la hidroelectricidad, de la cual dependía una amplia gama de los procesos de utilización de recursos y la actividad manufacturera. Para movilizar la economía, el estado canadiense dispuso el represamiento de ríos, la interconexión de sistemas y la racionalización de las demandas industrial y del consumidor doméstico. Tomando prestados datos de una amplia variedad de archivos, examino las políticas de explotación de recursos en tiempos de guerra, la estrategia centralista del gobierno federal y el papel de otros gobiernos para conformar la agenda del desarrollo. La guerra mundial obligó a cambiar las reglas de aprovechamiento de los ríos y su potencial hidroeléctrico, creó la oportunidad para el advenimiento de nuevos poderes del estado y debilitó la postura de los críticos del desarrollo de los ríos. El caso nos permite considerar los efectos directos e indirectos de la actividad militar global, para caracterizar el desarrollo de recursos en una variedad de escalas geográficas y temporales. *Palabras clave: Canadá, hidroelectricidad, la guerra y los recursos, Segunda Guerra Mundial.*

At the New York World's Fair in 1939, the Canadian pavilion contained a mural depicting the country as a panorama of modern resources (Figure 1; Canadian Government Exhibition Commission 1939). "Canada at Work," by Edwin Holgate and Albert Cloutier, included industries, people, and commodities, bound together by a narrative of material progress (Mosquin 2003, 273). In the foreground a man

Figure 1. The Holgate–Cloutier mural. Source: Canadian Government Exhibition Commission, 1939. Interior View of Canadian Pavilion at the New York World's Fair in 1939. Library and Archives Canada. Accession 1968–169. Image PA-195875.

with a horse and plough stood for agrarian tradition and at the center lay a hydro-electric dam, in front of which hung a white horse, symbolizing power. Clustered around this dam and connected by transmission wires stood basic industries: pulp and paper, mining, and manufacturing. The mural bore no military content, and implied no military vision of Canada, but did position hydro power at the country's metaphorical center. Several months after the pavilion opened, Canada declared war on Germany.

Five years later, in a very different historical moment and exhibition context, the National Gallery of Canada mounted a traveling show of posters of Canadian resources commissioned by the Wartime Information Board and produced by some of the country's leading artists. One of the exhibit's most striking images, "This Is Our Strength—Electric Power," (Figure 2) by Marian Mildred Scott, rendered a river as a captured resource, with a masculine hand grasping its power (Scott 1939–1945; Trépanier 2000, 166). The image differs from the Holgate–Cloutier mural in its explicit military content, its claim that hydro power drives Canada's war, and its merging of human strength and primordial energy. If the Holgate–Cloutier mural placed hydro power at Canada's center, then Scott's composition asserted hydro as its military prime mover. These representations of a river of industry and a river of war, produced by artists for state projects at either end of World War II, provide analogical bookends for what follows. They speak to how war initiated a broad hydro-electric program in Canada and how the state practically and metaphorically mobilized rivers.

The centrality of hydro power in the Holgate–Cloutier mural was not simply an example of artistic license. In 1939, hydro-electricity accounted for more than 98 percent of all electric power generated in Canada (Dominion Bureau of Statistics 1941, 3). With the exception of Norway, few other countries depended so heavily on hydro generation. Canada's total hydro generation was also massive. Its output was second only to the United States and led the world per capita ("Power generation and utilization in Canada" 1941, 360). The central Canadian provinces accounted for most of this capacity, although British Columbia and Manitoba also had considerable installations.

Whereas hydro development ground to a halt in most states during the war owing to defense risks and the material and labor costs of development, in Canada installed hydro-generating capacity increased by 40 percent over six war years (Figure 3). According to a U.S. Geological Survey report compiled at the end of the conflict, only Canada and the United States substantially increased hydro generation between 1939 and 1945 (U.S. Geological Survey 1946). In frontline states, projects were difficult to develop and hydro systems were vulnerable to attack. During the war, major dam and hydro-electric plants were destroyed in Germany, the Soviet Union, and Norway (Milward 1972, 200; Muller 2003, 322–23). Outside of core regions, hydro development projects awaited the return of peacetime conditions.

Unlike the United States, Canada had no state-led program to develop hydro resources during the Great Depression. Whereas large U.S. federal projects rose on the Columbia and Tennessee Rivers in the late 1930s with importance for American wartime production, in Canada public and private utilities sought to follow rather than promote demand. Late in the 1930s, therefore, Canada's major power systems were in a poor position to meet surges in wartime demand. After the fall of France in June 1940, when Canada stood as the United Kingdom's primary ally and economic mobilization moved forward rapidly, there was serious concern that Canada's power supply would fail to meet industrial needs. At this point, the federal state moved to control this strategic energy source and initiate new developments.

This article contributes to recent debates about the environmental consequences of global warfare by examining how total war imposed political pressures on states, industries, and citizens that conditioned the resource development process and overrode preexisting constraints (Russell 2001; Tsutsui 2003; Laakkonen 2004; McNeill 2004; Tucker 2004; Tucker and Russell 2004).[1] I argue that the Canadian and Allied demand for aluminum to prosecute the air war drove a broader centralization of control over hydro-electric resources in Canada; that previous political obstacles to development were removed in the process; and that the balance of powers within the state system shifted to accommodate military priorities. These changes were temporary but bore a range of longer term consequences for the exercise of state power and the character of hydro-electric development. State actions in the war modeled possibilities for the postwar period and created future institutional and infrastructural boundaries of possibility. Like Tucker and Russell (2004), I argue that wartime resource development created new demands and markets that did not disappear in the conversion to peacetime but rather imposed path dependencies.

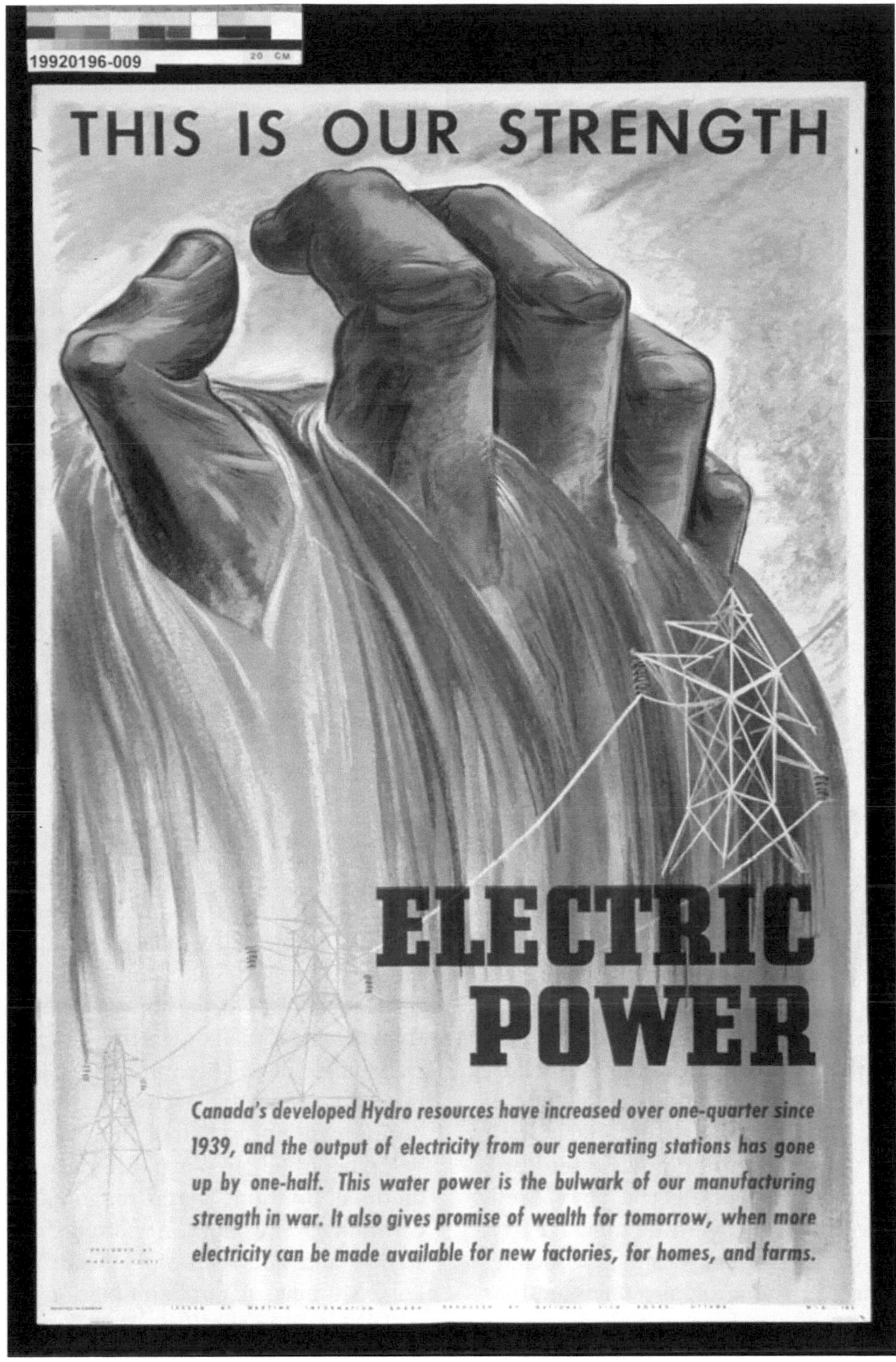

Figure 2. Marien Mildred Scott. 1939–1945. This Is Our Strength: Electric Power. Canadian War Museum, Artifact number 19920196-009.

Controlling Power

As the Battle of Britain began in the summer of 1940, the Canadian Department of Munitions and Supply, under Ontario minister C. D. Howe, was given sweeping authority to engage the Canadian economy for total war (de N. Kennedy 1950; Bothwell and Kilbourn 1979, 129–30; Bothwell 1981). Industrial controls served as

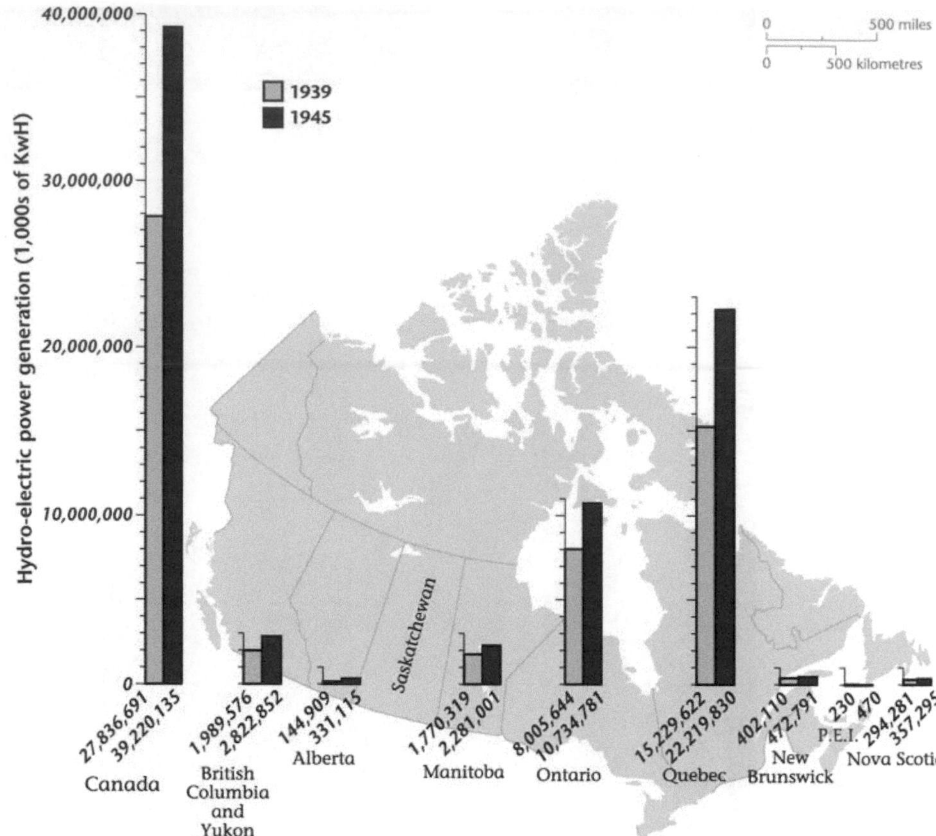

Figure 3. Hydro-electric energy generated, 1939 and 1945. Source: Dominion Bureau of Statistics (1941, 1947).

one of Howe's primary means of intervention. They assigned a single individual drawn from the private sector authority to regulate the production, distribution, and consumption of a strategic commodity, such as steel or timber. Power control was one of the first such controls established, and Herbert J. Symington, one of the first controllers.[2] A corporate lawyer, and general counsel of Royal Securities, one of Canada's most important investment houses, with primary holdings in utilities, Symington knew the business of electric power well. His strong connections to the governing Liberal Party, and particularly with his former business associate, C. D. Howe, meant that Symington also knew another side of the business of power (Bothwell and Kilbourn 1979, 98).[3]

The establishment of power control reconfigured the politics of Canadian hydro development. Before the war, power had been regulated by the provinces, except on international and navigable rivers. The War Measures Act and the Munitions and Supply Act shifted this authority to the federal center. To avoid potential provincial concerns, Symington persuaded the Ontario and Quebec governments to provide civil servants knowledgeable of the industry to support his activities.[4] At the corporate level, Symington also encouraged business leaders to pursue state policies voluntarily rather than passing formal orders that risked confrontations or noncompliance. This informality also allowed corporations considerable access to a regulator who favored business interests. Before the war, critics of river development had often exploited federal-provincial jurisdictional divisions (Armstrong and Nelles 1986). The centralization of authority after 1939 closed this possibility and one high-profile dam project, located on Lake Minnewanka in Banff National Park, went ahead in 1941 despite the Parks Act that forbade it and the protests of a preservationist lobby who had defended the lake since the 1920s against power development (Reichwein 1995).[5] By deftly enrolling federal and provincial politicians and using emergency wartime legislation, Calgary Power extracted an order-in-council from the federal cabinet setting aside the Parks Act and allowing for construction.[6] That Symington served as general counsel for and held stock in Calgary

Power reinforced the conviction of development critics that wartime necessity provided cover for corporate interests.

The Canadian state entered the power control field with no regulatory experience or expertise, which left Symington the considerable task of developing a coherent policy. Several factors shaped his approach. First, available power and imaginable sites of expansion were located in the central provinces, as were the majority of industries on military contracts. Second, Ontario faced a looming crisis because of sharply rising electricity demands. Third, the importance of the air war, and the shortage of aluminum in the United Kingdom and the United States, placed political pressure on Canada and the Aluminum Company of Canada (Alcan) to increase output massively, a task that would require a diversion of electricity from other industrial centers in southern Quebec and an expansion of hydroelectric facilities. By November 1940, after high-level meetings with U.S. officials who shared sensitive information about the potential shortage of aluminum and requested electricity exports from Canada, Symington developed a control policy centered around aluminum expansion in Quebec, realizing latent power in existing central Canadian systems through interconnection and increased water diversions, as well as conservation aimed at nonstrategic industries, commercial use, and domestic consumption.[7] The goal was to make aluminum while providing for a general expansion of wartime production across the industrial heartland of Ontario and Quebec.

Mobilizing Rivers for Aluminum

In 1939, the aluminum industry was a major power consumer in Canada. In the early 1920s, the Aluminum Company of America (Alcoa) had established a plant in Arvida, Quebec, on the Saguenay River and constructed a hydro dam at the outlet of Lac St. Jean. The location brought together a major waterpower site and access to a deepwater port where the Saguenay River joins the St Lawrence. Bauxite shipped from British Guiana provided the facility's ore (Massell 2000). After corporate reorganization separated Alcan from its parent company in 1927, it faced a difficult market in the 1930s and only recovered with the beginning of the war. The first substantial military contract signed with U.K. authorities in 1939 envisioned a major expansion in smelter and hydro facilities (Campbell 1985, 255).

Smelting aluminum became one of Canada's chief material contributions to the Allied war effort (Massell 2004). Alcan's production of primary aluminum rose from 75.2 thousand metric tons in 1939 to 1545.4 in 1945, a more than twentyfold increase (Campbell 1985, 251). As aircraft proved critical to most aspects of the conflict and aluminum alloys provided the majority of the material used in their construction, aluminum supply became strategically important to all belligerent states. Over six war years, Canada supplied more than 90 percent of U.K. and Commonwealth needs and 35 percent of total Allied needs (Campbell 1985, 251).

Canada's focus on aluminum was partly based on the vulnerability of the U.K.'s aluminum industry. Not only were the hydro facilities of the British Aluminum Company in Scotland badly exposed to potential air attack, but the company also lost access to its major bauxite supply in the French southwest. Added to this, Britain no longer had access to aluminum exports from Norway or Switzerland. Canadian aluminum thus became a central aspect of the U.K. aircraft production program, and Alcan received loans from the British government in the first half of 1940 totaling $40 million to increase capacity (Hall 1955, 18–19). U.S. demand also propelled Canadian production to offset shortages in the U.S. market caused by limited electricity near smelting sites and the difficulties of expanding aluminum smelting facilities in the early stages of the war (Nash 1990, 91–121; Koistinen 2004, 136).[8]

When U.S. President Roosevelt and Canadian Prime Minister Mackenzie King negotiated the Hyde Park Declaration (1941), which called for the two countries to integrate their economies for maximum efficiency in war production, Roosevelt insisted that aluminum be specifically mentioned in the diplomatic text (Granatstein 1979, 143). The U.K., U.S., and Australian governments all supported Alcan's expansion through advance contracts and payments (which amounted to credit-free loans) and the Canadian government provided substantial tax depreciation benefits (Campbell 1985, 255–64).

To support the aluminum production program, Symington pressured Alcan to expand its power generation facilities massively, despite the risks of postwar overcapacity, and sought to ensure sufficient power deliveries from neighboring systems until new hydro dams could be built. The expansion program had three phases: the construction of light structures in the upper Saguenay basin to retain the spring flood waters to create additional generating potential; the building of new dams on the Peribonka and Saguenay Rivers; and the completion of a new powerhouse at Shipshaw. The scale of the project and the challenges of construction were such that the new power would not be available until 1943. Until that time, Alcan and the power controller had to cope.

As had occurred in World War I, the prospect of power shortages drove a wave of interconnections among power systems (Hughes 1983). Quebec's major power utilities, the Montreal Light, Heat and Power Company (MLHP) and the Shawinigan Water and Power Company (SWP), built new transfer stations and ran a new high-voltage line under the island of Montreal to facilitate larger power transfers between the two systems (Farnham and Titus 1942; Way 1942). The new interconnections allowed MLHP to divert power east to SWP during the night and in the summer when Montreal loads were at their lowest levels. SWP, which served a mainly industrial market, could decrease generation in these periods and bank water in its reservoirs. By saving water, SWP gained the flexibility to increase generation when Alcan required and divert power east to Arvida.[9] Although the interconnection program did not avert some moments of crisis as Alcan's demand soared in 1941, it did realize the latent potential of separate systems, among business competitors (Evenden 2006).

Beyond Quebec

Solving the issue of power for the aluminum industry in Quebec raised other problems for the Canadian economy and for Symington's power strategy. Although aluminum was a crucial war material, the majority of war contracts had been let in Ontario, and providing electricity to the province's industrial southwest was also of crucial importance. The situations in Quebec and Ontario were closely connected because Ontario's public monopoly, the Hydro-Electric Power Commission (Ontario Hydro), depended substantially on imported electricity from Quebec. Although it operated its own hydro facilities on the Niagara River and in the provincial north, Ontario Hydro had shifted to an import strategy in the 1920s to avoid the risks of large capital investments. Private producers in Quebec therefore provided for Ontario's public system. This arrangement faltered once wartime drove up Ontario's demands. In December 1939, Ontario Hydro officials observed a rise in peak demand of 13 percent over the previous year; based on industrial growth projections, they estimated that demand would outstrip generating capacity by December 1941.[10] When Ontario Hydro officials approached Quebec companies to increase power deliveries, they were told that only existing contracts could be met.[11]

Without the ability to import electricity to alleviate its difficulties, Ontario Hydro planned major dam projects. In the 1930s, the Ontario government had resisted proposals to launch a St. Lawrence Seaway of integrated navigational works and hydro dams between Canada and the United States, owing to the belief that additional electricity would produce a massive surplus (Armstrong 1981, 166–70). Instead, the provincial monopoly installed diversion works in northern Ontario to increase flows at its Niagara plants. Owing to objections from the United States, under the terms of the International Waterways Treaty, however, they were not completed. Under wartime conditions, these earlier disputes took on a different complexion and Canada and the United States (with full support from Ontario) renegotiated a St. Lawrence Treaty in 1941, and Ontario completed its Great Lakes diversion projects to increase power generation potential on the Niagara with U.S. support.[12] By 1941, with demand climbing across the province, Ontario was fully committed to a St. Lawrence development program. When the U.S. Congress deemed the treaty impractical because of the war, the province was left with few options.

In the circumstances, Ontario Hydro moved quickly to rationalize its network of power systems and to launch development investigations on the Ottawa River. As in Quebec, the wartime shortage presented the opportunity to improve the transmission network of southern Ontario to allow for regional surpluses to be transferred between systems. Ontario Hydro also completed additional power lines reaching the Quebec border to improve transmission capacity. A dam was built on the Madawaska River and negotiations were opened with the Quebec government to determine the best means to develop the boundary Ottawa River. In the interwar period, various provincial governments and private firms had sought to develop the river, but none had succeeded owing to the claims of the federal government that it should hold regulatory authority because of the river's navigability, and therefore federal jurisdictional interest. The pressures of war cut through this development politics. Ontario and Quebec agreed to a site division agreement that reserved the upper river sites for Ontario and the lower ones, closer to Montreal, for Quebec.[13] Following the intervention of Symington and C. D. Howe with the federal cabinet and the prime minister, the federal government waived any former objections.[14] In this instance, the federal power controller acted in the interests of the provincial governments against the federal authority. Nevertheless, the Ottawa River Agreement (1943) came too late to serve Ontario Hydro's wartime needs. None of the major dams outlined in the agreement were completed until after 1945.

As a result, Ontario's rising wartime power demands had to be met by incremental additions, improving system efficiencies and a wide-ranging conservation program (Evenden 2005). Although power control held the authority to intervene to set rates, Symington determined that he would not follow this course unless disputes arose in the market.[15] Raising rates to reduce consumption posed other difficulties in terms of the federal government's price control program and the risk of stimulating inflation. Conservation had to operate according to time and behavioral changes in the market, not as a price discipline. One of Symington's first measures was to implement a Daylight Savings program in 1940 (de N. Kennedy 1950, 181–82).[16] He also argued for a shuffling of industrial timetables in both the central provinces to redistribute peak demands toward as steady a system load as possible. On several occasions he ordered pulp and paper mills to close, sometimes for weeks at a time, to reduce nonessential industrial demands (de N. Kennedy 1950). In December 1942, as the annual winter peak in power demand approached, he mandated a sweeping ban in Ontario and Quebec on evening commercial store lighting and a reduction in street lighting.[17]

Although Symington's conservation measures applied to both Ontario and Quebec, they were primarily crafted to meet the needs of Ontario. Whereas Quebec's power systems had been developed by private companies to meet the needs primarily of large industrial customers, Ontario's public monopoly had also held broad objectives to provide electricity cheaply to domestic and rural consumers. In the interwar period, Ontario Hydro had used stimulative pricing policies, sold load-building appliances cheaply, and pioneered rural extension programs. As a result, consumer demand accounted for a much more important segment of overall demand in Ontario than in Quebec (Dominion Bureau of Statistics 1947, 16). In step with Symington's 1942 light restrictions, Ontario Hydro launched a broad publicity campaign to educate consumers about the looming power shortages and the role that consumers could play to reduce power demand in the home (Evenden 2005). This conservation program targeted female consumers who were represented in advertising as wasteful consumers while wartime production went short. Although Cohen (2003, 75) argues that publicity campaigns in the United States that valorized the role of women in reducing consumption on the homefront gave agency to women as patriotic consumers and moral guides, it is difficult not also to recognize a strategy to provoke anxiety among women as selfish consumers in a political context that demanded selflessness.

As Ontario and Quebec pursued a mixture of strategies under the guidance of federal power control to meet pressing wartime needs, Symington intervened rarely in other parts of the country. He authorized a dam on the Kootenay River at Brilliant, British Columbia, to deliver electricity to Cominco's Trail smelter that processed copper and zinc, as well as heavy water for the Manhattan Project (Andrews 1971; Mouat 1997, 73–96). In Alberta, he allowed Calgary Power to dam Lake Minnewanka in Banff National Park; however, beyond these two western projects, Symington ensured that no other developments would occur outside Ontario and Quebec where there was insufficient strategic justification. When the British Columbia utility, BC Power, approached Symington in 1944 to dam a major Fraser River tributary in the face of looming shortages in Vancouver, he refused. The city faced frequent brownouts at the end of the war and depended on an intertie with the Bonneville Power Administration to avert more drastic problems (Evenden 2004, 119–48).

The wartime emergency legitimized the unprecedented centralization of regulatory authority over hydro-electricity in Canada in one federally appointed controller. This centralization of authority diminished provincial agency, although not ambition, and constrained corporate activity where no strategic merits could be justified; however, the power controller also acted on behalf of provinces when this supported broader national goals and promoted corporate interests when they coincided with his power strategy. More broadly, the federal power controller crafted his national policies with an eye to aluminum supply and a broader set of allied demands on Canadian economic resources. The central authority therefore triangulated Canadian policy within a British geopolitics and a continental military economy. As a result, the sharp centralization of state control over power contained within itself contradictions and pathways for the pursuit of diverse interests.

By 1944, with the end of the war in sight, the federal state began to unwind the wartime state, including power control. Provinces reassumed their jurisdictional primacy in the power field, but several redesigned the purpose and scope of that authority. The regulatory role of the war years and the capacity of state planning to meet critical needs paved the way for a series of new provincial corporations that nationalized private utilities, including Hydro Quebec (created out of MLHP in 1944) and the British Columbia Power Commission (formed from a series of utilities outside the urban southwest in 1945). The exercise of wartime power control modeled the advantages of state intervention in the power field and helped to produce a postwar model of provincial state building.

The expansion of power generation, the rationalization of transmission systems, and the implementation of conservation measures all contributed materially to Canada's war effort. No major power shortages occurred in the core industrial regions of Ontario and Quebec, and generation expanded to meet rising industrial needs, particularly for aluminum. The wartime power program not only met short-term needs and delivered a comparatively substantial increase in power generation on a world scale but laid the foundation for a host of postwar projects. The Ottawa River Agreement (1943) produced no new dams during the war but several in the late 1940s. The St. Lawrence Treaty (1941) negotiated with the United States at the height of the wartime emergency put in place a program and process for the St. Lawrence Seaway in the 1950s. The massive expansion of Alcan's power works on the Saguenay River converted this company into a global producer and led the corporation to continue its hydro and smelter expansion in British Columbia at Kemano and Kitimat starting in the late 1940s. Policies and projects struck during the war inspired postwar plans while also setting up path dependencies.

Although power control focused primarily on expanding power generation and improving distribution, its actions also incidentally produced delays to development outside the industrial heartland and introduced novel conservation strategies to limit consumer and commercial demands. After 1945, these aspects of wartime policy were rapidly overcome and forgotten. Outside the industrial heartland, private and provincial utilities alike launched hydro expansion programs. In British Columbia, for example, the BC Power Corporation immediately launched a new dam in the middle Fraser Basin and committed itself to a major expansion of facilities to counter rising political pressures for nationalization (Evenden 2004). Evidently, few consumers or commercial establishments sought to maintain conservation programs in the postwar period. Provincial regulators advocated a reduction in consumer power prices, consumer demand spiked, and electrical consumption became one marker of a bright future. A wartime narrative of limitation was followed, in Nye's (1997) terms, by a narrative of abundance. The environmental consequences of these narratives and actions were yet to be fully appreciated.

Acknowledgments

The research leading to this article was supported by a Standard Research Grant from the Social Sciences and Humanities Research Council of Canada, which I am pleased to acknowledge. My thanks to Graeme Wynn for reading an earlier version of this article, to Audrey Kobayashi for her editorial advice, and to Eric Leinberger for preparing the map.

Notes

1. Space constraints do not allow for a full discussion of this literature. Please consult Evenden (2005) and Evenden (2006).
2. Library and Archives Canada (LAC), RG 28, Department of Munitions and Supply, Vol. 54, file 1-1-98, 23 August 1940, P.C. 4129.
3. "Symington, Herbert James," in *The Canadian Who's Who* Vol. II , ed. Sir C. G. D. Roberts and A. L. Tunnell, 1048. Toronto: Murray Printing, 1937. On Symington's investments, see LAC, RG 28, Vol. 54, File 1-1-98 Power Control, "May 24, 1943 Question Mr Stokes, MP—Votes and proceedings #65, page 1."
4. For a description of his control strategy, see Archives of Ontario (AO), RG 3–10, Hepburn Papers, Box 324, File: Hydro Electric Power Commission, 1942, Symington to Mitchell, 17 March 1942, and LAC, RG 14 Accession 1997–98/628, Box 8, File: "War Expenditures Subcommittee 1—1943, Chairman's Copies, War Expenditure's Committee, Subcommittees No 2 and 3, Vol 24, November 5*th*, 1943" B-2, B5-B6.
5. LAC, RG 84, file B39-5, Edwards to Camsell, 10 December 1940; Aberhart to Crerar, 5 December 1940; Crerar to Gaherty, 16 December 1940; Crerar to Gaherty enclosing PC 7382, 13 December 1940. Protests arose from the Canadian Parks Association: LAC, RG 84, Vol. 502, File B39-5, Walker to Smart, 1 October 1940.
6. LAC, RG 84, Vol. 502, File B39-5, P.C. 7382, 13 December 1940. Legislation confirming this procedure followed in 1941: LAC, RG 84, Vol. 502, File B39-5, Bill 60 to Amend the Alberta Natural Resources Transfer Act, as Passed by the House of Commons, June 4, 1941, Second Session, Nineteenth Parliament, 4-5, George VI, 1940–41.
7. Symington refers to these meetings in LAC, RG 14, Accession 1997–98/628, Box 8, "War Expenditure's Committee, Subcommittees No 2 and 3, Vol 24, November 5*th*, 1943," pp. E-5- E-6
8. On aluminum production in the United States during the war, see United States National Archives (USNA), RG 179, Entry 40, Box 1, Elliott M. Helfgott, "Metals

and materials," in *Industrial mobilization for war (1940–1945), Vol. II, Metals and minerals*, ed. G. W. Auxier, War Production Board Historian, n.d.

9. Archives de Hydro Quebec (AHQ), Fonds Shawinigan Water and Power Company, Box 4109, F1/1190, "Distinctive Engineering Features of the Shawinigan System" (SWP, 1947), 22.
10. AO, RG 3, Mitchell Hepburn Papers, Box 306, File: Hydro Electric Power Company, 1940, Thomas Hogg, "Hydro and the War."
11. AO, RG 3–10, Hepburn Papers, Box 306, File: Hydro Electric Power Company, 1940, Hogg to Norris, President of Montreal Light, Heat and Power Co., 18 November 1939; Norris to Hogg, 24 November 1939.
12. AO, RG 3–10, Hepburn Papers, Box 301, Hogg to Premier, "Memorandum Relative to the St. Lawrence Development, December 18, 1939."
13. AO, RG 3–15, Premier Conant Papers, Box 423, Agreement, 2 January 1943 between the Province of Ontario and the Province of Québec, re: Apportionment of the Ottawa River; RG 3–17, Premier Drew Papers, Box 439, File 137-G Hydro-Electric Power Commission Des Joachims Development, Sessional Paper No. 41 Session of 1943, 10 February 1943, "Ontario-Qúebec Power Sites Allocation Agreement (Bill No. 25, An Act respecting the Water Powers of the River Ottawa)" (1943).
14. AO, RG 3–17, Premier Drew Papers, Box 439, File 137-G Hydro-Electric Power Commission Des Joachims Development, Order in Council, P.C. 651, 26 January 1943.
15. LAC, RG 28, Vol. 54, File 1-1-98 Memo re Conference on power, 12 September 1940.
16. LAC, RG 28, Vol. 54, File 1-1-98 Symington to Howe, 19 September 1940.
17. LAC, RG 28, Vol. 251, File 196-11-2, Department of Munitions and Supply, the Power Controller, Order No. P.C. 5 (Power Shortage Areas), 20 September 1942.

References

Andrews, C. D. 1971. Cominco and the Manhattan Project. *BC Studies* 11 (Autumn):51–62.

Armstrong, C. 1981. *The politics of federalism: Ontario's relations with the federal government, 1867–1942*. Toronto: University of Toronto Press.

Armstrong, C., and H. V. Nelles. 1986. *Monopoly's moment: The organization and regulation of Canadian utilities, 1830–1930*. Philadelphia: Temple University Press.

Bothwell, R. 1981. "Who's paying for anything these days?" War production in Canada, 1939–1945. In *Mobilization for total war: The Canadian, American and British experience*, ed. N. F. Dreisziger, 57–70. Waterloo, ON, Canada: Wilfred Laurier University Press.

Bothwell, R., and W. Kilbourn. 1979. *C. D. Howe: A biography*. Toronto: McClelland and Stewart.

Campbell, D. C. 1985. *Global mission: The story of Alcan*. Montreal: Alcan.

Canadian Government Exhibition Commission. 1939. Interior view of Canadian pavilion at the New York World's Fair in 1939. Library and Archives Canada. Accession 1968–169. Image PA-195875.

Cohen, L. 2003. *A consumers' republic: The politics of mass consumption in postwar America*. New York: Knopf.

de N. Kennedy, J. 1950. *History of the department of munitions and supply*. Ottawa, ON, Canada: King's Printer.

Dominion Bureau of Statistics, Transportation and Public Utilities Branch. 1941. Table 14: Electric energy generated. In *Central electric stations in Canada, 1939*, 36–37. Ottawa, ON, Canada: King's Printer.

———. 1947. Table 14: Electric energy generated. In *Central electric stations in Canada, 1945*, 36–37. Ottawa, ON, Canada: King's Printer.

Evenden, M. 2004. *Fish versus power: An environmental history of the Fraser River* Cambridge, UK: Cambridge University Press.

———. 2005. Lights out: Conserving electricity for war in the Canadian City. *Urban History Review* 34:88–99.

———. 2006. La mobilisation des rivières et des fleuves pendant la Seconde Guerre mondiale: Québec et l'hydroélectricité, 1939–1945 [Mobilizing rivers during the Second World War: Quebec and hydro-electricity]. *Revue d'histoire de l'Amerique Francaise* 60:125–62.

Farnham, D. M., and O. W. Titus. 1942. Design, manufacture, and installation of 120-Kv oil-filled cables in Canada. *Transactions of the American Institute of Electrical Engineers* 61:881–88.

Granatstein, J. L. 1979. *Canada's war: The politics of the Mackenzie King Government, 1939–1945*. Toronto: Oxford University Press.

Hall, H. D. 1955. *North American supply*. London: HM Stationary Office.

Hughes, T. P. 1983. *Networks of power: Electrification in Western societies, 1880–1930*. Baltimore: Johns Hopkins University Press.

Koistinen, P. A. C. 2004. *Arsenal of World War II: The political economy of American warfare, 1940–1945*. Lawrence: University of Kansas Press.

Laakkonen, S. 2004. War—An ecological alternative to peace? Indirect impacts of World War II on the Finnish environment. In *Natural enemy, natural ally: Toward an environmental history of war*, ed. R. Tucker and E. Russell, 175–94. Corvallis: Oregon State University Press.

Massell, D. 2000. *Amassing power: J. B. Duke and the Saguenay River, 1897–1927*. Montreal, PQ, and Kingston, ON, Canada: McGill-Queen's Press.

———. 2004. "As though there was no boundary": The Shipshaw project and continental integration. *American Review of Canadian Studies* 34:187–222.

McNeill, J. R. 2004. Woods and warfare. *Environmental History* 9 (3): 388–410.

Milward, A. S. 1972. *The fascist economy in Norway*. Oxford, UK: Clarendon.

Mosquin, A. 2003. Advertising Canada abroad: Canada on display at international exhibitions, 1920–1940. PhD dissertation, York University, Toronto.

Mouat, J. 1997. *The business of power: Hydro-electricity in south-eastern British Columbia, 1897–1997*. Victoria, BC, Canada: Sono Nis Press.

Muller, R.-D. 2003. Albert Speer and armaments policy in total war. In *Germany and the Second World War. Vol. V. Organization and mobilization of the German sphere of power: Part 2 Wartime administration, economy and manpower resources, 1942–1944/5*, ed. B. R. Kroener, R.-D. Muller, and H. Umbreit, 293–832. Oxford, UK: Clarendon.

Nash, G. D. 1990. *World War II and the West: Reshaping the economy*. Lincoln: University of Nebraska Press.

Nye, D. E. 1997. Energy narratives. In *Narratives and spaces: Technology and the construction of American culture*, 75–92. Exeter, UK: University of Exeter Press.

Power generation and utilization in Canada. 1941. *Canada yearbook 1940*. Ottawa, ON, Canada: King's Printer.

Reichwein, P. 1995. "Hands off our national parks": The Alpine Club of Canada and hydro-development controversies in the Canadian Rockies, 1922–1930. *Journal of the CHA/ Revue de la SHC* 6:129–55.

Russell, E. 2001. *War and nature: Fighting humans and insects with chemicals from World War I to Silent Spring*. New York: Cambridge University Press.

Scott, M. M. 1939–1945. This is our strength—Electric power. Canadian War Museum, Artifact number 19920196-009.

Trépanier, E. 2000. *Marian Dale Scott: Pioneer of modern art*. Montréal: Musée du Québec.

Tsutsui, W. 2003. Landscapes in the dark valley: Toward an environmental history of wartime Japan. *Environmental History* 8 (2): 294–311.

Tucker, R. 2004. The World Wars and the globalization of timber cutting. In *Natural enemy, natural ally: Toward an environmental history of war*, ed. R. P. Tucker and E. Russell, 110–41. Corvallis: Oregon State University Press.

Tucker, R., and E. Russell, eds. 2004. *Natural enemy, natural ally: Toward an environmental history of war*. Corvallis: Oregon State University Press.

U.S. Geological Survey. 1946. *Developed and potential water power of the world*. Unpublished manuscript, available from U.S. Geological Survey Library.

Way, W. R. 1942. Power-system interconnection in Québec. *Transactions of the American Institute of Electrical Engineers* 61:841–47.

Practicing Radical Geopolitics: Logics of Power and the Iranian Nuclear "Crisis"

Julien Mercille and Alun Jones

The theory of "radical geopolitics" is directly concerned with identifying the roots of U.S. foreign policy from a critical political economic perspective, seeking to determine the relative importance of political factors and economic forces in shaping foreign policy. It builds on Harvey's (2003) conceptualization of two logics of power and deploys a "geopolitical logic" and a "geoeconomic logic" to interpret political events. The former logic arises out of capitalism's tendency to expand geographically and the latter out of officials of statecraft's need to maintain their state's credibility internationally. Post-World War II U.S. foreign policy has largely followed the geoeconomic logic but has also been oriented (in sometimes divergent directions) by the geopolitical logic. A discussion of the Iranian nuclear "crisis" illustrates the radical geopolitics approach using national security documents outlining U.S. policy toward the Middle East and Iran, along with detailed interviews with senior diplomats from Iran, the United States, China, Russia, Germany, the United Kingdom, France, and the European Union. The Iranian crisis is a product of (1) American interest in the control of Iranian energy resources (a geoeconomic logic); and (2) American officials' need to reaffirm U.S. credibility in the face of Iranian defiance of U.S. hegemony in the Middle East (a geopolitical logic). *Key Words: critical geopolitics, Iran, oil, radical geopolitics, U.S. foreign policy.*

在"激进地缘政治学"理论中，直接涉及到从一个批判的政治经济角度确定美国外交政策的根源，寻求在制定外交政策的过程中政治和经济因素的相对重要性。它是建立在哈维（2003）对两个权力逻辑的概念化基础上，通过"地缘政治逻辑"和"地缘经济逻辑"来解释政治事件。前者的逻辑起因于资本主义地理上扩张的趋势，后者的逻辑起因于治国方略的需要，以维持其国际公信力。二战后的美国外交政策基本上遵循了地缘经济的逻辑，但也被地缘政治逻辑所导向（有时是在不同的方向）。本文对伊朗"核危机"的讨论展示了"激进地缘政治学"的方法，文中使用了国家安全文件中美国外交政策对中东和伊朗的阐述，以及对来自伊朗，美国，中国，俄罗斯，德国，英国，法国和欧盟等多位高级外交官的详细采访。伊朗危机是下述因素的产物：（1）美国对控制伊朗能源资源的利益（地缘经济逻辑）和（2）对伊朗无视美国的霸权，美国必须重申美国在中东的信誉（地缘政治逻辑）。*关键字：批判性地缘政治学，伊朗，石油，激进地缘政治学，美国的外交政策。*

La teoría de "geopolítica radical" se relaciona directamente con la identificación de las raíces de la política exterior de los EE.UU., desde una perspectiva político-económica crítica, buscando establecer la importancia relativa que tienen factores políticos y fuerzas económicas en el diseño de la política externa. Esta teoría elabora a partir de la conceptualización de Harvey (2003) sobre las dos lógicas del poder, y despliega una "lógica geopolítica" y una "lógica geoeconómica" para interpretar los eventos políticos. La primera de tales lógicas surge de la tendencia del capitalismo a expandirse geográficamente, y la segunda de la necesidad que tienen funcionarios oficiales por conservar internacionalmente la credibilidad de su estado. La política exterior de los EE.UU. después de la II Guerra Mundial en general ha seguido la lógica geoeconómica, aunque también ha sido orientada (algunas veces en direcciones divergentes) por la lógica geopolítica. Una discusión de la "crisis" nuclear iraní ilustra el enfoque geopolítico radical, con el suo de documentos de seguridad nacional que bosquejan la política americana hacia el Medio Oriente e Irán, más entrevistas detalladas con diplomáticos de alto nivel de Irán, los Estados Unidos, China, Rusia, Alemania, el Reino Unido, Francia y la Unión Europea. La crisis iraní es el producto de (1) el interés americano por controlar los recursos energéticos iraníes (una lógica geoeconómica); y (2) la necesidad de funcionarios del gobierno americano de reafirmar la credibilidad de los EE.UU. ante el reto iraní a la hegemonía americana en el Medio Oriente (una lógica geopolítica) *Palabras clave: geopolítica crítica, Irán, petróleo, geopolítica radical, política exterior de EE.UU.*

Radical geopolitics (Mercille 2008, 2009) provides the theoretical framework for this article, supplementing critical geopolitics by offering critical, political economic analyses to investigate the "why" (the causes) of policy and political events, without neglecting the "how" (the way they unfold). It examines the relative importance of the geopolitical and geoeconomic factors, or "logics," that drive foreign policy and addresses questions that have been neglected by current approaches in geopolitics, a number of which fail to incorporate political economy or to identify the fundamental reasons behind foreign policy. Inspired by revisionist cold war historiography, it modifies and reformulates David Harvey's logics of power into a "geoeconomic logic" and a "geopolitical logic" through which postwar U.S. foreign policy may be interpreted. The former logic arises out of capitalism's tendency to expand geographically and the latter out of officials of statecraft's need to maintain credibility internationally. A discussion of the current Iranian nuclear "crisis" illustrates the approach, clarifies empirically the nature and respective roles of the two logics of power, and highlights its usefulness in understanding contemporary and historical geopolitical events. The analysis presented here is one part of a much larger empirical and theoretical project and for lack of space must be restricted to a few key issues and ignore a number of other relevant ones.

Radical geopolitics recognizes that critical geopolitics is a diverse school of thought that is difficult to characterize in general terms but identifies two of its blind spots. First, critical geopolitics has tended to ignore the causes of policy and political events, instead focusing on the ways in which they unfold and are represented. Second, its engagement with political economy could arguably be deepened. Some studies emphasize discourses and representations (Dalby 1990; Toal and Agnew 1992; Dodds 2005; Jones 2006); others integrate political economy, but are not radical, as they emphasize the institutional affiliation of elite groups but neglect the workings of the political economic systems that shape policymaking, insufficiently emphasize the geoeconomic factors underpinning policy (Campbell 1992; Agnew and Corbridge 1995; Toal 2003; Toal, Dalby, and Routledge 2006), or both. Similar observations can be made about feminist geopolitics (Hyndman 2005) and the geopolitics of Lacoste (2006), which, important contributions notwithstanding, have not focused on the political economic drivers of foreign policy. The work of Flint and Taylor (2007) has reconfigured political geography in terms of world systems theory. Rooted in radical political economy, it shares many of this article's goals and methodology; however, it tends to be nomothetic, fitting complex political events into rigid models although its focus on large spatial and temporal scales makes it difficult to connect them to the specificities of particular geopolitical events. We prefer contextual interpretations more attentive to the agency of political actors and actual policy analysis.

Recent critical geopolitics scholarship on U.S. intervention in the Middle East reflects these contentions (Mercille 2009). For example, attention has focused on the imaginative geographies associated with the "War on Terror" (Gregory 2004; Sparke 2005; Dalby 2007); the conceptualization of Bush foreign policy in terms of empire or hegemony (Agnew 2005) and the gendered nature of its discourses (Hyndman 2005); 11 September 2001 as "somatic marker" (Toal 2003); and Arab media (Falah, Flint, and Mamadouh 2006); but political economy and Middle Eastern oil supply and procurement have been neglected. One unfortunate result is that Sidaway's (2003) call to geographers to investigate how U.S. involvement in the Middle East relates to the key commodity of oil has not received the attention it deserves.

Geoeconomic and Geopolitical Logics

Radical geopolitics draws principally on Harvey's (1982, 2003, 2006) work and specifies a geoeconomic logic and a geopolitical logic that refer, respectively, to the geoeconomic and geopolitical drivers behind U.S. foreign policy. The geoeconomic logic and its impact on state policy may be understood through the concept of the "spatial fix," which refers to the physical fixation of capital in places or to the spatial expansion of capitalist activities. The latter is closely associated with our geoeconomic logic, as the "outer" fix resolves (although only temporarily) the tendential overaccumulation of capital and labor power that threatens the devaluation of capital. We use the term *geoeconomic logic*—in lieu of reference to Harvey's capitalist logic—because we focus on the broad political economic aspects of capitalist expansion (Agnew and Corbridge 1989), and particularly the role of oil, rather than on detailed economic analytics or the geoeconomic discourses related to the economic imperatives of the global economy (Sparke 2005, 2007).

The geopolitical logic captures the U.S. need to maintain international credibility, a symbolic process whereby U.S. officials of statecraft signal to others that challenges to U.S. hegemony will be resisted. Failing to respond decisively even to isolated instances of defiance could embolden challenges elsewhere. The diplomatic record (declassified and public) of postwar U.S. foreign policy contains many references to such concerns on the part of policymakers, articulated in terms of "falling dominos," "apples in a barrel infected by one rotten one," or a growing "cancer"—all referring to the potential for spreading "instability" if place-specific challenges are not checked effectively (McMahon 1991).

The relative importance of and relationship between geopolitical and geoeconomic logics derive from our conception of the capitalist state (for details see Mercille 2008), a subject extensively debated by (neo) Marxists and (neo) Weberians, among others, the former emphasizing economic factors and the instrumental nature of the state and the latter countering that power is distributed more or less equally among various types (ideological, political, and economic; e.g., Poulantzas 1973; Skocpol 1977; Mann 1984; Toal 2006). Aware of those various debates, we follow the argument developed by Ashman and Callinicos (2006), Harman (1991), Block (1987), and Miliband (1983), among others, that the geoeconomic logic has predominated in orienting the direction of U.S. foreign policy since World War II, but also that the capitalist state enjoys a significant degree of relative autonomy, resulting in the geopolitical logic playing a significant role as well. The two logics can orient policy in the same direction or along divergent and even contradictory paths (this important issue falls outside the scope of this article and has been treated elsewhere; see Mercille 2008).

In short, our reference to geoeconomic logic is mostly to the need felt by business and state officials to control Middle East oil as a key aspect of the U.S. regulation of and hegemony over the world economy. Our reference to geopolitical logic is to officials' need to respond forcefully to the Iranian challenge to prevent it from spreading elsewhere. We contend that the geoeconomic logic is the main driver of U.S. foreign policy because if the United States did not seek hegemony over the world economy in the first instance, the need to maintain credibility would not arise, at least not to the same extent (this follows revisionist Cold War historiography, but a full discussion is outside the scope of this article).

Methodologically, we rely on declassified U.S. national security planning documents outlining policy toward the Middle East and Iran since 1945 and a series of highly detailed interviews conducted by the authors with diplomats at the cutting edge of this nuclear diplomacy. All eleven diplomats interviewed represented the governments involved in the crisis and were often themselves its key movers and shapers: Some were in charge of drafting sanctions; others were directors of their countries' foreign ministries' nuclear nonproliferation departments, personal representatives of the highest officials involved in the negotiations, or officers in the Iran or nuclear desks of their countries' foreign ministries. Semistructured and open-ended interviews were conducted in London, Paris, Brussels, Berlin, and Dublin from May 2008 to August 2008, lasted one to two hours, and were fully transcribed. Due to the very sensitive nature of the subject, the names and positions of the interviewees must remain confidential; most consented to be quoted in full on a nonattribution basis. The Israeli diplomatic corps was unwilling to agree to an interview. Interviews probed three question sets: officials' interpretation of the crisis, their governments' positions and policies in the crisis, and their perception of the principal states' roles and policies in it.

Interpreting the Iranian Nuclear Crisis Through Radical Geopolitics

In this section, we set out, albeit in a preliminary manner, the contours of a radical geopolitical interpretation of the Iranian nuclear crisis, an event largely instigated and maintained by the U.S. government and mainstream media. Nevertheless, the crisis constitutes a serious threat to world peace—a U.S. or Israeli military strike on Iran could inflame the Middle East. Yet, surprisingly, its prospect has been little discussed by political geographers.

U.S.–Iran relations have undergone important shifts since the 1950s. In 1953, a U.S.-backed coup removed Iran's nationalist leader Mohammad Mosaddeq and returned the Shah to power, who ruled the country until 1979 in close alliance with the United States (Keddie 2006). The 1979 Islamic revolution deposed the Shah and installed a theocratic regime that has ruled to this day and is openly antagonistic and challenging to the United States. For example, from 1979 to 1981 Iranian militants stormed the U.S. embassy and held fifty-two Americans captive for 444 days. The United States responded by attempting to isolate Tehran from 1979 to this day, for example, through sanctions that have been progressively tightened over time. The current round of sanctions and its nuclear crisis pretext should be seen as the latest chapter in American attempts to discipline Iran.

Some academic and journalistic accounts argue that Iran is not at fault in this nuclear diplomatic crisis (Beeman 2005; Ritter 2007) and others hold the opposite viewpoint (Clawson and Rubin 2005; Pollack 2005); however, to date, no credible evidence has been discovered supporting the conclusion that Iran has a military nuclear program, as underlined in a number of International Atomic Energy Agency (IAEA) reports (e.g., IAEA 2008, 5)—although it is possible that Tehran may be seeking to develop nuclear weapons in reaction to Israel's nuclear arsenal.

One of the most important issues illuminating the dynamics of the crisis is the failure of the United States and the European Union (EU) to engage Iran constructively. For example, in 2005, the EU reneged on the 2004 Paris Agreement, a bargain previously made with Iran in which the EU had pledged to provide Tehran with "firm guarantees on nuclear, technological and economic cooperation and firm commitments on security issues" (IAEA 2004, 4) on the condition that Iran suspend enrichment and provide assurances that its program was civilian, which it did, as verified by the IAEA. In its 2005 package, however, the EU failed to provide concrete security guarantees (e.g., it failed to persuade the United States to take the military option "off the table") and technological cooperation offers were limited at best (Varadarajan 2005). Of course, the official story related in the media and by some diplomats in interviews blames Iran for its "huge mistake" in rejecting "so bluntly" Europe's "quite generous" offer in 2005 and hence allegedly breaking the 2004 deal (German and French diplomats). A senior EU official, however, conceded that "security guarantees are weak in the offers, even now," and an American official, when asked whether a good step toward providing Iran with security guarantees could be to eliminate Israel's nuclear weapons, refused to comment. Moreover, the Nuclear Non-Proliferation Treaty, which attempts to limit the spread of nuclear weapons globally, obligates signatories who possess nuclear weapons to eliminate them, which could have been a helpful step in providing security guarantees. When asked for their views on this matter, however, Western diplomats brushed aside their obligations: "Yeah but there is no deadline, I'm sorry" (senior EU diplomat).

Russia and China have acted to inhibit the West's drive toward sanctioning Iran. A senior Chinese diplomat admitted that "we never believed that sanctions . . . could be effective in bringing Iran on board." Further, although the West refuses to allow Iran to conduct enrichment on its territory, a Russian official endorsed doing so, saying this approach "could be one of the solutions if at the same time there could be strict mechanisms of control." China and Russia are also interested in Iran's resources, as demonstrated by recent negotiations over energy contracts but do not push for sanctions as forcefully as does the United States, as similar credibility issues do not arise in their relations with Iran. Nevertheless, they support Iranian sanctions to preserve their important economic and political relationships with the West.

Geoeconomic and Geopolitical Logics Driving the Crisis

We now discuss the ways in which Iran's energy resources motivate U.S. interest in seeing a friendly regime emerge in Tehran (geoeconomic logic). We will also see that state officials perceive the need to enact sanctions to maintain credibility in the face of Iranian defiance (geopolitical logic).

The Middle East accounts for approximately two thirds of world energy reserves, and Iran holds the world's second largest reserves of gas and third largest of oil. Only a few geographers have discussed the role of oil in U.S. intervention in the Middle East (Harvey 2003; Smith 2003, 2005; Jhaveri 2004; Le Billon and El Khatib 2004). This account clarifies and adds some important points. The fundamental motivation for U.S. intervention in the Middle East is to *control* the amount of oil released on the world market, its geographical allocation, and the destination of petrodollars (where they will be invested; Bromley 2006; Stokes 2007). Access to oil for American consumption and potential profits for U.S. oil companies in the Middle East are two additional motivations but relatively incidental compared to the control of oil—even if it is often to those two motivations that critics of the Iraq invasion refer through the slogan "No Blood for Oil" (Rutledge 2005).

Declassified U.S. government planning documents support these claims. During the cold war, one main objective was to control Middle Eastern oil to allocate it to the "free world" (Japan and Western Europe in particular) and deny it to the Soviet bloc, in addition to recycling the petrodollars from the Middle East to the West, either through investments in New York and London banks or through the purchase of Western military weapons. For example, in 1951, in the face of rising Iranian nationalism under Mosaddeq, the U.S. National Security Council stated that "The loss of Iran by default or by Soviet intervention would . . . deny the free world access to Iranian oil. . . . These developments would seriously affect Western economic and military interests in peace or war in view of the great dependence on Western Europe on Iranian oil." Truman's Secretary of Defense James Forrestal noted that "whoever sits on the valve of Middle East oil may control the destiny of Europe" and George Kennan argued that controlling Japan's oil imports would mean a "veto power" over its industrial development; however, the need to access and use Middle East oil for American consumption and to generate profits for American companies receive comparatively little attention in the declassified record. Indeed, in the early postwar period, the United States used virtually no oil from the Middle East for its own domestic consumption—in 1948 it accounted for less than 1 percent of total U.S. consumption and in 1970 for only roughly 4 percent—but still intervened in the Middle East to direct its oil toward Western Europe and Japan to fuel their postwar recoveries along capitalist lines (NSC 1951; Forrestal quoted in Shaffer 1983, 143; Kennan quoted in State Department 1949; Painter 2002).

Planning documents outlining the thinking of the recent Bush administration have not been declassified yet, but the parallels are readily apparent (Brzezinski 1997; Klare 2006). Middle East oil now needs to be preserved for the Western world and denied to a Eurasian,[1] as opposed to Soviet, bloc. Indeed, Eurasian powers seek to develop an Asian energy grid to manage the continent's energy resources, into which Iran could be integrated. This would sustain Eurasian military and political cooperation through the Shanghai Cooperation Organization founded in 2001 between Russia, China, and other central Asian countries—and perhaps soon Iran—an ominous sign for the West. Also, although Americans are relying increasingly on Middle Eastern oil and this dependence is projected to rise in the future—which adds to interest in Iranian oil resources—the Middle East still only accounts for about 13 percent of total American consumption (National Energy Policy 2001). In short, the United States could live without Iranian oil and accessing it for American consumption must therefore be seen as an incidental motivation behind the crisis. Arguments denying that oil motivates American intervention in the Middle East such as the 2003 invasion of Iraq because oil could simply be bought on the open market without invasion (Agnew 2003) are misplaced: Iraqi oil could indeed have been obtained on the market, but this would not have translated into control over Iraqi oil, the United States' primary goal. This control can, however, potentially be achieved through an invasion leading to regime change, if a U.S.-friendly government emerges in Baghdad.

Therefore, we contend that U.S. (and earlier British) desire to control Iran's vast energy reserves is a fundamental factor that pushes U.S. officials to seek to maintain a friendly regime in power in Tehran, a geoeconomic logic. This motivates policymakers to react forcefully when Iran challenges U.S. hegemony, lest the "instability" spread elsewhere in the strategic region that is the Middle East, a geopolitical logic. This geopolitical logic is explicit in declassified documents; for example, those produced when Tehran moved toward nationalizing its then British-controlled oil resources in the early 1950s. The British Ministry of Fuel warned the U.S. State Department that "the security of the free world is dependent on large quantities of oil from Middle Eastern sources. If the attitude in Iran spreads to Saudi Arabia or Iraq, the whole structure may break down along with our ability to defend ourselves" (quoted in Abrahamian 2001, 188). Britain and the United States then backed a coup that removed Mosaddeq and brought back to power the Shah, who proceeded with an agreement leaving effective control of oil in the hands of Western oil companies. The *New York Times* (1954) editorialized, "Underdeveloped countries with rich resources now have an object lesson in the heavy cost that must be paid by one of their number which goes berserk with fanatical nationalism" ("The Iranian Accord" 1954, 16).

Similarly, former National Security Council director for Gulf Affairs Ken Pollack noted that the Islamic revolution and 1979 to 1981 Iranian hostage crisis "made the United States look weak in the eyes of the world, and weakness invites challenge" (Pollack 2005, 176). As a result, the United States imposed sanctions on Iran and tightened them over time. Today, the need to preserve U.S. credibility in dealings with Iran is sometimes voiced through the United Nations Security Council, which enacts sanctions in the nuclear crisis. For example, former Undersecretary of State Nicholas Burns recently declared that new sanctions were needed to keep the mullahs in check and warned that "the credibility of the Security Council is at stake" (Gollust 2007). From the interviews we conducted with senior diplomats, the role of credibility in provoking the crisis is apparent. For example, a European official stated that "there was a sense, when the Iran nuclear issue really came to prominence in 2002–2003, that this would be a test for the credibility of European diplomacy." When asked, "What would be the consequences of not responding to Iran's defiant attitude?" an American official answered that "the Security Council would lose its ability to be the force of last resort in countries' eyes." An Iranian official seemed to be aware of such motives when he stated that "because a country [Iran] is independent and is brave enough to announce that the major source of the regional problem is America, they are not happy with us."

Conclusion

Iran's importance to the United States has been significant historically due to its location in the region holding most of the world's energy resources and its own large reserves of oil and gas. For this reason, it is crucial for U.S. officials to attempt to install a friendly, pro-West regime in Tehran, explaining the intensity of the Iranian crisis that in effect has been going on in various forms since the 1979 Iranian revolution, a challenge to U.S. hegemony. U.S. policymakers have tried to reassert U.S. credibility by isolating Iran internationally to force it to change its attitude. This policy's ultimate goal, historically and today, is to control Iran's large energy resources, following a geoeconomic logic.

By providing critical, political economic, and contextualized interpretations, the radical geopolitics sketched in this article furnishes a promising platform to interpret geopolitical events, such as the Vietnam War (Mercille 2008) and the 2003 Iraq War (Mercille 2009). Although we have focused on American policy, the argument could certainly be extended to other cases exploring variations in the balance and tensions between geopolitical and geoeconomic logics shaping the foreign policies of other capitalist states. We would expect our theory of the state to be similar for nonhegemonic states, although given their smaller economic power, the geoeconomic logic would be less powerful in pushing them to intervene abroad, and the need to maintain credibility would correspondingly exert a weaker influence in inducing policymakers to repress challenges to their policies.

Acknowledgments

The authors wish to thank Audrey Kobayashi, John Agnew, James Sidaway, and three anonymous reviewers for helpful and constructive comments.

Note

1. Eurasia refers to China, Russia, and Central Asian states.

References

Abrahamian, E. 2001. The 1953 coup in Iran. *Science & Society* 65 (2): 182–215.

Agnew, J. 2003. Commentary III: Learning from the War on Iraq. *The Arab World Geographer* 6 (1). http://users.fmg.uva.nl/vmamadouh/awg/ (last accessed 14 September 2008).

———. 2005. *Hegemony: The new shape of global power*. Philadelphia: Temple University Press.

Agnew, J., and S. Corbridge. 1989. The new geopolitics: The dynamics of geopolitical disorder. In *The world in "crisis": Geographical perspectives*. 2nd ed., ed. R. Johnston and P. Taylor, 266–88. Oxford, UK: Basil Blackwell.

———. 1995. *Mastering space*. London and New York: Routledge.

Ashman, S., and A. Callinicos. 2006. Capital accumulation and the state system: Assessing David Harvey's *The new imperialism*. *Historical Materialism* 14 (4): 107–31.

Beeman, W. 2005. *The "great satan" vs. the "mad Mullahs."* Westport, CT: Praeger.

Block, F. 1987. *Revising state theory*. Philadelphia: Temple University Press.

Bromley, I. 2006. Blood for oil? *New Political Economy* 11 (3): 419–34.

Brzezinski, Z. 1997. *The grand chessboard*. New York: Basic Books.

Campbell, D. 1992. *Writing security: United States foreign policy and the politics of identity*. Minneapolis: University of Minnesota Press.

Clawson, P., and M. Rubin. 2005. *Eternal Iran*. New York: Palgrave Macmillan.

———. 2007. Regions, strategies and empire in the global War on Terror. *Geopolitics* 12 (14): 586–606.

Dalby, S. 1990. *Creating the second cold war*. New York: Guilford.

Dodds, K. 2005. Screening geopolitics: James Bond and the early cold war films (1962–1967). *Geopolitics* 10 (2): 266–89.

Falah, G. W., C. Flint, and V. Mamadouh. 2006. Just war and extraterritoriality: The popular geopolitics of the United States' War on Iraq as reflected in newspapers of the Arab world. *Annals of the Association of American Geographers* 96 (1): 142–64.

Flint, C., and P. Taylor. 2007. *Political geography: World economy, nation-state and locality*. 5th ed. New York: Pearson/Prentice Hall.

Gollust, D. 2007. US says UN credibility at stake on Iran nuclear issue. *Voice of America* 24 September. http://www.voanews.com/english/archive/2007-09/2007-09-24-voa67.cfm (last accessed 16 May 2009).

Gregory, D. 2004. *The colonial present*. Malden, MA: Blackwell.

Harman, C. 1991. The state and capitalism today. *International Socialism* 2 (51): 3–54.

Harvey, D. 1982. *The limits to capital*. Oxford, UK: Blackwell.

———. 2003. *The new imperialism*. Oxford, UK: Oxford University Press.

———. 2006. *Spaces of global capitalism: Towards a theory of uneven geographical development*. London: Verso.

Hyndman, J. 2005. Feminist geopolitics and September 11. In *A companion to feminist geography*, ed. L. Nelson and J. Seager, 550–64. Malden, MA: Blackwell.

International Atomic Enegy Agency (IAEA). 2004. INFCIRC/637 ("Paris Agreement"). http://www.iaea.org/Publications/Documents/Infcircs/2004/infcirc637.pdf (last accessed 12 September 2008).

———. 2008. May 2008 report. Implementation of the NPT Safeguards Agreement and relevant provisions of security council resolutions 1737 (2006), 1747 (2007), and 1803 (2008) in the Islamic Republic of Iran. http://www.iaea.org/Publications/Documents/Board/2008/gov2008–15.pdf (last accessed 12 September 2008).

"The Iranian Accord." 1954. *The New York Times* 6 August: 16.

Jhaveri, N. 2004. Petroimperialism: US oil interests and the Iraq War. *Antipode* 36 (1): 2–11.

Jones, A. 2006. Narrative-based production of state spaces for international region building: Europeanization and the Mediteranean. *Annals of the Association of American Geographers* 96 (2): 415–31.

Keddie, N. 2006. *Modern Iran: Roots and results of revolution*. New Haven, CT: Yale University Press.

Klare, M. 2006. The tripolar chessboard: Putting Iran in Great Power context. http://www.tomdispatch.com (last accessed 10 August 2008).

Lacoste, Y. 2006. *Géopolitique: La longue histoire d'aujourd'hui*. Paris: Larousse.

Le Billon, P., and F. El Khatib. 2004. From free oil to "freedom oil": Terrorism, war and US geopolitics in the Persian Gulf. *Geopolitics* 9 (1): 109–37.

Mann, M. 1984. The autonomous power of the state: Its origins, mechanisms and results. *Archives Européennes de Sociology* 25: 185–213.

McMahon, R. 1991. Credibility and world power: Exploring the psychological dimension in postwar American diplomacy. *Diplomatic History* 15 (4): 455–71.

Mercille, J. 2008. The radical geopolitics of US foreign policy: Geopolitical and geoeconomic logics of power. *Political Geography* 27 (5): 570–86.

———. 2009. The radical geopolitics of US foreign policy: The 2003 Iraq War. *GeoJournal*. http://www.springerlink.com/content/j5103751j2872873/fulltext.pdf (last accessed 31 August 2009).

Miliband, R. 1983. State power and class interests. *New Left Review* 138:57–68.

National Energy Policy. 2001. Washington, DC: U.S. Government Printing Office.

National Security Council (NSC). 1951. NSC 107/2 (declassified): The position of the United States with respect to Iran, 27 June 1951.

Painter, D. 2002. Oil. In *Encyclopedia of American foreign policy*. 2nd ed., Vol. 3, ed. R. DeConde, D. Burns, and F. Logevall, 1–20. New York: Scribner.

Pollack, K. 2005. *The Persian puzzle*. New York: Random House.

Poulantzas, N. 1973. *Political power and social classes*. London: New Left Books.

Ritter, S. 2007. *Target Iran*. New York: Nation Books.

Rutledge, I. 2005. *Addicted to oil: America's relentless drive for energy security*. London: I. B. Tauris.

Shaffer, E. 1983. *The United States and the control of world oil*. London: Croom Helm.

Sidaway, J. 2003. Decentering political geographies. *The Arab World Geographer* 6 (1). http://users.fmg.uva.nl/vmamadouh/awg/ (last accessed 9 August 2008).

Skocpol, T. 1977. Wallerstein's world system: A theoretical and historical critique. *American Journal of Sociology* 82: 1075–90.

Smith, N. 2003. *American empire: Roosevelt's geographer and the prelude to globalization*. Berkeley: University of California Press.

———. 2005. *The endgame of globalization*. London and New York: Routledge.

Sparke, M. 2005. *In the space of theory*. Minneapolis: University of Minnesota Press.

———. 2007. Geopolitical fears, geoeconomic hopes, and the responsibilities of geography. *Annals of the Association of American Geographers* 97 (2): 337–48.

State Department. 1949. Transcript of round table discussion on American policy toward China, 6, 7, 8 October 1949 (unclassified).

Stokes, D. 2007. Blood for oil? Global capital, counterinsurgency and the dual logic of American energy security. *Review of International Studies* 33: 245–64.

Toal, G. 1996. *Critical geopolitics: The politics of writing global space*. London and New York: Routledge.

———. 2003. "Just out looking for a fight": American affect and the invasion of Iraq. *Antipode* 35 (5): 856–70.

———. 2006. General introduction; introduction to part 2; introduction to part 3. In *The geopolitics reader*, 2nd ed., ed. G. Toal, S. Dalby, and P. Routledge, 1–14, 59–74, 119–34. London and New York: Routledge.

Toal, G., and J. Agnew. 1992. Geopolitics and discourse: Practical geopolitical reasoning in American foreign policy. *Political Geography* 11 (2): 190–204.

Toal, G., S. Dalby, and P. Routledge, eds. 2006. *The geopolitics reader*. 2nd ed. London and New York: Routledge.

Varadarajan, S. 2005. The Persian puzzle. *The Hindu* 21 September 2005. http://www.hindu.com/2005/09/21/stories/2005092105231000.htm (last accessed 27 August 2009).

"A Microscopic Insurgent": Militarization, Health, and Critical Geographies of Violence

Jenna M. Loyd

Wars do not maim with bullets and bombs alone but cause economic and environmental destruction that leave enduring bodily harms. Preparations for war-making also cause negative health effects, from toxic waste to the redirection of social wealth from investment in social needs. The commonsense juxtaposition of exceptional war to normal peace makes it difficult to recognize processes of militarization, the violent continuities between war and peace, and geographic ties binding spaces of relative health with spaces of harms. This article advances a critical geographic analysis of violence to analyze the ways in which militarization and structural violence reinforce one another. A 2007 cholera epidemic in Iraq was militarized through material and discursive geographies of cholera and violence. Humanitarian claims to cure cholera rested on this dualistic geopolitical imagination, distorting the agents of violence and erasing the grave effects of peacetime and wartime structural violence. By situating cholera within a broader historical and geographic context that shows links between "wartime" and "peacetime" places also suffering premature deaths from the destruction or abandonment of necessary infrastructures, a critical human geography can contribute to struggles for peace and justice. *Key Words: critical geographies of violence, health, militarization, peace, structural violence.*

战争带来的不仅仅是子弹和炸弹造成的身体伤害，它所造成的经济和环境破坏是持久而具体的。对战争的准备同样会造成不利的健康后果，例如有毒废物，社会财富的重定向，对社会需要的投资变化等等。常识将特殊的战争与正常的和平并列对待，这使得识别下述三种情况变得困难：军事化进程，战争与和平之间的暴力连续性，连接相对健康的地区和有危害的地区的地理联系。本文对暴力进行了进一步的批判地理学分析，研究了军事化和结构暴力是如何相辅相成的。发生在 2007 年伊拉克的霍乱疫情，在实质性和推论性的霍乱和暴力区域被军事化了。人道主义者声称要治愈停留在这个二元的地缘政治空想上的霍乱，并歪曲暴力机构，和消除和平时期与战时结构性暴力的严重影响。将霍乱置于更广泛的历史和地理背景下，通过其所展示的同样受困于基础设施被破坏和遗弃所造成的夭亡的"战争时期"与"和平时期" 的联系，批判的人文地理学则有助于争取和平与公义的斗争。*关键词：关于暴力的批判地理学，卫生，军事化，和平，结构暴力。*

Las guerras no mutilan solo con balas y bombas sino que causan destrucción económica y ambiental que deja heridas perdurables. Las preparaciones guerreristas también causan efectos negativos a la salud, desde la proliferación de desperdicios tóxicos hasta la exclusión de la salubridad social del rubro de inversión en necesidades sociales. La obvia yuxtaposición de la guerra excepcional a la paz normal dificulta el reconocimiento de los procesos de militarización, las continuidades violentas entre guerra y paz, y los lazos geográficos que ligan espacios relativamente saludables con espacios dañados. Este artículo se adentra en un análisis geográfico crítico de la violencia para analizar la manera como la militarización y la violencia estructural se refuerzan entre sí. En 2007 una epidemia de cólera en Irak se militarizó por medio de geografías materiales y discursivas del cólera y la violencia. Las demandas humanitarias para que se curara el cólera descansaban en esta imaginación geopolítica dualista, distorsionando los agentes de violencia y borrando los graves efectos de la violencia estructural en tiempos de paz y de guerra. Al situar el cólera dentro de un contexto histórico y geográfico más amplio que muestra lazos entre lugares de "tiempos de guerra" y de "tiempos de paz," que también sufren de muertes prematuras por la destrucción o abandono de infraestructuras necesarias, una geografía humana crítica puede contribuir a los esfuerzos en pro de la paz y la justicia. *Palabras clave: geografías críticas de la violencia, salud, militarización, paz, violencia estructural.*

> Cholera is a grave threat for the American project in Iraq, but also an opportunity to capture the hearts and minds of the population. The average Iraqi will feel truly secure only when the vicious disease-poverty-insurgent feedback loop is snapped. As we plan the post-surge phase of American operations, our leaders must bear in mind that healthy people make healthy decisions that serve as the bedrock for healthy societies.
>
> —Drapeau (2007)

In late 2007 the United Nations (UN) warned that Baghdad faced a cholera epidemic. Cholera is a severe diarrheal disease with symptoms that include massive fluid loss from diarrhea and vomiting, leading to painful body cramps. Left untreated, cholera can kill within hours to days of onset. Spread by the consumption of contaminated water, cholera signals a grave absence or breakdown of infrastructures for clean water and sanitation, electricity included. At a moment in which debate raged over the effectiveness of the U.S. military escalation in Iraq, cholera signaled the marked failure of U.S. reconstruction.

In a *New York Times* opinion piece, however, Mark Drapeau, a fellow at the United States' National Defense University,[1] interpreted the humanitarian crisis as an opportunity for success of the "American project." Because "Iraqi areas with the filthiest water and most raw sewage are breeding grounds for both *V. cholerae* and insurgents," remedying cholera would improve security conditions and thereby strengthen the newly democratic Iraqi nation. To dramatize the stakes and asymmetric threat that this "microscopic insurgent" posed to the world's largest military power, he warned: "[S]uicide bombers don't call in sick" (Drapeau 2007).

Savvy readers no doubt understand that Drapeau also was trying to shape the hearts and minds of the American public. Although his opinion hardly represents official U.S. health or military policy, its rhetorical power leveraged in and by the *New York Times* draws on Drapeau's position as a state expert. Close analysis of this piece alongside official U.S. and UN documents, nongovernmental reports, and scholarly and journalistic sources offers one illustration of how geographies of war are not just about battlegrounds and home fronts but are mediated through competing efforts to shape representations and hegemonic geographic imaginations of war (Woodward 2005; Bialasiewicz et al. 2007; Graham 2008).

Drapeau's invocation of the commonsense imaginary of a world divided between war-torn, diseased, "overcrowded communities of poor countries" and spaces of health, wealth, democracy, and peace makes it difficult to see how health no less than democracy is being militarized. Unlike controversy over the "politicized" science of the war's civilian casualties (Hyndman 2007), the authoritative objectivity of biomedicine works to naturalize this dualistic geopolitical imagination and obscure the role of U.S. military occupation. Treating a microbe as an agent of violence turns "curing" violence into an apparently apolitical, technocratic exercise that erases historical and contemporary violence in Iraq under the mantle of health and democracy promotion. Most important, this relatively unremarkable foray into the debate over the war in Iraq bears close scrutiny for registering a geopolitical moment in which humanitarian claims are used contentiously to legitimate military interventions and the geopolitics of disease increasingly is securitized and militarized (Ingram 2005; Mamdani 2008).

Precisely because modern war so thoroughly blurs times and spaces of peace and war, this dualism dangerously facilitates war-making. Indeed, feminist and peace scholars have long argued that such dichotomization is a militarized ideology that produces a series of dangerous occlusions. First, treating war as exceptional in relation to the liberal presumption of peace ignores the "normalcy of war" (Cowen and Gilbert 2007, 6) within the "peacetime" political-economic order. Second, the false localization and temporization of "war" obscures the broader geographies and histories of militarization: a "contradictory and tense social process in which civil society organizes itself for the production of violence" (Geyer 1989, 79, cited in Lutz 2002, 723). Militarization orients political economies and institutions to military projects and discursively shapes social relations, values, and geographic understandings in ways that prioritize military goals, promote martial values, and legitimate the use of force to solve social problems. Third, this dichotomy obscures everyday state and structural violence, including the extent to which militarization rests on, naturalizes, and sustains racial, gender, and sexual hierarchies (Gilmore 1998; A. Smith 2005; Alexander 2006; Incite! 2006; Puar 2007; Loyd 2009).

Because the war–peace dualism hides "the indistinguishability and interdependence of [direct] physical and structural violence" (Lutz 2002, 725), scholars have worked to conceptualize wars as part of a continuity of violence (C. Cockburn 2004; Scheper-Hughes and Bourgois 2004). The concepts of structural violence and militarization enable useful analytic distinctions that can inform a critical geography of violence. In this article I ask a set of interrelated questions: How do structural violence and militarization sustain each other and through what material-discursive geographies?

Rather than framing health in opposition to violence, I begin by theorizing how health is entwined with violence. To illustrate, I situate Iraq's cholera epidemic and Drapeau's representation of it within a broader historical and geographic context of structural violence and militarization. I conclude with a discussion of how a critical geography of violence can contribute to efforts for peace and social justice.

Health, Premature Death, and Continuities of Violence

Although "social forces ranging from poverty to racism become *embodied* as individual experience" (Farmer 2003, 30), it is not at the scale of the body that one can understand the social production of bodily harms (Hayes 1999; Nguyen and Peschard 2003; Curtis 2004). Premature deaths are the visceral marks of structural violence and militarization alike and offer a clear way of tracing a critical geography of violence. Representations and imaginative geographies of health and violence are crucial for understanding how violence is naturalized and legitimated.

The concept of structural violence disturbs hegemonic liberal theories of war as exceptional and legitimate violence as the sole province of the state. Such understandings facilitate state violence during war and peace and ratify everyday social injustice (Scheper-Hughes and Bourgois 2004; Skurski and Coronil 2006), making structural violence a prickly concept because it breaks from liberal conceptualizations of power that neatly demarcate power and violence. Rather, Gilmore (2002) explains, violence is structural precisely because, "the application of *violence*—the cause of premature deaths—*produces* political power in a vicious cycle" (16, italics added).

By the late 1960s, when peace scholar Johan Galtung and liberation theologians were theorizing structural violence, fascism, colonialism, and nuclearism had already undermined this liberal commonsense. Anticolonial and liberation movements' worldwide challenge to Western domination and violence of Enlightenment reason further spurred political theorists and actors to rethink the relations between power and violence and to establish new ways of distinguishing legitimate from illegitimate violence. To that end, we can situate structural violence against Arendt's contemporaneous recovery of the liberal distinction between power and violence and alongside Fanon and Foucault, for whom power and violence were deeply entwined and extend beyond state institutions. Fanon's work on colonial power relations emphasized the "violent peace" of the Cold War, whereas Foucault devoted his attention to the violence of apparently legitimate state institutions and to technologies of power threading through social institutions and practices of daily life (Hanssen 2000).

Galtung (1969) honed in on the embodied traces of structural inequities. For him, violence is not just about direct, intentional harm, incapacitation, or deprivation. Rather, societies are violent to the extent that "*human beings are being influenced so that their actual somatic and mental realizations are below their potential realizations*" (168). Power relations that create and sustain the uneven distribution of life necessities constitute a daily "violence [that] is built into the structure and shows up as unequal power and consequently unequal life chances" (171). Thus, structural violence results in avoidable, or premature, deaths. Peace, defined as the absence of violence and presence of social justice, would be marked by the flourishing of human capabilities.[2]

Medical anthropologist Paul Farmer's (2003) call for a "geographically broad" and "historically deep" (42) analysis of deadly inequalities is perhaps the best known elaboration of this theoretical perspective. He shows how the *longue durée* of slavery, French colonialism, and American imperialism, including their backing of military dictatorships, are responsible for Haiti's devastated urban and rural economies and ecologies, and hence its contemporary health crises. Farmer's ethnographic descriptions vividly demonstrate how the "*simultaneous* consideration of various social 'axes' [including "gender, ethnicity ('race'), and socioeconomic status"] is imperative in efforts to discern a political economy of brutality" (43).

Yet, a gap opens between Farmer's (2003) analysis and his political-medical practice when he settles on "the primacy of the economic" to decide whose "suffering needs to be taken care of first" (49, 50). Arguably, it is such a veer toward economism that has received the greatest criticism from colleagues who otherwise regard structural violence as a critical tool for exposing the hidden violence of everyday social inequities (Farmer 2004). Because structural violence captures the simultaneous destructiveness and productivity of categorical power relations, it allows us "to see how the violent destruction of landscape (and livelihood) in one place can redound very much to the benefit of landscapes (and people) in other places" (Mitchell 2003, 791). Indeed, to understand who is made most vulnerable where and how socially produced harms are naturalized discursively and materially, it is necessary to theorize specific economic, political, and social relations of oppression and domination and how they articulate (or intersect) in particular historical, geographic moments (Hall 1980; Hart 2006; Loyd 2009).

It is important to analyze how health becomes, to paraphrase Hall, one of the modalities through which structural violence is lived.[3] Take cholera, "the classic disease of social inequality" (Briggs 2004, 166). Historically, the Western colonial treatment of cholera localized blame by fusing racial difference and "a perceived lack of medical modernity" (Briggs 2004, 166). Yet deadly epidemics resulted not from the lack of modern economic development but precisely because of vulnerabilities produced through colonial wealth extraction, including capitalist agricultural projects and food regimes (Watts 1999; Davis 2001). Health and medical discourses that focus on narrowly proximate or cultural explanations for disease are complicit in obscuring the "pathogenic roles of [broader scale] social inequalities" (Farmer 1999, 5). Such reifying geopolitical imaginations have the effect of reinforcing individualistic biomedical and coercive public health interventions, thereby reproducing inequities rather than enabling changes that would create conditions for healthy living.

Harms of Militarization

War kills and maims not with bullets and bombs alone. Nor do wars singularly or even primarily harm combatants; however, the full scope of harms of militarization is obscured through hegemonic categorizations and institutions of war. The reification of direct violence fails to capture military strategies that target infrastructures for living, avoids the contradictions of militarization, and obscures the structural violence engendered by privileging of militarized priorities.

The sheer scope of the health and environmental effects of militarization makes it virtually impossible to separate the direct from enduring harms of war and war-making (Seager 1999). Broadly speaking, war disrupts or destroys productive economic practices and means of livelihood, which undermines food security and other basic needs. The devastation of infrastructures—water and water treatment; public health services; medical services, medicines, and medical facilities; transportation; and telecommunications—has immediate and long-term health effects and enduring economic effects. Mass displacement frequently disrupts social networks that enable mutual aid. Interpersonal violence becomes more common in the short and long term, and women become more vulnerable to sexual violence within and outside the home. State spending is often diverted from basic needs to the military during conflict and reconstruction. Finally, people are forced to live in less healthy ecologies, whether due to the breakdown of public health infrastructures (such as malaria eradication efforts), relocation to refugee camps, or exposure to military toxics and hazards, including cluster bombs, land mines, Agent Orange, and depleted uranium (Hewitt 1987; Levy and Sidel 1997; Kalipeni and Oppong 1998; Smallman-Raynor and Cliff 2004; Graham 2005).

These harms of war and war-making are not experienced evenly, which is to say that militarization exacerbates and reshapes existing social inequalities. Since World War II, wars have been concentrated in the global South and over this time civilian deaths increased from 67 percent to 90 percent of all war-related deaths (Sidel 1995). War has exacerbated poverty in these places, and poverty exacerbates the harms of war (Johnston, O'Loughlin, and Taylor 1987; Zwi 1991; Stewart, FitzGerald, and Associates 2001). Thus, "high rates of civilian mortality are determined more by the *pre-existing fragility* of the affected population than the intensity of the conflict" (Guha-Sapir and van Panhui 2003, 2127, italics added).

Likewise, preparations for war create uneven geographies of harm. In the abstract, nuclearism poses a universal threat to humanity, but Native peoples' lands specifically have been appropriated and used for nuclear weapons development, and only the indirect effects of testing have been shared, if still unevenly (Kuletz 2001; Loyd 2009). The use of social wealth to build and sustain the military rather than fulfilling universal needs like clean water, good food, housing, and education entrenches socially produced scarcities and social hierarchies (Lutz 2002). These needs, in turn, become militarized to the extent that they are more readily available in exchange for military service than from other state institutions (Enloe 2000). In this way, militarization sustains and necessitates structural violence at multiple scales.

A Militarized Geography of War and Health

At the outset I showed how Drapeau's (2007) "A Microscopic Insurgent" militarized cholera by attaching its cure to military priorities. Health can also be militarized to the extent that it is used to legitimate state violence. For example, Drapeau marshals the "actively responsible" citizen-subject produced through hegemonic health discourses (Briggs and Hallin 2007) to position his American audience as concerned humanitarian healers and promoters of healthy social bodies. In this section I analyze how Drapeau's militarized geographic imagination of war and health extends Western colonial imaginaries and works to elide histories of violent Western interventions and citizens' complicity for violence on their behalf.

Drapeau (2007) criticizes the inadequacy of the Iraqi government's personal hygiene campaign and rightly suggests that "the best solution is a clean-water program and better management of waste." But who is responsible for the terrible condition of Iraq's water, sewage, and electricity infrastructures? At one moment Drapeau attributes the destruction of Iraq's infrastructure narrowly to "war" and in the next displaces it to underdevelopment and to Saddam Hussein's rule. This leads to a series of historical inaccuracies and distortions of meaning. If "war" alone were to blame, the United States and its allies would be responsible for reconstruction under international humanitarian law. But he turns this legal obligation into an unsolicited charitable mission, which contravenes the definition of humanitarianism. "To be humanitarian, assistance must be in response to a crisis that was not caused by the provider of aid" (Terry 2002, 241). What is more, depicting the war as an "American project" to help this "developing nation" build a healthy society is a fantastic revision of the initial justification of the United States and United Kingdom for invasion: regime change and disarmament.

Although Hussein bears responsibility for egregious ethnic and sectarian oppression, the ease with which Drapeau transforms a military mission into a humanitarian one rests on an orientalist geopolitical imagination that enables him to obscure the U.S. Cold War backing of Hussein and recent history of U.S. military violence and to erase twelve years of Western-imposed sanctions. Like wars, economic sanctions impede access to basic goods and damage infrastructures such as transportation, water purification, and disposal, all of which work to harm health and well-being. Indeed, one would not know from reading his piece that before the 2003 invasion it was common knowledge—among health professionals, antiwar activists, and U.S. postwar planners alike—that this violence already had undermined Iraq's water and health infrastructures.

The cumulative effect of this violence was a reversal of Iraq's comprehensive health index to 1950s levels. UNICEF estimated that some 500,000 Iraqi children died during the 1990s alone (Arnove [2000] 2002; Daponte and Garfield 2000; Rawaf 2005).[4] Between 1990 and 2005, Iraq's child mortality rate worsened more than 6 percent each year—the world's most severe deterioration—partially due to high rates of (chronic) malnutrition (De Belder 2007; World Health Organization 2008b). In late 2007, the time of Drapeau's writing, some 70 percent of Iraqis did not have access to clean water; only 20 percent of homes outside of Baghdad had sewage connections; and Baghdad's sewage treatment plants were operating at only 17 percent of capacity (D. Smith 2007).

By the end of the 2007 rainy season, cholera cases had been identified in half of Iraq's provinces, but the major outbreaks were concentrated in Kurdistan (in the predominantly Kurdish city of Sulaimaniya and the ethnically mixed city of Kirkuk) and in Baghdad (particularly the Shia neighborhood of Sadr City). Yet, Drapeau's (2007) ahistorical and environmental determinist explanations for the spread of cholera fail to account for this geographic unevenness. He asserts that "population density" is responsible for cholera's spread, but this claim is patently insufficient; relatively robust water and health infrastructures make dense cities such as New York relatively invulnerable to such epidemics. What is more, only two cases of cholera were confirmed in Basra, Iraq's second largest city (although the city did experience an outbreak after the 2003 invasion). "Refugee camps" are also an unsatisfying explanation. According to the UN, more than 4.7 million Iraqis have been displaced: Some 2.7 million people remain within the country and more than 2 million have become international refugees. There were reports of people crowding into camps near Sulaimaniya, but this does not capture the overarching geography of displacement. The International Rescue Committee (2008) found that Iraqi refugees "are not huddled together in a camp" (i) but had mostly moved into existing neighborhoods within and outside the country.

The linchpin of Drapeau's (2007) argument—"Iraqi areas with the filthiest water and most raw sewage are breeding grounds for *V. cholerae* and insurgents"—is even more misleading. Considering the scope of long- and short-term devastation of Iraq's infrastructure, it is hard to say where Iraq's water problems were the worst. Nonetheless, the places where the epidemic was concentrated were not the "breeding grounds" of the "insurgency." Drapeau's equation of cholera with insurgency displays either a knowing dissimulation of the facts or ignorance of ethnic, class, and sectarian divisions in Iraq at the time. In Iraq war discourse, the "insurgency" referred to armed Sunni opposition to U.S. military occupation. Sunni groups also began directing attacks against the Shia majority population, which had gained historic political power in the U.S.-created Iraqi government. In early 2006, Shia militias began returning Sunni violence, resulting in a cycle of "Sunni-Shia sectarian slaughter" (P. Cockburn 2008) that fueled displacement and the segregation of previously ethnically mixed neighborhoods. In early 2007, the United States increased its military presence in a bid to stem the violence and create the possibility for political negotiations. But the United States was simultaneously fostering sectarian divisions and the military escalation only solidified newly homogeneous geographies, leading to still more killings (Glanz and Farrell 2007; Agnew, Gillespie, and Min 2008; P. Cockburn 2008; World Health Organization 2008b).

Drapeau's misrepresentations matter not simply because they fail to explain Iraq's geography of cholera but because they severely distort the violence in Iraq and the U.S. role there. Such misrepresentations are made possible and plausible by their consonance with dominant Western geopolitical imaginations. For example, Drapeau's (2007) juxtaposition of his sensible, comprehensive solution to "the general ineffectiveness of Baghdad's government" rests on and upholds a world divided between war ("failed" and "rogue" states) and peace (free market democracies), but the U.S. funding of Iraq's response disturbs his imperial narrative (Kaplow 2007).[5] Likewise, Drapeau draws on the Third World iconography of dense cities and the refugee camp but otherwise erases the existence of millions of displaced Iraqis. Indeed, his only discussion of cross-border movement invokes an orientalist, neoconservative discourse indicting Iran, Syria, and other "Islamist" agitators for Iraq's turmoil. This geographic imagination of war and health enables Drapeau to frame his "counterinsurgency" tactic as a "humanitarian" impulse from a helpful, but beleaguered, partner nation, rather than the violent redux of a similar U.S. medico-military operation in the Philippines at the turn of the last century (Cooter 2003).

By late 2007, however, the comprehensive reconstruction Drapeau recommended had been rejected by the U.S. military as a counterinsurgency tactic. Irony, however, fails to capture the extent to which Drapeau's appeal to U.S. benevolence and superior infrastructural know-how distorts what was happening on the ground. According to Stuart Bowen, the Special Inspector General for Iraq Reconstruction, postwar planners knew that sanctions had weakened Iraq's infrastructure, and they were split on how much rebuilding the United States should commit its (and Iraq's) money to. A "no nation-building" camp (favored by Donald Rumsfeld and the White House) advocated remediating only the damage from the 2003 invasion. The other (including the United States Agency for International Development) favored more extensive reconstruction, which they claimed would enable economic development and a "democratic transition" (Bowen 2008, 10, 29). The former won out, but even the sums allocated to rebuilding to pre-2003 levels were insufficient.

By the time Drapeau's piece was written, the Government Accountability Office had concluded that U.S. reconstruction efforts fell far short of goals (Glanz 2007). Outside official circles it was widely criticized for corruption, no-bid corporate contracts, lack of oversight, and failure to employ Iraqis. Reconstruction of Iraq's health care system was among the most controversial and failed programs (Rawaf 2005; De Belder 2007; Bowen 2008). This failure is surpassed by the U.S. military's counterinsurgency strategy of redirecting money from comprehensive rebuilding toward immediate, high-profile projects. Bowen (2008, 320) found that between June 2004 and June 2005 funds for electricity and water infrastructures were reduced by 21 percent and 49 percent, respectively. The effects of this shift for Iraqi well-being were not insignificant: "The water sector lagged most of all [reconstruction projects]. U.S.-funded projects were on track to bring access to potable water to 8.2 million Iraqis [in September 2006]. Although this met the goal set . . . *after* funds had mostly been reprogrammed to security, only one in three Iraqis had regular access to fresh water" (Bowen 2008, 506, italics added).

Iraq's 2007 cholera epidemic was fueled not by the "insurgency" but by the U.S. invasion and occupation and by sectarian violence; high rates of unemployment and U.S. free market policies imposed heavier harms among Iraqis unable to afford to buy bottled water or the soaring cost of cooking fuel (Kaplow 2007; Sparke 2007). These immediate causes must be situated within particular histories of direct and structural violence. Further research could confirm the extent to which Hussein's oppression of Kurdistan and Sadr City resulted in poor infrastructural investment and hence their greater vulnerability to cholera years later. On the flip side, by virtue of its importance as a port within an oil-rich region, Basra's relative imperviousness to cholera in 2007 might be due to the concerted international response to the city's 2003 epidemic (Valenciano et al. 2003). (Indeed, even as most Iraqis had no access to clean water, it was being was pumped into oil fields near Basra for oil extraction; Bowen 2008, 336.)

How Does the War–Health Dualism Sustain Militarization and Structural Violence?

The dualistic geopolitical imagination of war and violence rests on and sustains the reification of direct violence, but the violence of wars and everyday structural violence foster similar conditions for premature deaths, including coerced migration, disruption of livelihoods and social ties, and infrastructure destruction. Peacetime outbreaks of cholera in Venezuela in 1992 and 1993 claimed some 500 lives, and 250 people were killed in South Africa between 2000 and 2002. Each of these epidemics claimed more lives than cholera did in wartime Iraq, where one had a better chance of survival than in "peacetime" Niger or Senegal (World Health Organization 2008a).

Drapeau's criticism of band-aid solutions to cholera should be taken seriously the world over. Yet Iraq's epidemic echoed South Africa's experience just a few years prior when the implementation of free market policies exacerbated historic vulnerabilities (Mbali 2002). Further, Iraq did not depart from the World Health Organization and the World Bank's advice for immediate response. These organizations also recommend longer term infrastructural development, yet this basic remedy to predictable and preventable deaths is undercut significantly by the dominance of neoliberal regimes that prioritize privatization and marketization over universal provision of water, sewage infrastructure, and health care (Graham and Marvin 2001; Briggs 2009). Iraq no less than South Africa epitomizes "Western" water provision.

Drawing attention to the violent policies linking Iraq and South Africa is not intended to lessen the socially produced deaths in either place, but it is intended to question the ways in which dominant categorizations obscure relations and harms of violence. Health and medicine, of course, are not solely Western concerns (half of Iraq's doctors were displaced and thousands have been killed), but health itself became a means of violence when Drapeau (2007) allied it to Western values and peace while simultaneously erasing Iraq's well-developed health record. What relations of geopolitical and social power, including institutions of health, are sustained by maintaining hegemonic distinction between health and violence, peace and war? Drapeau's warning that "suicide bombers don't call in sick" rests on the apparent commonsense distinction between legitimate and illegitimate violence, which works to erase Iraqi suffering and negate the legitimacy of opposition to U.S. military occupation. This distinction is underscored by his invocation of the humane Western institution of "sick days," not shared by the unhealthy global South. Yet, this obscures the structural violence of the U.S. failure to guarantee sick days or provide health care for its citizens. The ideological equation between sickness and illegitimate violence thereby displaces the harms of structural and U.S. state violence.

Demilitarizing Geographies of War and Health

There have been significant geopolitical shifts since the Cold War moment in which Galtung theorized universal human flourishing as the mark of peace. At a time in which deepening structural violence is called "globalization," a critical geography of violence is imperative for distinguishing between interventions that entrench more than remedy the conditions for premature death. Theorizing health as entwined with violence offers a compelling entrée into contemporary debates over war, so-called humanitarian military interventions, and the geopolitics of disease alike. Drapeau's unofficial entry into the U.S. and U.K.'s revisionist justifications for invasion illustrates one convergence between discourses of health and sovereign military violence. This suggests that the geopolitics of disease and humanitarianism are important sites for contributing to theoretical debates over the threshold between the biopolitical production of life and necropolitics (Mbembe 2003; Rabinow and Rose 2006; Bhungalia 2009).

As we have seen, geographic imaginations that sharply separate health from violence and war from peace contribute to the naturalization and legitimation of violence. The reification of violence and disease within discretely bound places is an ideological process of localization that severs these places from broader scale processes that concentrate well-being and harms in different, but viscerally and violently related, places. In this process of political inversion, the people and places harmed by structural and wartime violence are alchemized into agents of their own violent deaths, into exceptional sources of violence, or both. Rendered as security risks, such threats are then used to justify violent interventionism. Despite the marked failure of neoliberal economic and health policies, the inertia of such policies and militarization of "biosecurity" threaten to reinforce a violent logic of intervention, including coercive policing powers of public health, rather than opening up an imagination of mutual care and responsibility (Briggs 2009).

A critical geography of violence situates all of these harms within broader histories and geographies of colonialism and contemporary structural violence. Refusing to draw sharp distinctions between categories of violence means that complicity, and hence responsibility, is more broadly shared, which suggests a role for "internationals"; however, "humanitarian intervention" has become one of the virtuous ways in which violence is naturalized and structures of power are maintained. The imagined peacefulness of the global North justifies and, in this logic, necessitates their humanitarian intervention, doubly erasing international culpability and relations of structural violence (Orford 2003; Mamdani 2007).

Struggles over health are also struggles over definitions of violence and ultimately over who has the power to organize and legitimate the use of violence. In contemporary ideological battles over war, the legal and ethical distinction between intentional, deliberate violence and unintentional, "collateral" damage is inevitably invoked. Yet, contemporary warfare—including the strategic targeting of "dual-use" infrastructures—makes such distinctions materially and analytically moot. This ideological distinction fosters the legitimacy of war (T. W. Smith 2002) but is possible only by erasing the *inevitable* enduring, indirect, and direct harms of war. We can see these harms in Iraq no less than Gaza (Bhungalia 2009).

While there is social injustice, health and violence cannot be understood as mutually exclusive. A too narrow focus on direct violence in antiwar organizing capitulates to state legitimation efforts and to militarized distinctions between soldiers and civilians, which situates civilians in the already prepolitical category of humanitarianism; however, situating antiwar arguments within a broader antiviolence frame shifts attention to justice and to enlivening discourses that imagine health as a mark of human flourishing. This shift in turn can enable relationships of mutuality and solidarity in place of imperial, humanitarian relationships of protection (Koopman 2008; Mamdani 2008).

A critical geography of violence engenders skepticism about whose material privileges are served and what hierarchies of economic, social, and cultural power are maintained through attention to apparently aberrant spaces of violence and disease to the exclusion of broader structures. How often are calls for international intervention in cholera-ravaged Zimbabwe accompanied by proposals to dismantle global military capacities—the vast majority of which are maintained by the world's wealthiest countries—and redistribute this social wealth? How might reinvesting this wealth make living in South Africa or Iraq or New Orleans safer and freer? Tracing connections among places that are being disorganized, abandoned, and destroyed during war and peace illustrates how building peace means situating demilitarization efforts within longer histories and shared geographies of racial violence, imperialism, and capitalist exploitation. In such places, no less than Iraq, (re)building safe and clean water infrastructures for all will not result from humanitarian largesse but be the mark of struggles for justice and peace.

Acknowledgments

This article would not be possible without conversations with Jennifer Casolo, Lisa Bhungalia, and Eddie Yuen. Thank you also to Don Mitchell and Chris Niedt for their comments on an early version of this article, to anonymous reviewers for their useful suggestions and criticisms, and to Audrey Kobayashi for her clear editorial guidance.

Notes

1. According to his blog, Drapeau is trained as an evolutionary biologist and studies the national security applications of the life sciences at NDU's Center for Technology and National Security Policy.
2. Galtung's later work elaborates the importance of cultural violence, or meaning, to naturalizing and sustaining structural violence. This is important, as Scheper-Hughes and Bourgois remind us, because violence "goes beyond physicality to include assaults on self-respect and personhood. The social and cultural dimensions of violence are what give it its force and meaning" (Farmer 2004, 318). We might usefully understand Galtung's focus on physicality in relation to Amartya Sen and Martha Nussbaum's "human capabilities" approach to development, freedom, and justice (Confortini 2006).
3. Space prevents discussion of Foucault's (1978) richly elaborated theory of biopower, which informs this analysis (see Rabinow and Rose 2006).
4. Drapeau echoes Secretary of State Madeline Albright, who, in response to medical doctors' criticisms of the deaths caused by the sanctions, resolved her vacillation between whether sanctions are "like force" or an "alternative to force" by attributing the harms to Saddam Hussein: "There is no greater enemy to public health in Iraq than he" (Albright 2000, 155–56).
5. Drapeau invokes the hegemonic narrative of the West's progressive, rational sanitary reform to criticize Iraq's personal hygiene campaign. Around the world, however,

infrastructures have been built unevenly as part of economic development and nation state-building projects and as a result of political and professional mobilization to improve living conditions (Graham and Marvin 2001).

References

Agnew, J., T. W. Gillespie, and B. Min. 2008. Baghdad nights: Evaluating the US military "surge" using nighttime light signatures. *Environment and Planning A* 40:2285–95.

Albright, M. 2000. Economic sanctions and public health: A view from the Department of State. *Annals of Internal Medicine* 132 (2): 155–57.

Alexander, M. J. 2006. Not just (any)*body* can be a patriot: "Homeland" security as empire building. In *Interrogating imperialism: Conversations on gender, race, and war*, ed. R. L. Riley and N. Inayatullah, 207–40. New York: Palgrave.

Arnove, A., ed. [2000] 2002. *Iraq under siege: The deadly impact of sanctions and war*. Boston: South End Press.

Bhungalia, L. 2009. A liminal territory: Gaza, executive discretion, and sanctions turned humanitarian. *GeoJournal*. http://springerlink.com/content/q7h6g20280604h15/fulltext.html (last accessed 12 September 2009).

Bialasiewicz, L., D. Campbell, S. Elden, S. Graham, A. Jeffrey, and A. J. Williams. 2007. Performing security: The imaginative geographies of current US strategy. *Political Geography* 26:405–22.

Bowen, S. W. J. 2008. *Hard lessons: The Iraq reconstruction experience*. http://graphics8.nytimes.com/packages/images/world/20081213_RECONSTRUCTION_DOC/original.pdf (last accessed 28 May 2009).

Briggs, C. 2004. Theorizing modernity conspiratorially: Science, scale, and the political economy of public discourse in explanations of a cholera epidemic. *American Ethnologist* 31 (2): 164–87.

———. 2009. Politics of the flu (interview). *Against the Grain* 4 May. http://www.againstthegrain.org/program/183/id/191420/mon-5-04-09-politics-flu (last accessed 28 May 2009).

Briggs, C., and D. C. Hallin. 2007. Biocommunicability: The neoliberal subject and its contradictions in news coverage of health issues. *Social Text* 25 (4): 43–66.

Cockburn, C. 2004. The continuum of violence: A gender perspective on war and peace. In *Sites of violence: Gender and conflict zones*, ed. W. Giles and J. Hyndman, 24–44. Berkeley: University of California Press.

Cockburn, P. 2008. Violence is down—But not because of America's "surge." *The Independent* 14 September. http://www.independent.co.uk/news/world/middle-east/ iraq-violence-is-down-ndash-but-not-because-of-americas-surge-929896.html (last accessed 28 May 2009).

Confortini, C. C. 2006. Galtung, violence, and gender: The case for a Peace Studies/feminism alliance. *Peace & Change* 31 (3): 333–67.

Cooter, R. 2003. Of war and epidemics: Unnatural couplings, problematic conceptions. *Social History of Medicine* 16 (2): 283–302.

Cowen, D., and E. Gilbert. 2007. The politics of war, citizenship, territory. In *War, citizenship, territory*, ed. D. Cowen and E. Gilbert, 1–30. London and New York: Routledge.

Curtis, S. 2004. *Health and inequality: Geographical perspectives*. London: Sage.

Daponte, R. O., and R. Garfield. 2000. The effects of economic sanctions on the mortality of Iraqi children prior to the 1991 Persian Gulf War. *American Journal of Public Health* 90:546–52.

Davis, M. 2001. *Late Victorian holocausts: El niño famines and the making of the third world*. London: Verso.

De Belder, B. 2007. Four years into the occupation: No health for Iraq. *The Brussels Tribunal*. http://www.brusselstribunal.org/Health200307.htm (last accessed 28 May 2009).

Drapeau, M. 2007. A microscopic insurgent. *New York Times* 4 December. http://www.nytimes.com/2007/12/04/opinion/04drapeau.html (last accessed 21 May 2009).

Enloe, C. 2000. *Maneuvers: The international politics of militarizing women's lives*. Berkeley: University of California Press.

Farmer, P. 1999. *Infections and inequalities: The modern plagues*. Berkeley: University of California Press.

———. 2003. *Pathologies of power: Health, human rights, and the new war on the poor*. Berkeley: University of California Press.

Farmer, P., with comments by P. Bourgois and N. Scheper-Hughes, D. Fassin, L. Green, H. K. Heggenhouen, L. Kirmayer, and L. Wacquant. 2004. An anthropology of structural violence. *Current Anthropology* 45 (3): 305–25.

Foucault, M. 1978. *The history of sexuality: Volume I*. New York: Vintage Books.

Galtung, J. 1969. Violence, peace, and peace research. *Journal of Peace Research* 6 (3): 167–91.

Geyer, M. 1989. The militarization of Europe, 1914–1945. In *The militarization of the Western world*, ed. J. R. Gillis, 65–102. New Brunswick, NJ: Rutgers University Press.

Gilmore, R. W. 1998. Globalisation and US prison growth: From military Keynesianism to post-Keynesian militarism. *Race & Class* 40 (2–3): 171–88.

———. 2002. Fatal couplings of power and difference: Notes on racism and geography. *The Professional Geographer* 54 (1): 15–24.

Glanz, J. 2007. In report to Congress, oversight officials say Iraqi rebuilding falls short of goals. *New York Times* 31 October. http://www.nytimes.com/2007/10/31/world/middleeast/31reconstruct.html (last accessed 28 May 2009).

Glanz, J., and S. Farrell. 2007. More Iraqis said to flee since troop increase. *New York Times* 24 August. http://www.nytimes.com/2007/08/24/world/middleeast/24displaced.html (last accessed 28 May 2009).

Graham, S. 2005. Switching cities off: Urban infrastructure and US air power. *City* 9 (2): 169–93.

———. 2008. Cities and the "War on Terror." In *Indefensible space: The architecture of the national insecurity state*, ed. M. Sorkin, 1–28. London and New York: Routledge.

Graham, S., and S. Marvin. 2001. *Splintering urbanism: Networked infrastructures, technological mobilities and the urban condition*. London and New York: Routledge.

Guha-Sapir, D., and W. G. van Panhui. 2003. The importance of conflict-related mortality in civilian populations. *The Lancet* 361:2126–28.

Hall, S. 1980. Race, articulation, and societies structured in dominance. In *Sociological theories: Race and colonialism*, 305–45. Paris: UNESCO.

Hanssen, B. 2000. *Critique of violence: Between poststructuralism and critical theory*. London and New York: Routledge.

Hart, G. 2006. Denaturalizing dispossession: Critical ethnography in the age of resurgent imperialism. *Antipode* 38 (4): 977–1004.

Hayes, M. 1999. "Man, disease and environmental associations": From medical geography to health inequalities. *Progress in Human Geography* 23 (2): 289–96.

Hewitt, K. 1987. The social space of terror: Towards a civil interpretation of total war. *Environment and Planning D: Society and Space* 5:445–74.

Hyndman, J. 2007. Feminist geopolitics revisited: Body counts in Iraq. *The Professional Geographer* 59 (1): 35–46.

Incite! Women of Color Against Violence, ed. 2006. *Color of violence: The Incite! anthology*. Cambridge, MA: South End Press.

Ingram, A. 2005. The new geopolitics of disease: Between global health and global security. *Geopolitics* 10: 522–45.

International Rescue Committee. 2008. *Five years later, a hidden crisis: Report of the IRC Commission on Iraqi refugees*. New York: International Rescue Committee. http://www.theirc.org/resources/2008/iraq_report.pdf (last accessed 28 May 2009)

Johnston, R. J., J. O'Loughlin, and P. J. Taylor. 1987. The geography of violence and premature death: A world-systems approach. In *The quest for peace: Transcending collective violence and war among societies, cultures and states*, ed. R. Väyrynen, D. Senghaas, and C. Schmidt, 241–59. London: Sage.

Kalipeni, E., and J. Oppong. 1998. The refugee crisis in Africa and implications for health and disease: A political ecology approach. *Social Science & Medicine* 46 (12): 1637–53.

Kaplow, L. 2007. Another tribulation. *Newsweek* 28 September. http://www.newsweek.com/id/41683 (last accessed 21 May 2009).

Koopman, S. 2008. Imperialism within: Can the master's tools bring down empire? ACME 7 (2): 283–307. http://www.acme-journal.org/v017/SKo.pdf (last accessed 7 July 2009).

Kuletz, V. 2001. Invisible spaces, violent places: Cold War nuclear and militarized landscapes. In *Violent environments*, ed. N. L. Peluso and M. Watts, 237–60. Ithaca, NY: Cornell University Press.

Levy, B. S., and V. W. Sidel, eds. 1997. *War and public health*. New York: Oxford University Press.

Loyd, J. M. 2009. War is not healthy for children and other living things. *Environment and Planning D: Society and Space* 27 (3): 403–24.

Lutz, C. 2002. Making war at home in the United States: Militarization and the current crisis. *American Anthropologist* 104 (3): 723–35.

Mamdani, M. 2007. The politics of naming: Genocide, civil war, insurgency. *London Review of Books* 8 March. http://www.lrb.co.uk/v29/n05/mamd01_.html (last accessed 27 May 2009).

———. 2008. The new humanitarian order. *The Nation* 29 September. http://www.thenation.com/doc/20080929/mamdani (last accessed 27 May 2009).

Mbali, M. 2002. "A bit of soap and water and some jik": Historical and feminist critiques of an exclusively individualising understanding of cholera prevention in discourse around neoliberal water policy. University of Natal. http://www.ukzn.ac.za/ccs/files/cholera_msp3.pdf (last accessed 28 May 2009).

Mbembe, A. 2003. Necropolitics. *Public Culture* 15 (1): 11–40.

Mitchell, D. 2003. Cultural landscapes: Just landscapes or landscapes of justice? *Progress in Human Geography* 27 (6): 787–96.

Nguyen, V.-K., and K. Peschard. 2003. Anthropology, inequality, and disease: A review. *Annual Review of Anthropology* 32:447–74.

Orford, A. 2003. *Reading humanitarian intervention: Human rights and the use of force in international law*. Cambridge, UK: Cambridge University Press.

Puar, J. K. 2007. *Terrorist assemblages: Homonationalism in queer times*. Durham, NC: Duke University Press.

Rabinow, P., and N. Rose 2006. Biopower today. *BioSocieties* 1:195–217.

Rawaf, S. 2005. The health crisis in Iraq. *Critical Public Health* 15 (2): 181–88.

Scheper-Hughes, N., and P. Bourgois, eds. 2004. *Violence in war and peace*. Malden, MA: Blackwell.

Seager, J. 1999. Patriarchal vandalism: Militaries and the environment. In *Dangerous intersections: Feminist perspectives on population, environment, and development*, ed. J. Silliman and Y. King, 163–88. Cambridge, MA: South End Press.

Sidel, V. W. 1995. The international arms trade and its impact on health. *British Medical Journal* 311:1677–80.

Skurski, J., and F. Coronil. 2006. Introduction: States of violence and the violence of states. In *States of violence*, ed. J. Skurski and F. Coronil, 1–31. Ann Arbor: The University of Michigan Press.

Smallman-Raynor, M. R., and A. D. Cliff. 2004. *War epidemics: An historical geography of infectious diseases in military conflict and civil strife, 1850–2000*. Oxford, UK: Oxford University Press.

Smith, A. 2005. *Conquest: Sexual violence and American Indian genocide*. Cambridge, MA: South End Press.

Smith, D. 2007. Cholera crisis hits Baghdad. *The Observer* 2 December. http://www.guardian.co.uk/world/2007/dec/02/iraq.davidsmith (last accessed 28 May 2009).

Smith, T. W. 2002. The new law of war: Legitimizing hi-tech and infrastructural violence. *International Studies Quarterly* 46:355–74.

Sparke, M. 2007. Geopolitical fears, geoeconomic hopes, and the responsibilities of geography. *Annals of the Association of American Geographers* 97 (2): 338–49.

Stewart, F., V. FitzGerald, and Associates, eds. 2001. *War and underdevelopment*. Oxford, UK: Oxford University Press.

Terry, F. 2002. *Condemned to repeat?: The paradoxes of humanitarian action*. Ithaca, NY: Cornell University Press.

Valenciano, M., D. Coulombier, B. Lopes Cardozo, A. Colombo, M. Alla, S. Samson, and M. Connolly. 2003. Challenges for communicable disease surveillance and control in southern Iraq, April–June 2003. JAMA: *Journal of the American Medical Association* 290 (5): 654–58.

Watts, S. 1999. *Epidemics and history: Disease, power and imperialism*. New Haven, CT: Yale University Press.

Woodward, R. 2005. From military geography to militarism's geographies: Disciplinary engagements with the geographies of militarism and military activities. *Progress in Human Geography* 29 (6): 718–40.

World Health Organization. 2008a. *Countries reporting cholera cases in 2007*. http://gamapserver.who.int/mapLibrary/Files/Maps/Global_Cholera(WER)_2007.png (last accessed 28 May 2009).

World Health Organization. 2008b. *Social determinants of health in countries in conflict: A perspective from the Eastern Mediterranean*. World Health Organization, Regional Office for the Eastern Mediterranean. http://www.emro.who.int/dsaf/dsa955.pdf (last accessed 28 May 2009).

Zwi, A. B. 1991. Militarism, militarization, health and the third world. *Medicine and War* 7:262–68.

The Political Utility of the Nonpolitical Child in Sri Lanka's Armed Conflict

Margo Kleinfeld

During Sri Lanka's protracted civil conflict, linkages between ethnicity and homeland have been under considerable strain. Subsequently, alternative rationales for territorial claims have emerged, including each belligerent's human rights record and moral fitness to govern. To this end, child-based narratives such as the child as a zone of peace have been used strategically by each side to rearticulate conflict relations and spaces into moral terms. The case of the Days of Tranquility (DOTs), a humanitarian ceasefire implemented between 1995 and 2001 to immunize children against the polio virus in Sri Lanka's conflict-affected areas, is examined here. Based on fieldwork and an analysis of texts that represent the DOTs, the research reveals how child-centered tropes and humanitarian spaces for children can be politically useful to parties to a conflict as well as to the United Nations agencies that support them. *Key Words: armed conflict, child, humanitarian space, Sri Lanka, territory.*

在斯里兰卡的长期内战期间，种族和国家之间的联系承受了相当大的压力。随后而出现了领土要求的其他的理由，包括对每个交战方的人权记录和道义水平的管理。为此，冲突各方开始策略性地使用基于儿童的对冲突的叙述，例如利用和平区的儿童来重新描述冲突关系和空间，并将此上升到道德的高度。本文使用了和平日个例（DOTs），对此进行了分析研究。和平日是 1995 年至 2001 年在斯里兰卡的冲突地区所实施的一个人道主义停火计划，为该地区的儿童进行脊髓灰质炎病毒的防疫保护。根据实地调查和分析代表和平日项目的文本，本研究揭示出以儿童为中心的叙述和在人道主义空间的儿童对冲突各方以及支持他们的联合国机构是如何具有政治性帮助的。*关键词：武装冲突，儿童，人道主义空间，斯里兰卡，领土。*

Durante el prolongado conflicto civil de Sri Lanka, los lazos entre etnicidad y patria han estado sometidos a considerable tensión. Subsiguientemente, han surgido justificaciones alternativas sobre reclamaciones territoriales, incluyendo el historial sobre derechos humanos y la idoneidad moral para gobernar de cada beligerante. Con tal fin, cada bando ha utilizado estratégicamente narrativas de origen infantil, tales como las que adoptan al niño como motivo central de paz, con el propósito de rearticular relaciones de conflicto y espacios en términos morales. Al respecto, aquí se examina el caso del Día de la Tranquilidad (DT), un cese al fuego humanitario que se implementó entre 1995 y 2001 para permitir la vacunación de los niños contra el virus del polio en áreas de Sri Lanka afectadas por el conflicto. Con base en trabajo de campo y un análisis de textos relacionados con el DT, la investigación revela cómo los tropos centrados en niños y los espacios humanitarios asignados para niños pueden ser políticamente útiles tanto para las partes en conflicto como para las agencias de las Naciones Unidas en que aquellos programas se apoyan. *Palabras clave: conflicto armado, niños, espacio humanitario, Sri Lanka, territorio.*

In Sri Lanka, tropes of the child have been a fixture in conflict relations between the Government of Sri Lanka (GoSL) and the Liberation Tigers of Tamil Eelam (LTTE), who have been fighting for territorial control of their respective homelands since 1983. Narratives invoking the protection given or harm done to children are used by both sides in the conflict to depoliticize conflict relations and spaces by rearticulating them in moral terms and locating belligerents outside of politics. The case examined here is the Days of Tranquility (DOTs), a humanitarian operation managed by the United Nations Children's Fund (UNICEF), an international child advocacy organization, which took place in Sri Lanka's conflict zone between 1995 and 2001 as part of a global polio eradication effort. I observed the DOTs directly in 2001 while working in the UNICEF country office in Colombo, Sri Lanka (hereafter UNICEF–SL).

As an intern, I worked on a variety of projects for UNICEF–SL's communications section between August 2001 and July 2002, during which time I created an archive of materials related to the DOTs comprised of field and technical reports, interview data, and direct observations of the operation in September and October 2001. I concluded that the parties to the conflict, who gave their consent to refrain from military operations and create a "nonpolitical" arena for four days each year, benefitted politically from the DOTs in a number of ways: by legitimizing themselves as morally fit territorial authorities and their adversaries as unfit ones; by using the DOTs as evidence of territorial control; by enjoying positive international publicity; and, possibly, by using the DOTs as part of their military calculus. UNICEF, which depends on its good relations with belligerents to maintain access to children in emergencies and which has developed a powerful communications and fundraising machinery based on the visibility of its work, perhaps unwittingly facilitated the belligerents' utility of the nonpolitical child in their contest for political and territorial legitimacy in Sri Lanka.

The Nonpolitical Child

In popular discourse, the child is imagined as innocent and prepolitical, occupying a privileged position as a nonagent that fundamentally links the child to caretakers and pathologizes those that do not serve the child's best interests (Berlant 1997; Edelman 1998). Because helping or harming a child is typically articulated as a moral issue, invoking the child puts a moral narrative into play that often distracts from political contest or social and economic realities (see Holland 1992; Burman 1995; Valentine 1996; Boyden 1997; Roberts 1998; Katz 2004). Humanitarian projects for children are a good example of this process. Reminiscent of Ferguson's (1994) antipolitics machine whereby development projects in Lesotho, articulated in technical and nonpolitical terms, masked power relations and an expanding state bureaucracy, good works for children can be similarly put to work for political purposes (for example, Nieuwenhuys 1998). Praised by relief and advocacy organizations, the United Nations (UN) agencies, and a public eager to support projects for suffering children during times of war, critical assessment of child-centered humanitarian efforts is often lacking. This has political utility for territorial authorities who can identify their participation as evidence of moral fitness and the right to govern as well as for advocacy and service-delivery organizations that can assert their own capacities in armed struggles or complex emergencies.

The unique identity of the child is especially effective in discourses on the depoliticization of space. As Natter and Jones (1997) have argued, attempts to stabilize any representation of space—humanitarian or otherwise—are always political projects. The child is an especially effective trope in efforts to "fix" or naturalize nonpolitical space not only because its moral appeal distracts from the political but, like other moral tropes, the child orders and gives meaning to events, thereby creating narrative closure and the appearance of stability (White 1987). In this case, the believability of the DOTs as a humanitarian, nonpolitical enterprise within Sri Lanka's conflict zone is achieved through citationality; that is, by creating the appearance of a "believable object" through its recitation in texts and among actors, even where a material referent is absent (Certeau 1984, 185–89; see also Foucault 1972; Laclau and Mouffe 1985; Butler 1993; Campbell 1998). Efforts to stabilize nonpolitical spaces for children derive to some extent from the concept of "children as a zone of peace," which became a key strategy for UNICEF protection activities, globally and in Sri Lanka.

The Child as a Zone of Peace

Children as Zones of Peace (CZoP) was originally coined in 1983 by Nils Thedin, head of the Swedish Save the Children organization, and promoted the right of children to always live in a conflict-free zone (Vittachi 1993). Humanitarian organizations were equally empowered through this principle to deliver assistance to children wherever they were geographically or in whatever kind of circumstance (Nylund 1999; UNICEF 2000), but like all humanitarian spaces, consent of belligerents is required to create a zone of peace within a war zone (Landgren 1995; van Brabant 1998; Hyndman 2003). The CZoP mandate was vigorously promoted by UNICEF's James Grant during the 1980s and became the rationale for corridors, zones, bubbles, and days of peace or tranquility to immunize children, deliver food or supplies, and evacuate civilians during armed conflict (Vittachi 1993; Walker 1993; UNICEF 1996; CZoP 1998). Many came to believe that humanitarian spaces for children could be uniquely effective in depoliticizing a conflict environment and used as confidence-building measures that might set warring parties on a path to peace (Hay 1990; Centre for Days of Peace 1991; Peters 1997; Bush 2000; Ebersole 2000).

The DOTs in Sri Lanka grew from the CZoP idea, taking the form of a four-day humanitarian ceasefire in eight conflict-affected districts between 1995 and 2001 to facilitate the implementation of national immunization days (NIDs), a World Health Organization strategy to vaccinate children on a single day as part of its polio eradication initiative (Figure 1). The NIDs occurred countrywide in Sri Lanka between 1995 and 1999 and in those areas that received low immunization coverage between 2000 and 2003, also known as the *Sub-National*

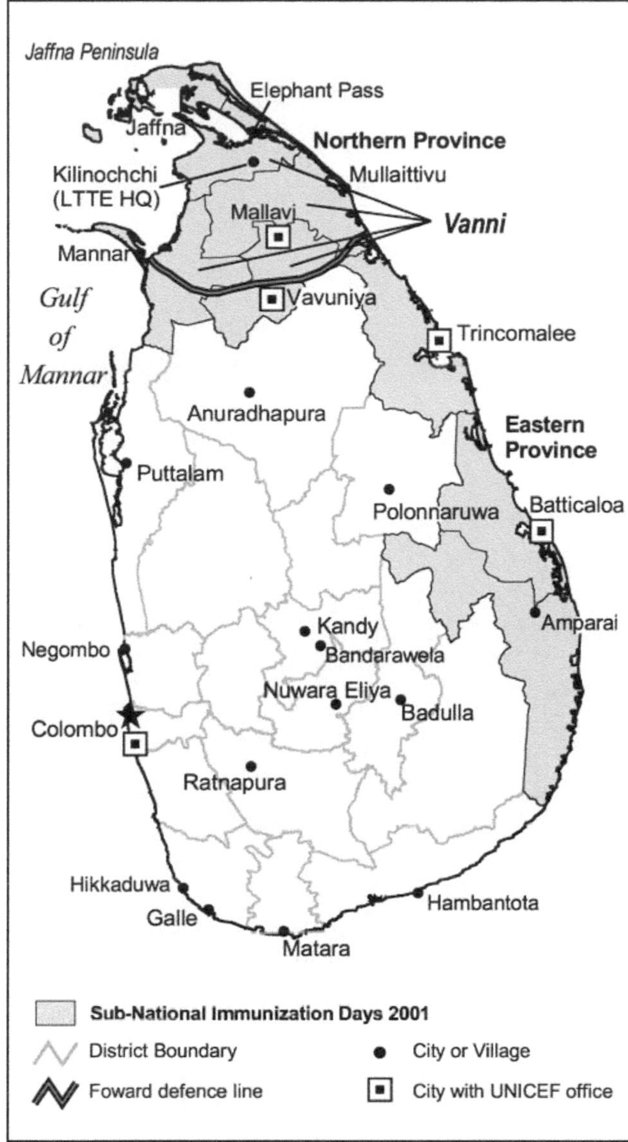

Figure 1. Map of the 2001 Sub-National Immunization Days (S/NIDs) and Days of Tranquility. The government of Sri Lanka and the Liberation Tigers of Tamil Eelam agreed to refrain from military operations while children under five were immunized against the poliovirus. Cartography: Steven Bolssen.

Immunization Days (S/NIDs). The DOTs "corridor of peace," as UNICEF (1997) called it, consisted of two rounds, two days each, one month apart at about 1,500 immunization sites in the conflict zone. On the first day of each round, vaccine and equipment were transported to immunization centers; on the second day, oral polio drops were administered to children by volunteers (Figure 2). The DOTs continued until 2001, after which a more permanent ceasefire agreement was signed between the GoSL and the LTTE in February 2002.[1]

It is perhaps noteworthy that although Sri Lanka had been polio free since 1993, it was thought that refugee movements between polio-endemic India and Sri Lanka could be a possible source of importation.[2] A UNICEF health official also explained that James Grant had been a frequent visitor to the office in Sri Lanka and sold the idea to UNICEF–SL staff, who had already conducted island-wide immunization campaigns through a media blitz involving sports and other celebrities. In 1995, when UNICEF–SL first approached the GoSL and the LTTE to consider the possibility of a humanitarian zone for children, they described zones of peace in El Salvador and Sudan and argued that the operation would be followed with interest by domestic and international media and could be a win–win intervention for children as well as the political parties that supported them (Figure 3).

Background to the Conflict

Today's war between the GoSL and the LTTE can be traced to events leading up to Ceylon's independence in 1948, when power-sharing arrangements between the majority Sinhalese and minority Tamils floundered. Although the concentration of Tamil speakers living in the northern and eastern parts of the island has been used as evidence of a vital Tamil homeland and nation (Arasaratnam 1998), the so-called Tamils are actually composed of several distinct groups including the northern Tamils, Moors, and "Upcountry" Tamils, differentiated by class, caste, religion, and geography. Eventually, the relationship between these groups and successive Sinhalese governments who had failed to create a multiethnic polity or to devolve power from the center escalated into two distinct nationalisms that culminated in the Tamil demand for a separate state (Tamil United Liberation Front 1988). Whereas Tamil nationalists have defended their right to territorial sovereignty based on the existence of an ancient homeland in the north and east, Sinhalese nationalists claim the entire island as primordial homeland and a sacred repository of Buddhism, evidence for the Tamils of an entrenched Sinhalese chauvinism and inability to rule a multiethnic state (Balasingham 1983; Little 1994). The Sinhalese have argued that Tamil homeland claims have no historical basis and that the division of the island for any reason will spell the end of the Sinhalese people (De Silva 1987; Kapferer 1988; Uyangoda 1994).

The militarization of Tamil groups dates to the early 1970s when Ceylon's new constitution reaffirmed Sinhala as the official state language as well as the state's

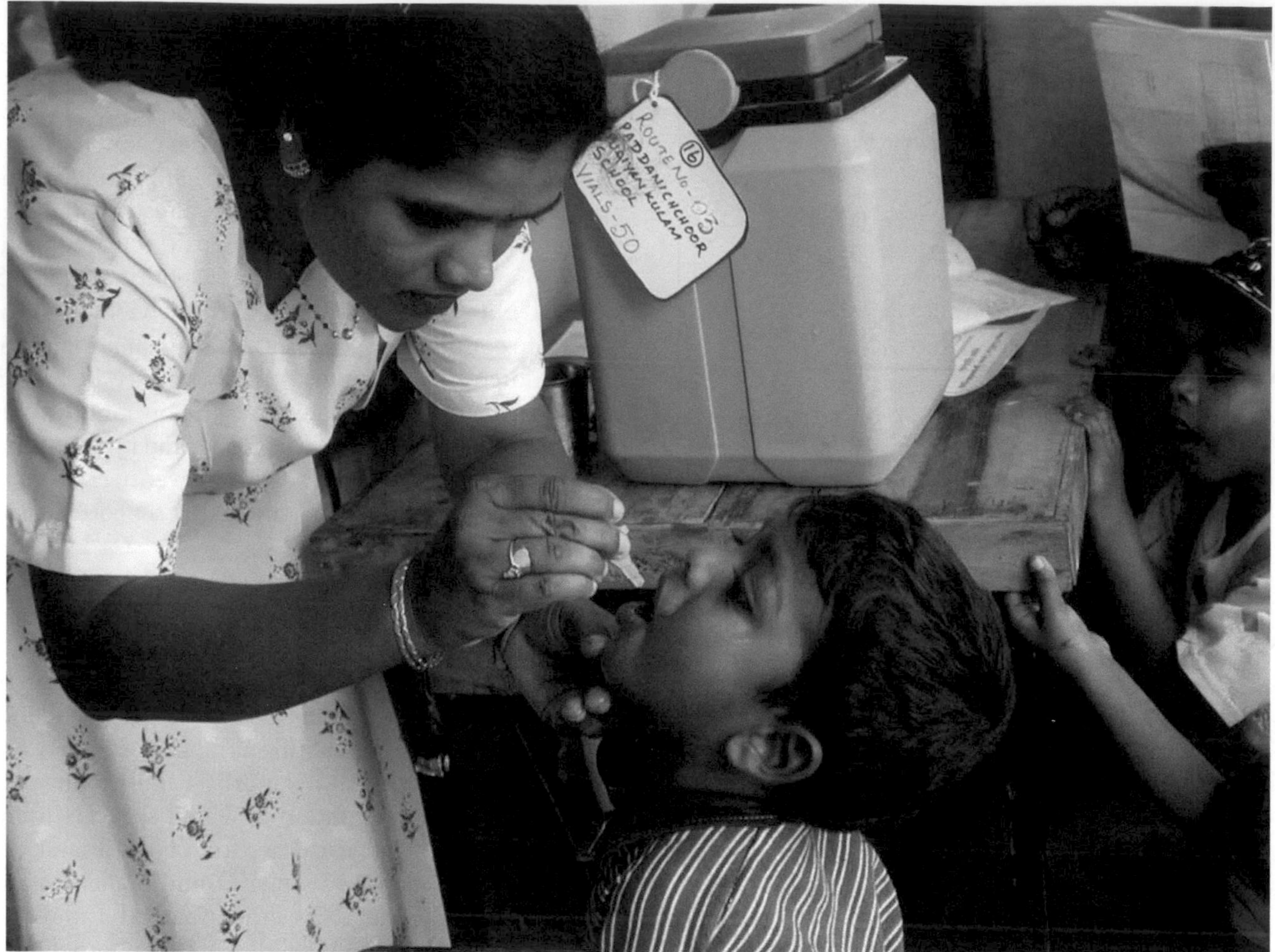

Figure 2. Oral polio vaccine. A volunteer gives a child polio vaccine in drop form during the 2001 Days of Tranquility in Sri Lanka.

duty to protect and foster Buddhism. This, along with rising underemployment among Tamil youth, a discriminatory quota system for university admissions, and the lack of development assistance in Tamil areas, led to calls for regional autonomy (Ponnambalam 1983; Manogaran 1987). Tamil youth formed into groups, including the LTTE, that began to take up arms against state oppression and ultimately in support of secession. Most scholars, however, date the war to July 1983 when anti-Tamil riots instigated by the government resulted in 2,000 to 4,000 deaths (Tambiah 1986; Senaratne 1997) and the creation of a separate Tamil state, Tamil Eelam, was considered the most attractive option even by moderate Tamils (Bose 1994; Wilson 2000).

In the twenty-five years between 1983 and 2008, 70,000 people were killed and both parties, the GoSL and the LTTE, were accused of using dirty war and terrorist tactics by independent sources. Charges of human rights abuses have also featured prominently in each side's propaganda, particularly since the 1980s when scholarly and political deconstructions and demystifications of territorial claims based on ethnicity and the historical record required alternative justifications for each side's national project (Kleinfeld 2005b). Allegations that belligerents were violating children's rights—through the employ of child soldiers or by ignoring chronic child malnutrition, the growth of the child sex tourism industry, and children in forced labor—were frequently used to suggest that one or the other was unfit to govern. These charges received a great deal of attention, partly because child victims are often the most compelling (Burman 1994; Richardson 2001) but also because evidence of child abuse at any scale is enormously useful in establishing the "nonlegitimacy" of an adversary (see Jackman 1993; Kane 2001, 15). At the same time, helping children can be used to establish a political regime as a moral agent and, therefore,

Figure 3. TV cameras surround a child being immunized, while several "internationals" are called to administer the drops.

worthy of support (see Eckstein 1971). The political utility of the child in the parties' contest over legitimacy in Sri Lanka took a notable international turn after the advent of the Convention on the Rights of the Child.

The Visibility of War-Affected Children

Contemporary protections for children were first codified in the Geneva Conventions of 1949, but the 1989 Convention on the Rights of the Child (CRC) created a new international rights-based political infrastructure for children that included monitoring and reporting mechanisms. It also dramatically repositioned the state as the child's chief guarantor of rights. Within two years of its ratification and every five years thereafter, state parties to the CRC are required to file a detailed report with the eighteen-member international Committee on the Rights of the Child describing the state's efforts to implement child rights protections. Unforeseen was the degree to which the CRC would become politically useful to states embroiled in armed struggles and the "legitimate basis for shaming warring parties over their disregard of children's well being" (Black 1996, 273).

In 1994, the Committee on the Rights of the Child commissioned a global study of children affected by war that produced the influential report, "Impact of Armed Conflict on Children," known as the Machel Report after its author Graça Machel (United Nations General Assembly 1996). The Machel Report has raised the profile of war-affected children in a number of ways: It recommended the creation of a Special Representative to the Secretary General (SRSG) on Children and Armed Conflict; it was used by UNICEF as evidence in its successful 1999 effort to place war-affected children on the UN Security Council's agenda; and it led to the Secretary General's "Naming and Shaming Report" (United Nations Security Council 2002), which identified parties that recruited or used children in conflict.

Sri Lanka capitalized on the heightened international publicity associated with the CRC and the Machel Report. In 1997, Foreign Minister Lakshman Kadirgamar[3] invited Olara Otunnu, the newly appointed SRSG on Children and Armed Conflict, to the country to extract promises from the LTTE on their use of children in the conflict. During the highly publicized trip, the LTTE admitted to the practice and promised to discontinue it.[4] Otunnu's visit encouraged Kadirgamar to highlight the LTTE's use of child soldiers in a subsequent offensive to stem the flow of foreign funds from the Tamil diaspora. Speeches given to the UN General Assembly, to foreign parliaments, and at international events such as the Winnipeg Conference on War-Affected Children in 2000, likely contributed to the proscription of the LTTE by the United States, the United Kingdom, and Australia. In its 2002 report to the Committee on the Child (Committee on the Rights of the Child 2002, 36–37), the GoSL devoted about one third of the fifty-eight-page report describing LTTE maltreatment of children:

> Their violent campaign has directly and indirectly affected the lives of children. . . . Many children have been killed, disabled and orphaned; more have been subjected to the trauma of witnessing extreme forms of brutality and violence unleashed by terrorism. Being rendered homeless or separated from families is often compounded by grief at the loss of loves ones. Children of all ethnic communities—Sinhala, Tamil and Muslim—have been affected.

The LTTE attempted to defend itself against these charges in press releases and statements from supporters such as the Ilankai Tamil Sangam USA, the Tamil Centre for Human Rights, and the editorial page of the *Tamil Guardian*, and at the same time also became adept at publicizing the GoSL's human rights violations and its abuse of children (Kleinfeld 2003, 2005a).

In 1998, UNICEF–SL increased the visibility of Sri Lanka's war-affected children when it set up a special unit, Children Affected by Armed Conflict (CAAC) in Colombo, and established field offices in the northern and eastern conflict areas. Despite UNICEF's global-level policies known as the Martigny Consultations advocating the mainstreaming of emergency programs into regular development work, UNICEF–SL could not pass up the fundraising boon that CAAC provided. Whereas funding for UNICEF regular resources around the world had been flat or declining since 1990, contributions for emergencies had increased by nearly 50 percent (UNICEF 2001). In Sri Lanka, emergency work in the conflict area attracted a disproportionate amount of donor interest, some years more than triple all other Sri Lanka programs combined. The CAAC field offices gave UNICEF–SL a presence on both sides of the conflict, an example of the child as zones of peace principle, and helped staff keep track of donor contributions, coordinate and monitor programs, and make projects such as the DOTs more accessible to donors and the press (Kleinfeld 2005a).

Repoliticizing Conflict Space

UNICEF managed the DOTs' consent process by requesting that belligerents refrain from military operations for four days in a letter sent to each side about six weeks before the ceasefire. The permissions process was valuable UN-endorsed evidence of the LTTE's de facto territorial authority in the northern region known as the Vanni and the government's ability to coordinate an island-wide operation through its ministry of health. Providing consent also established the parties' intents to protect children, futurity, and future space in front of interested international audiences. In its letters, UNICEF stressed that the parties' involvements would be a "practical demonstration of 'children as zones of peace' in a way that people can easily understand." Participation also affirmed the parties' capacity to work with the international community for humanitarian purposes, important criteria in establishing external legitimacy and territorial sovereignty (Anderson 1999; Terry 2002; Kleinfeld 2005b).

Neither side, however, allowed the spatial claims or display of humanitarian good works by their adversary to go unchallenged. After the first two DOTs in 1995 and 1996, both sides denounced the other in press statements reporting breaches and violations of the DOTs. Headlines such as "Lanka Says Rebels Violate DOTs," "Hospital Bombed; SL Violates Ceasefire," and "LTTE Terrorists Violate UNICEF Brokered 'Days of Tranquility'" made clear that the other would sacrifice children for military gain.

UNICEF's promotion of the DOTs as a demonstration of the child as a zone of peace contradicted internal UNICEF documents and ministry of health field reports from the first to the last DOTs indicating that shelling was often heard during the ceasefire and that security problems including active operations and multiple displacements prevented children from reaching immunization centers. Because ceasefire violations

Figure 4. Nelukkulam Welfare Centre, 2001. Many children were dressed in their finest clothes on what was commonly referred to as "polio day."

might have occurred for any number of reasons, it was nearly impossible to determine whose shells were fired or if a breach was a provocative tactic to illicit a response for later publicity. Staffers from other international organizations that I interviewed also witnessed fighting during the DOTs. UN security staff suggested to me that the DOTs presented opportunities for the parties to conduct covert operations and move troops and equipment. As long as tranquility existed precisely where kids were being immunized, however, the operation was considered a success by the UN team on the lookout for anything problematic. My own archive indicated that it was not unusual for a new military operation to commence directly after a round of immunizations, although it would be very difficult to prove such tactics. Reports from other locations also suggest that ceasefires for children might facilitate military activities (U.S. Agency for International Development 2002).

I traveled to Vavuniya and Trincomalee in 2001 to observe the DOTs and interview a variety of stakeholders. I found that conflict-affected children were immunized through the regular health infrastructure by dedicated GoSL health officials, UNICEF staff, volunteers, and international organizations such as Médecins sans Frontières and the International Committee of the Red Cross who provided vaccines to children living in areas under LTTE control, just as they did on other days at their mobile clinics. Although immunization was "scaled up" to vaccinate many children on a single day, I observed neither a geographic nor temporal space of "tranquility." Even though "polio day," as it was called locally, was a life-affirming event for families (Figure 4), parents and health workers I spoke with complained about security issues including shelling, military fencing, travel restrictions, and usual problems at military checkpoints that prevented them from reaching immunization sites. Government medical officers, who were made responsible for the immunization tally, offered technical solutions to some of these problems, but chronic low immunization rates persisted in the same places year after year, and the base count of children relied on an outdated census. Military operations affecting immunization counts were noted in the many reports I reviewed from local health officials, but there was little done to address it. According to official instructions from the ministry of health to medical officers organizing the Sub-National Immunization Days, security was not identified as a problem that might be encountered during the operation.

UNICEF–SL's positive public comments can be considered part of its desire to avoid politicizing the DOTs or antagonizing the parties to the conflict who had shut down or threatened to shut down humanitarian programs when confronted with bad publicity. At most they might note that the DOTs were a success despite reports of a few incidents. Behind the scenes, however, gentle reminders were sent to the Ministry of Defence, the LTTE, and the President of Sri Lanka, urging them to maintain their commitments to the DOTs and CZoP principle. Aware of the international visibility trained on the DOTs and war-affected children in general, UNICEF–SL maintained what one official termed a "quiet approach," much to the consternation of other international child rights organizations in the country that wanted UNICEF to use its clout to expose both GoSL and LTTE child rights abuses. Even so, UNICEF–SL was often accused by the GoSL of being "pro-LTTE" for either not denouncing the LTTE sufficiently or being "too friendly" with them while they coordinated programs for children in the conflict zone.

In a few cases, UNICEF did exert political pressure by reporting specific instances of LTTE underage recruitment or the use of schools by GoSL security forces (see Kleinfeld 2005a). When it chose to do so, UNICEF also made use of tropes of the nonpolitical child by shaming one or the other party into action. But UNICEF necessarily walked a fine line between ensuring humanitarian access to children and demonstrating that it had a firm hand if faced with violations of child rights. Most often, however, it found ways to avoid criticizing either the GoSL or the LTTE and, conceivably, enhanced both parties legitimacy as a result.

Conclusion

Representations of the DOTs as well as their value to the war-affected children in Sri Lanka are complex and contradictory, but the degree to which the DOTs have been politicized is rarely described. During the course of this research, I observed that by using tropes of the child and working with child advocacy organizations, parties to the conflict could provide evidence of their moral fitness, spatial stability, and thereby garner positive international support. UNICEF and its partners also benefited by publicizing the creation of humanitarian space for children in a conflict zone, and by minimizing potential and actual problems, could demonstrate the UN's capacity to make a difference during times of war.

Compliance with or violation of the DOTs in Sri Lanka was one example of how the category of the child was used by the GoSL and the LTTE to claim legitimacy and undermine its challengers—effectively repoliticizing the territorial contest from one rooted in historical linkages between ethnicity and space to one based on protecting or harming children. Providing consent to create a nonpolitical space within each side's territorial domain not only provided the belligerents with evidence to support their spatial claims but also signified that they had the interest and capacity to use spaces under their control for a greater good. That the child and its moral appeal were also inserted into advocacy narratives used by UNICEF and the UN suggests that the political utility of the child can be applied to a multitude of purposes.

Despite the positive publicity, enhanced legitimacy, and other rewards reaped from the DOTs, both parties continued to employ children in operations, did not meet commitments to refrain from fighting during the DOTs, ignored children's particular vulnerabilities such as their psychosocial or educational needs, and refused to fund teachers and pediatricians within the war zone where they were in desperately short supply. Like Ferguson's (1994) examples, the DOTs in Sri Lanka were repeated year after year without addressing these and other problems, most notably persistent low immunization rates in some areas due to insecurity. UNICEF, for its part, worked hard on programs and projects for Sri Lanka's children throughout the year, but by avoiding public critique of the parties or a critical evaluation of how their own work might have assisted belligerents, operations such as the DOTs were inevitably politicized. The degree to which UNICEF or any other relief organization can actually improve the quality of life for war-affected children remains an open question. The humanitarian impulse might best be served by making space for a political solution.

Notes

1. The ceasefire ended in January 2008.
2. Currently, polio is endemic in India, Afghanistan, Pakistan, and Nigeria, although the disease has been "imported" into several additional countries.
3. Lakshman Kadirgamar was Sri Lanka's foreign minister between 1994 and 2001. In 2004, he had started a second term when he was assassinated, allegedly by the LTTE.
4. The LTTE has maintained that children have joined the organization voluntarily and that the reported number of child soldiers is greatly exaggerated. Human Rights Watch has accused both the LTTE and the GoSL of using minors in their forces or militias.

References

Anderson, M. B. 1999. *Do no harm: How aid can support peace—Or war*. Boulder, CO: Lynne Rienner.

Arasaratnam, S. 1998. Nationalism in Sri Lanka and the Tamils. In *Sri Lanka: Collective identities revisited*, ed. M. Roberts, 295–313. Colombo, Sri Lanka: Marga Institute.

Balasingham, A. S. 1983. *Liberation Tigers and Tamil Eelam freedom struggle*. Madras, Sri Lanka: Political Committee, Liberation Tigers and Tamil Eelam.

Berlant, L. 1997. *The queen of America goes to Washington: Essays on sex and citizenship*. Durham, NC: Duke University Press.

Black, M. 1996. *Children first: The story of UNICEF, past and present*. Oxford, UK: Oxford University Press for UNICEF.

Bose, S. 1994. *States, nations, sovereignty: Sri Lanka, India and the Tamil Eelam movement*. New Delhi, India: Sage.

Boyden, J. 1997. Childhood and the policy makers: A comparative perspective on the globalization of childhood. In *Constructing and reconstructing childhood: Contemporary issues in the sociological study of childhood*, ed. A. James and A. Prout, 190–229. London: Falmer.

Burman, E. 1994. Innocents abroad: Western fantasies of childhood and the iconography of emergencies. *Disasters* 18:238–53.

———. 1995. The abnormal distribution of development: Policies for Southern women and children. *Gender, Place and Culture* 2 (1): 21–36.

Bush, K. 2000. Polio, war and peace. *Bulletin of the World Health Organization* 78 (3): 281–82.

Butler, J. 1993. *Bodies that matter*. London and New York: Routledge.

Campbell, D. 1998. *National deconstruction: Violence, identity, and justice in Bosnia*. Minneapolis: University of Minnesota Press.

Centre for Days of Peace. 1991. *Humanitarian ceasefires: Peacebuilding for children*. Ottawa, ON, Canada: Centre for Days of Peace.

Certeau, Michel de. 1984. *The practice of everyday life*. Trans. S. Rendall. Berkeley: University of California Press.

Children as Zones of Peace Contact Group (CZoP). 1998. *Children: Zones of peace, a call to action*. Colombo, Sri Lanka: UNICEF.

Committee on the Rights of the Child. 2002. Second periodic report submitted by the government of Sri Lanka. State party report submitted under article 44 of the Convention on the Rights of the Child.

De Silva, K. M. 1987. *The "traditional homelands" of the Tamils of Sri Lanka: A historical appraisal*. Colombo, Sri Lanka: International Centre for Ethnic Studies.

Ebersole, J. M. 2000. Health, peace and humanitarian ceasefires. *Health in Emergencies* 8 (1): 7–8.

Eckstein, H. 1971. *The evaluation of political performance: Problems and dimensions*. Beverly Hills, CA: Sage.

Edelman, L. 1998. The future is kid stuff: Queer theory, disidentification, and the death drive. *Narrative* 6:18–30.

Ferguson, J. 1994. *The anti-politics machine: "Development," depoliticization, and bureaucratic power in Lesotho*. Minneapolis: University of Minnesota Press.

Foucault, M. 1972. *The archaeology of knowledge and the discourse on language*. Trans. A. M. Sheridan Smith. New York: Pantheon.

Hay, R. 1990. *Humanitarian ceasefires: An examination of their potential contribution to the resolution of conflict*. Ottawa: Canadian Institute for International Peace and Security.

Holland, P. 1992. *What is a child? Popular images of childhood*. London: Virago Press.

Hyndman, J. 2003. Preventive, palliative, or punitive? Safe spaces in Bosnia-Herzegovina, Somalia, and Sri Lanka. *Journal of Refugee Studies* 16:167–85.

Jackman, R. W. 1993. *Power without force: The political capacity of nation-states*. Ann Arbor: University of Michigan Press.

Kane, J. 2001. *The politics of moral capital*. Cambridge, UK: Cambridge University Press.

Kapferer, B. 1988. *Legends of people, myths of state: Violence, intolerance, and political culture in Sri Lanka and Australia*. Washington, DC: Smithsonian Institution Press.

Katz, C. 2004. *Growing up global: Economic restructuring and children's everyday lives*. Minneapolis: University of Minnesota Press.

Kleinfeld, M. 2003. Strategic troping in Sri Lanka: September eleventh and the consolidation of political position. *Geopolitics* 8 (3): 105–26.

———. 2005a. Depoliticizing space in Sri Lanka: The discursive utility of the child during times of war. PhD thesis, University of Kentucky.

———. 2005b. Destabilizing the identity–territory nexus: Rights-based discourse in Sri Lanka's new political geography. *GeoJournal* 64:287–95.

Laclau, E., and C. Mouffe. 1985. *Hegemony and socialist strategy: Towards a radical democratic politics*. London: Verso.

Landgren, K. 1995. Safety zones and international protection: A dark grey area. *International Journal of Refugee Law* 7 (3): 436–58.

Little, D. 1994. *Sri Lanka: The invention of enmity*. Washington, DC: United States Institute of Peace.

Manogaran, C. 1987. *Ethnic conflict and reconciliation in Sri Lanka*. Honolulu: University of Hawaii Press.

Natter, W., and J. P. Jones III. 1997. Identity, space, and other uncertainties. In *Space and social theory*, ed. G. Benko and U. Strohmayer, 141–61. London: Blackwell.

Nieuwenhuys, O. 1998. Global childhood and the politics of contempt. *Alternatives* 23:267–89.

Nylund, B. V. 1999. UNICEF's experience working on both sides of conflict for the protection of internally displaced women and children. *Refugee Survey Quarterly* 18:116–25.

Peters, M. A., ed. 1997. *A health-to-peace handbook*. Hamilton, ON, Canada: McMaster University.

Ponnambalam, S. 1983. *Sri Lanka: The national question and the Tamil liberation struggle*. London: Zed Books.

Richardson, J. 2001. *The role of the communications officer in unstable situations in times of war: A communications strategy for emergencies*. Report prepared for UNICEF Division of Communication. New York: UNICEF Office of Emergency Programs.

Roberts, S. M. 1998. Commentary: What about the children? *Environment and Planning A* 30:3–18.

Senaratne, J. P. 1997. *Political violence in Sri Lanka, 1977–1990: Riots, insurrections, counter-insurgencies, foreign intervention*. Amsterdam: VU (Vrije Universiteit) University Press.

Tambiah, S. J. 1986. *Sri Lanka—Ethnic fratricide and the dismantling of democracy*. Delhi, India: Oxford University Press.

Tamil United Liberation Front (TULF). 1988. *Towards devolution of power in Sri Lanka: Main documents, August 1983 to October 1987*. Madras, Sri Lanka: TULF.

Terry, F. 2002. *Condemned to repeat? The paradox of humanitarian action*. Ithaca, NY: Cornell University Press.

United Nations Children's Fund (UNICEF). 1996. *The state of the world's children 1996*. New York: UNICEF.

———. 2000. *First global meeting of emergencies capacity building focal points: Proceedings and recommendations*. New York: UNICEF.

———. 2001. *An overview of UNICEF's humanitarian mandate and activities*. New York: EMOPS UNICEF.

United Nations General Assembly. 1996. Fifty-first session. Impact of armed conflict on children. Report of the expert of the secretary-general, Ms. Grac'a Machel, submitted pursuant to general assembly resolution 48/157.

United Nations Security Council. 2002. Report of the secretary-general on children and armed conflict. Report submitted pursuant to paragraph 15 of Security Council resolution 1379 (2001).

U.S. Agency for International Development. 2002. *Foreign aid in the national interest: Promoting freedom security, and opportunity*. Washington, DC: USAID.

Uyangoda, J. 1994. The state and the process of devolution in Sri Lanka. In *Devolution and development in Sri Lanka*, ed. S. Bastien, 83–120. Colombo, Sri Lanka: International Centre for Ethnic Studies in association with Konark Publishers.

Valentine, G. 1996. Children should be seen and not heard: The production and transgression of adults' public space. *Urban Geography* 17:205–20.

van Brabant, K. 1998. Security and humanitarian space: Perspective of an aid agency. *Humanitaires Volkerrecht* 11:14–24.

Vittachi, V. T. 1993. *Between the guns: Children as a zone of peace*. London: Hodder and Stoughton.

Walker, J. R. 1993. *Orphans of the storm: Peace-building for children of war*. Toronto: Between the Lines.

White, H. 1987. *The content of the form: Narrative discourse and historical representation*. Baltimore: The Johns Hopkins University Press.

Wilson, A. J. 2000. *Sri Lankan Tamil nationalism: Its origins and development in the nineteenth and twentieth centuries*. New Delhi, India: Penguin.

Terror, Territory, and Deterritorialization: Landscapes of Terror and the Unmaking of State Power in the Mozambican "Civil" War

Elizabeth Lunstrum

During the Mozambican "civil" war, residents across large areas of the countryside were terrorized out of their villages by the South African-backed Mozambican rebel organization Renamo. Drawing on the deterritorialization debates and investigations into the relation between territory and terror—literatures that have rarely engaged with one another—and bringing them together with interviews with survivors of the conflict, I show how Renamo unmade state power through a terror-induced deterritorialization. As the newly independent Mozambican state had attempted to build a new nation–state through communal villages as a particular ordering of space, Renamo used tactics of profound terror to destroy the lived spaces of these villages to empty them of residents but also of citizen and state. Speaking to a gap in the deterritorialization debates, this case illustrates that terror is a powerful force in realizing deterritorialization. Yet these debates, in particular their insights concerning the necessary relation between de- and reterritorialization, help clarify that terror is more accurately linked to territorialization processes rather than territory simply as space. This is a valuable addition to the literature on territory and terror and key to understanding Renamo's achievements. The de/reterritorialization coupling furthermore sheds light on the equally spatial and temporal aspects of terror and, more concretely, helps clarify why Renamo's terror stands out as particularly disturbing. Namely, Renamo effected a suspended state of deterritorialization; although it did rebuild spaces and spatial relations, this reterritorialization was ultimately aimed at social, political, and spatial annihilation and hence at ensuring the villages remain indefinitely empty. *Key Words: deterritorialization, Mozambique, territory, terror, war.*

在莫桑比克"内战"中，广大农村地区的居民由于受到南非支持的叛乱组织（莫桑比克全国抵抗运动）的恐吓而被迫离开他们的村庄。基于有关非领土化的争论和对领土和恐怖活动之间关系的研究（相关文献很少同时涉及二者），本文作者将二者放到一起，对冲突幸存者进行了采访调查，揭示了莫桑比克全国抵抗运动是如何通过恐怖活动造成的非领土化来瓦解国家权力的。当新独立的莫桑比克国曾经试图通过把社区村庄作为特定的空间构序而建立一个新的民族国家的时候，莫桑比克全国抵抗组织用高深的恐怖活动摧毁了这些村庄的生活空间，不但清空了它们的居民，而且清空了它们的公民和国家。本例强调在领地辩论中的空白，说明恐怖活动是实现领地的强大力量。然而，这些辩论，特别是他们关于领地和再区域化之间的必然联系的见解，有助于澄清恐怖活动是更准确地与地方化过程而不是仅仅作为空间的领土相关联。这是对领土和恐怖以及理解莫桑比克全国抵抗运动之成就关键的相关文献上的一笔有价值的贡献。领地和再区域化耦合更进一步揭示了恐怖活动同样的空间和时间方面的特性，更具体地说，有助于澄清为何莫桑比克全国抵抗运动很突出的特别令人不安。也就是说，莫桑比克全国抵抗运动影响了领地的悬浮状态。虽然它确实重建了空间和空间的关系，这种再区域化的最终目的是社会，政治和空间的湮没，从而确保这些村庄保持永久空置。*关键词：非领地化，莫桑比克，领土，恐怖活动，战争。*

Durante la guerra "civil" de Mozambique, los aterrorizados residentes de vastas áreas rurales fueron expulsados de sus aldeas por la organización mozambicana rebelde denominada Renamo, con apoyo sudafricano. Con datos extraídos de los debates sobre desterritorialización e investigaciones en el ámbito de la relación entre territorio y terror—literaturas que raramente se han involucrado la una con la otra—e integrándolas por medio de entrevistas practicadas a sobrevivientes del conflicto, muestro cómo Renamo deshizo el poder estatal por desterritorialización inducida por el terror. Mientras el nuevo estado independiente de Mozambique había intentado contruir una nueva nación-estado a partir de aldeas comunales, como una particular organización del espacio, Renamo utilizó tácticas de profundo terror para destruir los espacios vitales de estas aldeas, desocupándolas de residentes y también de ciudadanos y estado. Hablando de una brecha en los debates sobre desterritorialización, el presente caso deja ver que el terror es una fuerza poderosa para causar la desterritorialización. Sin embargo, estos debates, en especial las observaciones en lo que concierne a la necesaria relación entre des-y reterritorialización, ayudan a aclarar que el terror está más claramente relacionado con los procesos de territorialización que con el territorio considerado simplemente como espacio. Esta es una valiosa adición a la literatura sobre territorio y terror, y clave para entender los logros de Renamo. Todavía más, el acoplamiento desterritorialización/reterritorialización arroja luces sobre los igualmente importantes aspectos espaciales y temporales del terror y, más en concreto, ayuda a aclarar por qué el terror de Renamo se destaca como particularmente perturbador. Es decir, Renamo efectuó un estado suspendido de desterritorialización; aunque de hecho reconstruyó espacios y relaciones espaciales, la reterritorialización inoportunamente apuntaba hacia la aniquilación social, política y espacial, y por ende a asegurar que las aldeas permanecieran desiertas de manera indefinida. *Palabras clave: desterritorialización, Mozambique, territorio, terror, guerra*

As I sat down with several women from the village of Canhane, Eliasse began to explain how the violence and brutality unleashed by Renamo troops during the Mozambican "civil" war had left Canhane entirely abandoned. I then asked, rather naïvely, if they had ever talked with the Renamo soldiers. Eliasse's friend Noémia replied in a heavy tone, "How would we have done that?... Facing them made you turn pale; if you were to face them, you would lose your breath" (Canhane 30 September 2004). If any single word could capture this emotion, surely it was *terror*. From the mid-1980s until 1992, residents across large areas of Mozambique had, like Eliasse and Noémia, been terrorized out of their villages by the South African-backed Mozambican rebel organization named Renamo, short for Resistência Nacional Moçambicana. Such terror tells us much about the relation between terror and territory and in particular the ways in which terror is deployed to unmake and remake space. Bringing the experiences of survivors of the conflict in Canhane and its neighbor Massingir Velho (Figure 1) together with two literatures that have rarely engaged with one another—the deterritorialization debates and investigations into the relation between territory and terror—I show how Renamo unmade state power across large areas of rural Mozambique through a terror-induced deterritorialization. As the newly independent Mozambican state attempted to build a new nation–state through communal villages, these became one of Renamo's prime targets as it worked to dissolve state power. Employing tactics of profound terror, the rebel organization destroyed the lived spaces of these villages to empty them of residents but also of citizen and state. This case speaks to a general gap in the deterritorialization literature by illustrating that terror is a powerful force in realizing deterritorialization. Yet more than this, recognition of the necessary relation between de- and reterritorialization emerging from the deterritorialization debates makes clear that terror is more accurately linked to territorialization processes rather than territory simply understood as space. This contributes to our understanding of the relation between territory and terror and, more concretely, Renamo's achievements. This de/reterritorialization coupling also helps clarify what is unique, and uniquely disturbing, about Renamo's brand of terror-induced deterritorialization and sheds light on the equally spatial and temporal aspects of terror. Namely, Renamo effected what I refer to as a suspended state of deterritorialization. The rebel organization did reterritorialize the landscape; yet this remaking of space was ironically aimed at social, political, and spatial annihilation and hence at ensuring the villages remain indefinitely empty.

Before turning to the scholarly debates and the concrete ways in which Renamo's terror unmade territory, a note of caution is in order. Given that the conflict played out within southern Africa along with the often invisible nature of apartheid South Africa's support for Renamo, there is a danger of this conflict being read as "yet another" example of "African," "tribal," or "ethnic" "black-on-black" violence. Such commonly held assumptions are peppered with varying degrees of historical amnesia and geographical nearsightedness. I ask readers to keep in mind that the causes of Mozambique's civil war transgressed the country's borders, as the conflict was actively promoted by the apartheid South African state, a point I return to later.

Territory, Deterritorialization, and Terror

Although the concept of territory remains undertheorized, it received a theoretical boost in the form of critical rejections of the deterritorialized- or globalized-world thesis. In the early 1990s, a handful of scholars and pundits began announcing that forces of globalization were eroding the power of the nation–state and in some cases bringing about the end of territory. Globalizing phenomena—such as migrating capital, bodies, and information—were apparently becoming detached from territory and territorial boundaries and creating

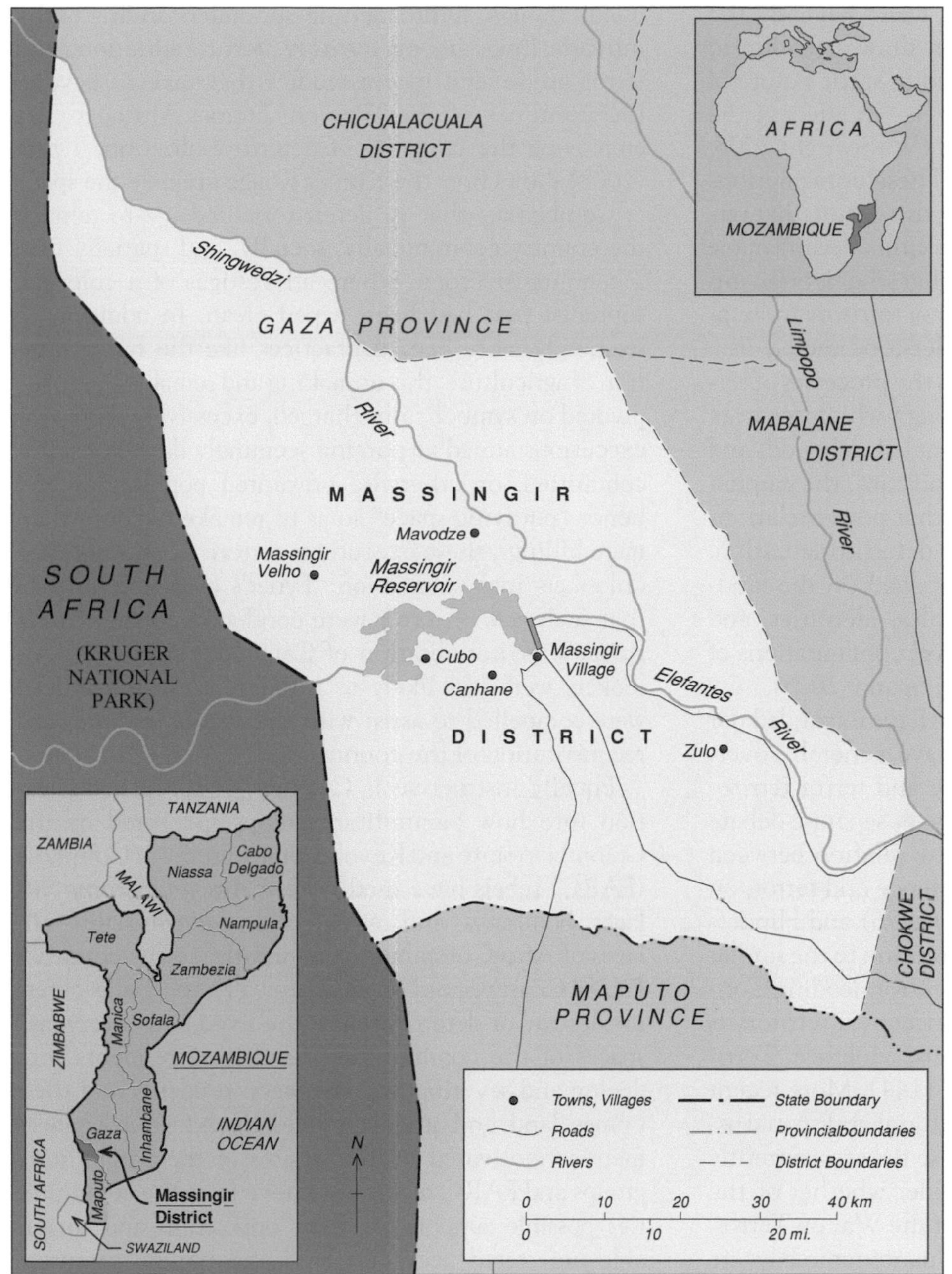

Figure 1. Mozambique's Massingir District.

a type of deterritorialized world in which territory was losing its significance (Ohmae 1995; Appadurai 1996). Many geographers and their spatially minded colleagues shot back, demonstrating that such alarmist predictions rested on theoretically and empirically weak understandings of state power, scale, and territory (Brenner 2004; Sparke 2005). In the process of laying out their criticisms, they had provided far more sophisticated theorizations of these concepts. Arguably the most important insight emerging from their critiques is that deterritorialization is never the endgame; it is always coupled with a *reterritorialization*, whether this is led by nongovernmental actors like corporations, supra- or substate actors, or the national state. Territory hence does not cease to be as deterritorialization (or globalization) takes place; it instead becomes something different as it is physically reconfigured or, in a word, reterritorialized. Ó Tuathail (1999) pushes this idea further in his insistence that "It is not simply that there is no de-territorialisation without re-territorialisation, but that both are parts of ongoing generalized processes of territorialization" (143).

Such insights have import well beyond the globalized-world debates. First, they draw attention to the dynamic and shifting as opposed to static nature of territory. It is not just that territory is "an effect of the practices that constitute it as such" (Wainwright 2008, 21), although it certainly is this. These contributions show, whether they explicitly assert this or not, that territory is actively and routinely made, unmade, and made again. Furthermore, they reinforce that what is often important in understanding the power of territory in shaping lives is not territory itself, understood merely as a container or inert space; rather, it is the processes of territorialization or the processes through which territory is fashioned, undone, reshaped, rescaled, rebound, and so on (Newman 2006). Equally significant, the emphasis on reterritorialization shows us that power relations are never erased in the process of deterritorialization. Rather, they are altered and reinscribed on the landscape, producing new markets, scales, identities, and forms of governance, as well as novel configurations of power and territory (Brenner 2004; Sparke 2005).

If the deterritorialization debates have expanded our understanding of territory, they have generally overlooked the ways in which violence and terror territorialize and thereby reshape territory. A separate debate has, however, begun to examine the relation between territory, on the one hand, and violence and terror, on the other. Scholars like Connolly (1996) and Hindess (2006) have, for instance, drawn attention to the similar etymological roots of territory and terror, leading Connolly (1996) to contend that "To occupy a territory is to receive sustenance and to exercise violence. Territory is land occupied by violence" (144). More recent scholarship has focused on the territorial and spatial aspects of the "War on Terror." Those debates currently receiving the most attention consider whether or the ways in which the defining spaces of the War on Terror, such as the U.S. Naval Station at Guantánamo Bay, are accurately characterized as Agambenian "spaces of exception" or legally ambiguous spaces or territories where the law is suspended and where state-orchestrated violence, including torture, plays out with impunity (e.g., Butler 2004; Gregory 2006; Comaroff 2007).

There has nonetheless been surprisingly little explicit engagement in these discussions with the deterritorialization debates or their central concepts, with a few notable exceptions. Elden (2007), for one, has critiqued claims that Al-Qaeda is a "deterritorialized" threat or a "network of networks" detached from or unconfined to any nation–state, showing there is a definite territorial basis to its coherence and operations. More useful, though, in broadening our understanding of the intricate links among territory, territorialization, and terror are several recent studies that take us beyond the frontlines of the War on Terror. Although not employing the language of deterritorialization, Tyner (2008) shows how the Khmer Rouge unmade the space of Cambodia—that is, deterritorialized it—to remake the country economically, socially, and spatially into a communist utopia where all vestiges of a colonial, capitalist past had been wiped clean. In addition to seemingly more benign practices like the rationalization of agriculture, this unmaking and remaking of space pivoted on symbolically charged, excessively grotesque executions aimed at purging seemingly dangerous, uncommitted, or otherwise unwanted populations and hence "purifying space" so as to remake it. More than mere killings, these executions terrorized or frightened onlookers into submission. Tyner's evidence suggests that such acts of terror were conducive to the Khmer Rouge's territorialization of Cambodia given that onlookers were less likely to interfere with and indeed were compelled to assist with the spatial undoing and reconstitution of the country.

Equally instructive is Oslender's (2007) investigation into how paramilitary groups sponsored by the Colombian state and Revolutionary Forces of Columbia (FARC) rebels have used violent threats, burning villages, bombings, and massacres to deterritorialize villages of Afro-Colombian communities in the country's Pacific Coast region. More specifically, these acts of terror destroy or deterritorialize the lived and "everyday" spaces of the communities, terrorizing residents into fleeing and severing links between residents and their homes, land, and other resources. Such forced displacement is motivated by the desires of the paramilitary groups and FARC rebels to remove from the communities possible support for their opposition and to enable unfettered access to land and natural resources once controlled by the now-displaced communities. Oslender is less explicit about reterritorialization, yet his analysis shows that this materializes in the form of both outsiders' consolidation of power over land and resources and networks of resistance spearheaded by displaced communities and their supporters. Both Oslender and Tyner are explicitly concerned with how violence and terror unmake or remake space, reflecting the dynamic understanding of territory that underpins rejections of the deterritorialized-world thesis. Yet they take one step further by showing several ways in which violence and terror transform territory. Building on and departing from these insights, I turn to examine the ways in which Renamo's terror deterritorialized communal villages as a means of dissolving state power, leaving these villages uninhabited and indefinitely uninhabitable.

Building a New Nation and State through Villagization

To understand how it was that Renamo's terror effectively deterritorialized rural villages, we first must take a step back to look at Frelimo's attempt to build a new nation–state. Upon independence in 1975, Frente de Libertação de Moçambique (Frelimo) came to power as the single-party head of the socialist state. Frelimo had inherited from the Portuguese a citizenry that was poor, overwhelmingly illiterate, and often physically difficult to reach and hence govern. The party additionally faced a nation only in the loosest sense of the term, as it lacked a common sense of identity, history, and even a common language. Frelimo in response set out to undertake a large-scale villagization project in which rural residents would be collected from their scattered distribution across the countryside and concentrated into rationally organized communal villages (Isaacman and Isaacman 1983; Newitt 1995). These would not only assist Frelimo in providing social services but also provide a space in which seemingly backward, "traditional" practices could be purged; national history taught; a sense of national unity forged; and modern, national subjects created. In short, it was through the communal villages that a modern, revolutionary, and unified nation was to be born (Lunstrum 2007).

The communal villages were also a vehicle of state formation. First, they enabled social control because it was easier to watch over and govern new national subjects if they lived near one another. The social and agricultural services enabled by the communal villages, farms, and trading networks, moreover, created a capillary system of patronage with nodal points in each village. In exchange for these services, Frelimo in theory would receive support and legitimacy from village residents. By repeating this pattern across the country, Frelimo could consolidate its power nationally as its patronage and influence would both span and unite the country (Lunstrum 2007). This new spatiality of village life in theory enabled the rise of a new nation and state under the guidance of Frelimo and amounted to Frelimo's attempt to territorialize the newly independent national landscape. Although the communal villages did not unfold exactly as planned due in part to varying degrees of peasant opposition, their implementation did reshape much of the countryside, including Canhane and Massingir Velho, whose residents were relocated into the new villages in the late 1970s.[1]

Renamo's Deterritorializing Tactics: Linking Terror, Territory, and State Unmaking

Although residents of the two villages were mostly supportive of Frelimo's vision of development, as was much of Gaza Province in the south of the country, resistance and resentment were swelling elsewhere, especially in areas of the center and north. Many residents there found Frelimo's reshaping of the countryside offensive, including the party's anti-"tradition" stance, its apparent commitment to developing Frelimo strongholds at the expense of the rest of the country, and especially villagization, which had grown progressively coercive and increasingly backed by state violence (Geffray and Pedersen 1986). Organizing around these grievances, a small number of Mozambicans united under the name Renamo. Yet from its earliest days, the group was actively shaped and encouraged by outsiders. After early support from southern Rhodesian and Portuguese agitators, Renamo became fiercely yet increasingly clandestinely backed by the anticommunist apartheid South African state, which was more than eager to train and support Renamo given the threat posed by a black-led, antiapartheid, and decidedly socialist Frelimo. Given the larger context of the Cold War, nominally anticommunist Renamo also received assistance from conservative factions in Portugal, the United States, and West Germany.

Generously supported, Renamo troops set out on a mission to destabilize Mozambique to bring down Frelimo and erase its imprint from the countryside. By the end of the war in 1992, there were 1 million casualties along with 5.5 million displaced civilians—over one third of the national population—who sought protection in the bush, larger villages and towns, and neighboring countries (Lunstrum 2007; United States Agency for International Development 2009). Although Frelimo troops were partly responsible, much of the violence throughout the country was directly attributable to Renamo insurgency and the actions of its 20,000 troops. Although it committed atrocities countrywide, Renamo reserved its most brutal acts for the south, where support for Frelimo was strongest, and where it was not possible simply to "pick off a handful of local party officials" (Hall 1990, 53).

Renamo's was a particular type of destabilization, indeed a deterritorialization, as it worked to undo Frelimo's power specifically by unraveling its territorial reordering of the countryside. Because it was through the communal villages that Frelimo had constituted space as state territory and individuals as state subjects, it was these spaces that required erasure. Renamo effected this erasure primarily through tactics of terror directed at once against the state, the peasantry, and the very structure of the communal villages. Renamo not only unmade the communal villages as spaces of daily life, similar to what we see in Oslender's analysis; it unmade these lived spaces in their concurrent role as state spaces.[2]

Within the communal villages, for example, Renamo routinely attacked local Frelimo party offices and the social services provided by the state (Vines 1991; Newitt 1995), thus denying Frelimo one of its most important sources of power, given that Frelimo gained legitimacy and patronage by providing these services. A hallmark of Renamo's tactics, its troops also performed acts of bodily mutilation that were often aimed at mocking Frelimo's power and presence in the villages. For example, drawing attention to Renamo's penchant for disfiguring and mutilating faces, Josué elaborated,

> Sometimes Renamo's men would cut out people's mouths and noses. When they cut your mouth out, your teeth would be exposed, which looked like you were laughing. So they used to say that you were laughing at the President. (Massingir Velho 1 February 2005)

While the mutilated body stood as a grotesque mockery of the President, the brutality of the act was at the same time a reminder of Frelimo's inability to protect citizens, let alone its supporters, and a reminder of just how dangerous Frelimo's national territory had become.

Yet what made Renamo's violence utterly terrorizing was that its tactics were often not aimed at (overt) signs of Frelimo or Frelimo supporters. Here is where much of its power to deterritorialize by terrorizing lay, as residents were terrorized out of the lived spaces and indeed the most intimate spaces of the communal villages, including their homes. Because Frelimo had radically restructured even the most personal spaces of community and family life through villagization, an attack on or within these spaces was simultaneously an attack on Frelimo and its attempt to modernize the countryside and peasantry. As such, Renamo soldiers actively and violently destroyed the material conditions necessary for social life and even basic existence within the villages. They systematically looted homes and stole livestock, destroyed household goods and crops and, before they moved on, set fire to houses and fields. As explained by Luciano, Renamo troops, "even burned the maize [in our fields] so that when the people thought about this, they would become discouraged and would prefer to stay where they were" (Canhane 28 September 2004).

Although the timing and targets of Renamo's violence were often unpredictable, one feature of the group's terror tactics that was predictable was that they were performed as spectacle, in many ways similar to the tactics used to unmake space in Cambodia and Colombia as highlighted by Oslender (2007) and Tyner (2008). In forcing people to witness the attacks or their aftermaths and as accounts of these atrocities spread, Renamo successfully transformed the villages into landscapes of terror and insecurity that residents were forced to abandon. For example, victims were often mutilated, having their ears, mouths, and noses cut out, and pregnant women were torn open because Renamo soldiers "wanted to know the sex of the baby" (multiple interviews, Massingir Velho and Canhane August 2004–February 2005). This violence, along with castrations and the killing of children, directly removed future support for Frelimo but also did so indirectly, as these acts terrorized survivors and witnesses into fleeing the villages. Similarly, bodies were placed on display to further terrorize populations. As Leandro recounted, "Once we were running away and Renamo soldiers killed a person and then buried him like a post" (Massingir Velho 1 February 2005). Understanding the spectacular nature of Renamo's violence also helps explain the fact that despite the fear these tactics generated, they were not altogether common. The numbers, depending on who is asked, range from three to thirteen deaths in Massingir Velho and four to five in Canhane.[3] Herein lies the "genius" of Renamo. What its troops accomplished was to terrorize and empty entire villages and entire areas of the country by unleashing violence that was strategically random, witnessed, and, above all, highly stylized and sadistically cruel. Wielding such terror, Renamo had degraded national territory into chaotic and unpredictable space, reduced citizens to refugees or internally displaced persons, and ultimately dissolved state power. In so doing, Renamo had fulfilled its mission of undoing Frelimo's territorializing of a new Mozambique. To be sure, this was a pattern reproduced throughout southern Mozambique. In fact, in Gaza Province's Massingir District (Figure 1), home to both Canhane and Massingir

Velho, Renamo's terror emptied over 90 percent of the district's twenty-one villages (Diriba, Leonhardt, and Cooke 1995).

This analysis differs from those of Oslender (2007) and Tyner (2008) in its focus on the ways in which Renamo worked to destroy the lived spaces of the communal villages as state spaces. There are nonetheless important parallels. First, the focus on the unmaking of space reflects a dynamic understanding of territory and the ways in which violence and terror do not merely play out in or occupy territory, as suggested by scholars like Connolly, but fundamentally transform it. In other words, violence and terror territorialize. Second, the terror-induced deterritorializations of Mozambique, Colombia, and Cambodia also reinforce what terror is: It amounts to acts of extreme and often symbolically charged violence used not necessarily to wipe out populations physically, as with genocide as strictly understood, but rather to scare its victims into compliance—for example, compliance in providing political endorsement, as in Cambodia, and compliance in vacating space and refusing political support for a particular political group, as in Mozambique and Colombia. These cases illustrate that terror, along with its spatial expressions and transformations, is above all a technology of governance (cf. Miller 2006), not merely a technology of destruction. Furthermore, given that acts of terror in all three contexts were directed primarily at civilians, they amount to unmistakable acts of terrorism, reminding us that the unmaking of space is a potent weapon in terrorism's arsenal.

Terror, Territory, and Renamo's Suspended State of Deterritorialization

Ending the analysis here, however, would overlook why residents not only abandoned their villages as Renamo's violence unfolded but why they stayed away until after the conflict ended in 1992. In fact, residents of Massingir Velho did not return to the village proper until several years later because they were "still afraid of the enemies, even though they didn't kill anymore" (Massingir Velho 2 February 2005). Unlike other examples of terror-induced deterritorialization, especially the Khmer Rouge's unmaking of Cambodia, which was aimed at rebuilding a new country and people, Renamo worked to ensure that its unmaking of space endured and that the villages stayed empty indefinitely. This was in effect a *suspended state of deterritorialization*. To grasp this and reflecting the insights of the deterritorialization debates, it is important to recognize that Renamo did indeed reterritorialize the landscape. With help from apartheid South Africa and others, Renamo built networks that crossed international borders and circulated money, supplies, weapons, reconnaissance, and other forms of knowledge. For instance, it developed a vast radio network that linked its bases with one another, with its headquarters in Gorongoza in central Mozambique, and with the communications exchange in Phalaborwa, South Africa. As a way of financing further attacks, moreover, Renamo developed networks to sell goods looted from the communal villages, illegally captured ivory, and so forth (Vines 1991). Also moving through these networks were soldiers and porters, many of whom had been kidnapped during Renamo raids. As Madalena recalled, Renamo soldiers found her brother working on his farm.

> They cut the sugarcane and gave it to my brother to carry. After a while, he asked to take a break. . . . They killed him to give him his last break. (Massingir Velho 5 February 2005)

In addition to underscoring Renamo's brutality, Madalena's description shows how Renamo reconstituted spatial relations and landscapes through its network of soldiers and porters and the labor they provided. It was ultimately networks like these that allowed Renamo to expand its operations and to cohere to undo Frelimo's spatial ordering and its own network of power.

This, however, proved to be a peculiar type of reterritorialization. For much of the war in the south, Renamo's tactics were aimed primarily at continuing the spatial destruction or ensuring that the communal villages remain empty. As with other reterritorializations, new markets, identities, and systems of governance emerged. One fundamental difference, though, lies in the fact that in villages like Massingir Velho and Canhane, reterritorializations were always secondary to the primary goal of spatial destruction; that is, of keeping these communal villages and Frelimo's power deterritorialized. This was a spatial unmaking in which Renamo worked to keep these spaces of the former communal villages empty, save for the weapons, soldiers, and so on, which periodically flowed through them. Renamo also left these villages empty of permanent settlement by its troops and supporters and of the possibility of constructing a new society under its direction.[4] This is illustrated by Josué's poignant observation, which also highlights the importance of cattle as a vital source of food, agricultural labor, cultural capital, and economic security: "Renamo's actions astonished me; how would they govern those people whose cattle they had stolen?" As Josué was well aware, Renamo never was interested in governing populations or even resources in much of the south beyond doing so as a means of social, economic, political, and spatial annihilation.

In other words, for most of the war in provinces like Gaza, neither Renamo nor apartheid South Africa worked to reterritorialize so as to create in any positive sense a state, nation, or territory or even a market or sense of community conducive to any type of long-term interest. In rendering the communal villages indefinitely homeless and stateless and keeping them empty of all people and activities—except for those bent on destruction—Renamo brought about a type of reterritorialization *as* deterritorialization or spatial fixity *as* unfixity. Highlighting both the spatial and temporal underpinnings of terror, this was a condition of suspended deterritorialization in which terror endured. Returning to the broader scholarly concerns, critics of the globalized-world thesis help clarify terror's aptitude for unmaking space by helping us understand that even the most destructive deterritorialization is coupled with a necessary reterritorialization. The Mozambican conflict, however, provides a profoundly disturbing relationship between the making and unmaking of territory—one effected through terror in which the reterritorialization refuses to move beyond the logic and ambition of spatial destruction.

To conclude, Connolly may be correct in his estimation that "Territory is land occupied by violence." In fact, Frelimo's implementation of the communal villages was sometimes backed by state violence, and certainly so were many of Portugal's prior colonial territorial configurations (see, for example, Isaacman and Isaacman 1983; Newitt 1995). Yet bringing together the insights of the critics of the globalized-world thesis and recent investigations into the links between territory and terror, the conflict in Mozambique demonstrates that if violence and terror underlie territory, so too do they unmake territory. In the Mozambican case, Renamo employed tactics of terror to dissolve a particular type of territory—state territory—through the destruction of the lived spaces of the communal villages. More broadly, though, it is not territory merely as a space that is associated with terror. Rather, it is this along with the practices through which territory is crafted, destroyed or undone, and ultimately built again; that is, it is the process of territorialization. Simply put, terror does not just occupy space, it radically transforms it, unmaking it and making it anew and into something terrifying. And, as I have attempted to illustrate, through Renamo's destruction of communal villages like Massingir Velho and Canhane, the connection between terror and territorialization proves particularly intimate in states of suspended deterritorialization, that is, when the goal of the territorialization is precisely to unmake territory and ensure that this spatial unmaking persists.

Acknowledgments

This article would not have been possible without the insights generously shared by residents of Massingir Velho and Canhane along with funding provided by the Fulbright Foundation, the University of Minnesota's Graduate School and Interdisciplinary Center for the Study of Global Change (ICGC), and York University's Faculty of Arts. The article has benefited greatly from the insightful critiques of Tricia Wood, Bruce Braun, three anonymous reviewers, and audiences at York University's Centre for International and Security Studies and the University of Minnesota's ICGC. All errors are my own.

Notes

1. The re-creation of Massingir Velho and Canhane into communal villages was made more urgent by the completion of the Massingir Dam in the late 1970s, which eventually would flood the two villages, requiring their relocation.
2. To comprehend this spatial undoing, it is helpful to have at least a rough understanding of what I mean by state territory: I see it as the space claimed and controlled (largely) by the state apparatus and claimed or occupied (largely) by or in the name of the nation or citizens, including the arrangements and agreements through which the space is calculated, organized, bounded, regulated, and transformed.
3. Although exact figures are difficult to attain, each village had approximately 800 to 1,000 residents in my estimation.
4. In areas of the country where Renamo had more support, it was able to develop systems of governance less bent on destruction and aimed at governing in a more positive sense (see, for example, Geffray and Pedersen 1986).

References

Appadurai, A. 1996. *Modernity at large: Cultural dimensions of globalization.* Minneapolis: University of Minnesota Press.

Brenner, N. 2004. *New state spaces: Urban governance and the rescaling of statehood.* Oxford, UK: Oxford University Press.

Butler, J. 2004. *Precarious life: The powers of mourning and violence*. London: Verso.

Comaroff, J. 2007. Terror and territory: Guantanamo and the space of contradiction. *Public Culture* 19 (2): 381–405.

Connolly, W. 1996. Tocqueville, territory, and violence. In *Challenging boundaries: Global flows, territorial identities*, ed. M. J. Shapiro and H. R. Alker, 141–64. Minneapolis: University of Minnesota Press.

Diriba, G., A. Leonhardt, and N. Cooke. 1995. Mozambique: Food security in a post-war economy: A rapid livelihood security assessment for Massingir District, Report 4. Maputo, Mozambique: Care International.

Elden, S. 2007. Terror and territory. *Antipode* 39 (5): 821–45.

Geffray, C., and M. Pedersen. 1986. Sobre a guerra na Província de Nampula [About the war in Nampula Province]. *Revista Internacional de Estudos Africanos* 4–5:303–18.

Gregory, D. 2006. The black flag: Guantanamo Bay and the space of exception. *Geografiska Annaler B* 88B (4): 405–27.

Hall, M. 1990. The Mozambican National Resistance Movement (Renamo): A study in the destruction of an African country. *Africa* 60 (1): 39–68.

Hindess, B. 2006. Terrortory. *Alternatives* 31 (3): 243–57.

Isaacman, A., and B. Isaacman. 1983. *Mozambique: From colonialism to revolution, 1900–1982*. Boulder, CO: Westview.

Lunstrum, E. 2007. The making and un-making of sovereign territory: From colonial extraction to postcolonial conservation in Mozambique's Massingir region. PhD dissertation, Department of Geography, University of Minnesota, Minneapolis.

Miller, B. 2006. The globalization of fear: Fear as a technology of governance. In *Globalization's contradictions: Geographies of discipline, destruction, and transformation*, ed. D. Conway and N. Heynen, 161–77. London and New York: Routledge.

Newitt, M. 1995. *A history of Mozambique*. Bloomington: Indiana University Press.

Newman, D. 2006. The resilience of territorial conflict in an era of globalization. In *Territoriality and conflict in an era of globalization*, ed. M. Kahler and B. Walter, 85–110. Cambridge, UK: Cambridge University Press.

Ó Tuathail, G. 1999. Borderless worlds? Problematising discourses of deterritorialisation. *Geopolitics* 4 (2): 139–54.

Ohmae, K. 1995. *The end of the nation state: The rise of regional economies*. New York: Free Press.

Oslender, U. 2007. Spaces of terror and fear on Colombia's Pacific coast: The armed conflict and forced displacement among black communities. In *Violent geographies: Fear, terror, and political violence*, ed. D. Gregory and A. Pred, 111–32. London and New York: Routledge.

Sparke, M. 2005. *In the space of theory: Postfoundational geographies of the nation-state*. Minneapolis: University of Minnesota Press.

Tyner, J. A. 2008. *The killing of Cambodia: Geography, genocide and the unmaking of space*. Burlington, VT: Ashgate.

United States Agency for International Development. 2009. Mozambique. http://www.usaid.gov/our_work/humanitarian_assistance/the_funds/galleries/csphotos/mozambique/index.html (last accessed 25 February 2009).

Vines, A. 1991. *Renamo: Terrorism in Mozambique*. Bloomington: Indiana University Press.

Wainwright, J. 2008. *Decolonizing development: Colonial power and the Maya*. Malden, MA: Blackwell.

The Geography of Conflict and Death in Belfast, Northern Ireland

Victor Mesev, Peter Shirlow, and Joni Downs

The conflict known as the "Troubles" in Northern Ireland began during the late 1960s and is defined by political and ethno-sectarian violence between state, pro-state, and anti-state forces. Reasons for the conflict are contested and complicated by social, religious, political, and cultural disputes, with much of the debate concerning the victims of violence hardened by competing propaganda-conditioning perspectives. This article introduces a database holding information on the location of individual fatalities connected with the contemporary Irish conflict. For each victim, it includes a demographic profile, home address, manner of death, and the organization responsible. Employing geographic information system (GIS) techniques, the database is used to measure, map, and analyze the spatial distribution of conflict-related deaths between 1966 and 2007 across Belfast, the capital city of Northern Ireland, with respect to levels of segregation, social and economic deprivation, and interfacing. The GIS analysis includes a kernel density estimator designed to generate smooth intensity surfaces of the conflict-related deaths by both incident and home locations. Neighborhoods with high-intensity surfaces of deaths were those with the highest levels of segregation (>90 percent Catholic or Protestant) and deprivation, and they were located near physical barriers, the so-called peacelines, between predominantly Catholic and predominantly Protestant communities. Finally, despite the onset of peace and the formation of a power-sharing and devolved administration (the Northern Ireland Assembly), disagreements remain over the responsibility and "commemoration" of victims, sentiments that still uphold division and atavistic attitudes between spatially divided Catholic and Protestant populations. *Key Words: Belfast, conflict, deaths, Northern Ireland, paramilitary, segregation*.

以“麻烦”为名的北爱尔兰冲突始于 20 世纪 60 年代后期，被定义为国家之间，支持国家，和反对国家的各种力量之间的政治和种族教派暴力。通过社会，宗教，政治和文化的争执，这场冲突的原因变得富有争议和复杂化，其中大部分关于暴力受害者的辩论由于竞争性的宣传角度而被强化。本文介绍了一个和当代爱尔兰冲突有关的个人死亡地点的信息数据库。对于每一个受害者，包括其人口普查基本状况，家庭住址，死亡方式，和对此负责的组织名称。运用地理信息系统（GIS）技术，该数据库被用于测量，测绘和分析 1966 年至 2007 年间发生于北爱尔兰的首都贝尔法斯特与冲突有关的死亡人数的空间分布，同时考虑到隔离水平，社会和经济匮乏程度以及它们之间的相互影响。地理信息系统的分析包括一个内核密度估计器，目的是为冲突相关的死亡，不论是事件或是家庭的地址，生成光滑的强度界面。具有高强度界面的街区往往是那些具有最高隔离程度（大于百分之九十为天主教或新教）和最贫困的地区，它们还多位于物理界线的附近，即那些所谓的“和平线”，分隔那些以压倒性多数的天主教或新教为主的社区。最后，尽管和平的到来和形成了权力分享和下放的管理机构（北爱尔兰议会），分歧依然存在于对受害者的责任和“纪念”上，即仍然继续分裂的情感和在空间上分隔的天主教和新教人口间的仇恨态度。*关键词：贝尔法斯特，冲突，死亡，北爱尔兰，准军事部队，隔离。*

El conflicto de Irlanda del Norte, también conocido como conflicto de "Los Líos," empezó a finales de los 1960s, y se le define en términos de violencia política y etno-sectaria entre fuerzas estatales, pro-estatales y anti-estatales. Las razones del conflicto son debatidas y complicadas a la luz de disputas sociales, religiosas, políticas y culturales, en las que gran parte de la discusión se concentra en las víctimas de una violencia endurecida por la competencia entre perspectivas condicionadas por la propaganda. Este artículo presenta una base de datos con información sobre la localización de muertes individuales conectadas con el conflicto irlandés contemporáneo. Por cada víctima, la información incluye un perfil demográfico, dirección residencial, modo de muerte y organización responsable. Por medio de técnicas de los sistemas de información geográfica (SIG), la base de datos se utiliza para medir, cartografiar y analizar la distribución espacial de muertes relacionadas con el conflicto entre 1966 y 2007 en Belfast, capital de Irlanda del Norte, con respecto a los niveles de segregación, privación social y económica, y proximidad a las barreras físicas que separan comunidades de diferente religión. El análisis del SIG incluye un estimador de la densidad kernel, diseñado para generar superficies de intensidad pareja de muertes relacionadas con el conflicto en localizaciones fortuitas y en el lugar de residencia. Los vecindarios con superficies de alta intensidad de muertes se corresponden con los que tienen altos niveles de segregación (>90 por ciento de católicos o protestantes) y privación, y localizados cerca de barreras físicas, las llamadas líneas de paz, situadas entre comunidades predominantemente católicas y comunidades predominantemente protestantes. Finalmente, pese a la implantación de la paz y la formación de una administración de poder compartido y restituído (la Asamblea de Irlanda del Norte), todavía subsisten desacuerdos en cuanto a la responsabilidad y "conmemoración" de víctimas, sentimientos que prolongan la división y actitudes atávicas entre las poblaciones católicas y protestantes espacialmente separadas. *Palabras clave: Belfast, conflicto, muertes, Irlanda del Norte, paramilitar, segregación.*

Since the 1960s the conflict in Northern Ireland has resulted in over 3,600 deaths, the maiming of at least 30,000 people, and the displacement of tens of thousands due to intimidation and political violence, which in turn has furthered residential segregation along ethno-religious boundaries. This conflict in Northern Ireland, commonly called the "Troubles," is the result of confrontation among three key sets of antagonists: the British state, which, through the use of force, aimed to maintain the majority's desire to remain within the United Kingdom; Irish Republicans, notably the Irish Republican Army (IRA), who sought to unite Ireland through physical force; and the loyalist paramilitaries, such as the Ulster Volunteer Force and Ulster Freedom Fighters, who as pro-state paramilitaries utilized violence to maintain the constitutional link with Britain.

To the world at large, the long-standing Northern Irish "problem" represents hatred and conflict between Catholics and Protestants. Indeed, despite high-profile media attention on the ideological motivations for violence by Catholics associated with republican paramilitaries, Protestants associated with loyalist paramilitaries, and the British state, the vast majority of victims were unarmed civilians (Shirlow and Murtagh 2006). To try to explain the continued political friction more fully demands appreciation of the multifaceted tangle of cross-community and within-community disputes, as well as during the postconflict period, the "commemoration" and "symbolism" of conflict, and the definition and legacy of victims and "victimhood" (Cosgrove and Daniels 1988; Verdery 1999; Aughey 2005; McEvoy 2007). This article will touch on some of these terms if only to demonstrate how research built primarily on social, cultural, and political narratives and interpretations has attempted to understand how space is contaminated by violence and how violence can lead to a moral reordering of power relationships; however, such debates on the Northern Irish conflict have traditionally lacked rigorous quantitative methodologies with which to measure and map violence. This is not to say that a conventional narrative approach lacks merit, but instead that a statistical approach can also be used to validate significance and, in so doing, challenge propaganda attached to the social messages that emerge concerning conflict (McDowell 2008).

The measurement and mapping of deaths by religion can assist a spatially sensitive analysis of inter- and intracommunity violence that is not always reproducible in current debates concerning sectarian violence and fatalities (cf. Schellenberg 1977; Murray 1982; Poole 1983; Douglas and Shirlow 1998; McPeake 1998). The analysis presented in this article utilizes a unique, spatially configured database that documents every known conflict-related death that occurred in Belfast, the capital city of Northern Ireland, during the Troubles. The database contains individual records of victims, including their demographic profile, home address, where they were killed, and the organization responsible. Geographic information systems (GIS) are used to analyze the database to complete the following objectives: (1) explore temporal trends in conflict-related deaths in terms of both victim groups and responsible party; (2) identify overall patterns and which areas were most impacted by the conflict; and (3) characterize spatial patterns of deaths specific to particular victim groups and relate these spatial patterns to socioeconomic characteristics of affected areas to identify which types of neighborhoods were disproportionally affected by the conflict. These objectives are designed to help unravel the intricate spatial patchwork of conflict-related deaths in Belfast and begin to appreciate the emotive symptoms of the deep social, cultural, and political struggles for which Northern Ireland has become notorious. Moreover, our approach aims to document a reasoned understanding of how the issues of conflict, interpretations of victimhood, and resolution of peace can be explored through the mapping of suffering combined with a balanced debate on the spatial location of victims.

Data and Methods

The database is built from a unique blend of information derived from official government sources, police records, media reports, and work undertaken by the second author with paramilitary and victim groups. Other more general sources based on human rights violations and journalistic accounts are also available, although they have not undertaken such a specific spatial analysis (examples include Amnesty data [Amnesty International 1994; Sutton 1994; Fay, Morrissey, and Smyth 1998; McKittrick et al. 2004]). Reported in Mesev et al. (2008), our work is a comprehensive store of information on conflict-related deaths that occurred during the Troubles in Northern Ireland, which is part of the United Kingdom (Figure 1), between 1967 and 2007. The database on Belfast holds details on 1,601 victims (Figure 2). Each victim is listed by name, gender, age, religion, postcode of home address, date and postcode of incident, and type of death (method and responsible party). Queries to the database produce maps of the spatial patterns of conflict-related deaths, with specific relation to levels of intracommunity violence, segregation, deprivation, and physical divisions.

Figure 1. Belfast, Northern Ireland.

Victim Groups and Responsible Parties

Conflict-related deaths are summarized by five major groups of victims: (1) republican paramilitary, (2) loyalist paramilitary, (3) state and local government, (4) Protestant civilian, and (5) Catholic civilian. The republican paramilitary group includes the Irish Republican Army (IRA), the Official Irish Republican Army (IRA O), Continuity (CIRA) and Real Irish Republican Army (RIRA), the Irish National Liberation Army (INLA), the Civilian Defence League (CDL), and Direct Action Against Drugs (DAAD; alleged as a front group for the IRA). Loyalist paramilitaries include the Ulster Volunteer Force (UVF), the Ulster Defence Association/Ulster Freedom Fighters (UDA/UFF), the Red Hand Commando (RHC), the Red Hand Defenders (RHD), the Protestant Action Group (PAG), and the Protestant Action Force (PAF). The third group is the state and local government, which includes the British Army (BA), Ulster Defence Regiment (UDR), and the Royal Ulster Constabulary (RUC), as well as judges, prisoner officers, and other civil servants. Persons not associated with any of these three groups were considered civilians, with Catholics and Protestants[1] treated separately. Maps illustrating temporal patterns of conflict-related deaths by each of these five groups are summarized by both counts of victims and responsible party.

Spatial Distribution of Deaths

The goal of the methodology is to both characterize overall trends in the geographic distribution of conflict-related deaths in Belfast as well as identify spatial patterns unique to each victim group. The initial stage involves the identification of areas across the city—in terms of both the locations of the incidents and the homes of the victims—that were most affected by the Troubles. Because incident locations and home addresses are georeferenced to U.K. postcodes (these are spatial units of approximately fourteen households), and because the impact of a death in a neighborhood would likely extend some distance beyond its exact position in space, a GIS data smoothing technique is deployed to quantify more accurately the impacted areas of the city compared to simply plotting the individual locations. The technique chosen is a kernel density estimation procedure (KDE), a data smoothing technique (Silverman 1986) commonly used by geographers to identify "hot spots" of point event locations (Fotheringham, Brunsdon, and Charlton 2000; Downs and Horner

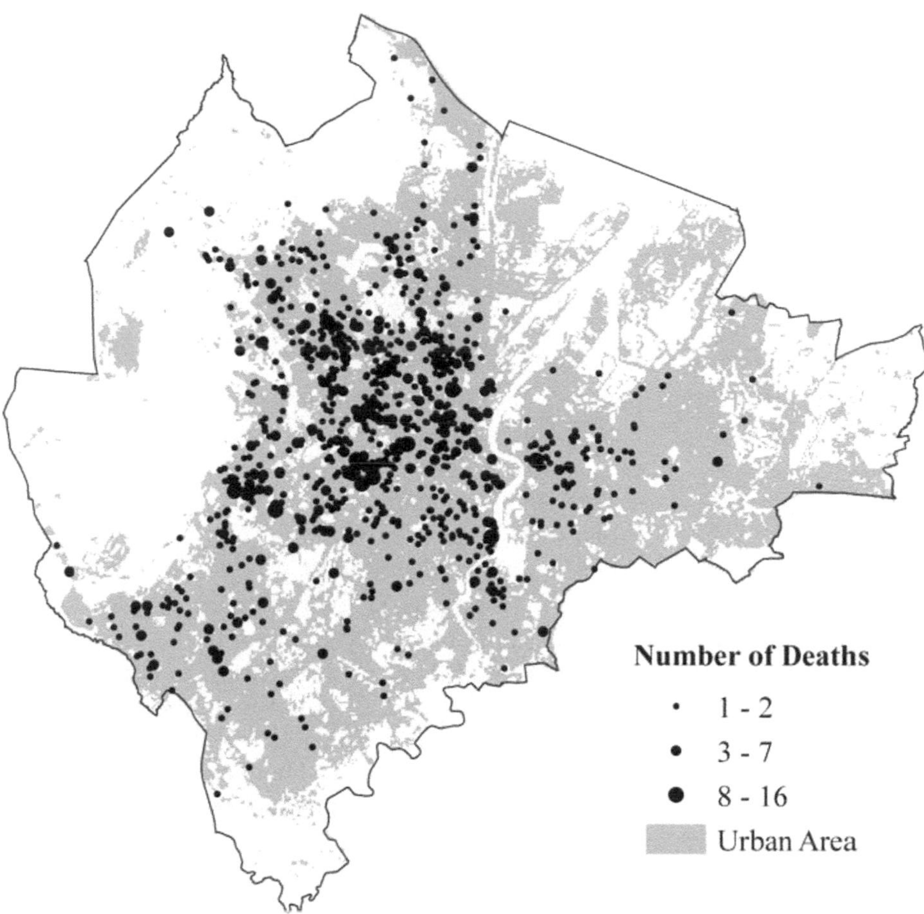

Figure 2. Distribution of conflict-related deaths in the Belfast urban area, 1966–2007.

2008). KDE smooths a pattern of event points using a distance-weighting function, or kernel, which creates a continuous intensity surface where higher intensities occur in areas of the map that are located close to a large number of events. The amount of smoothing produced by the kernel is controlled by its bandwidth, which effectively specifies how far the influence of each event extends into space. Here, KDE is used to characterize the spatial intensity of both incident locations and home addresses of victims by using a Gaussian kernel with bandwidths of 500 m. This bandwidth is selected based on the supposition that the highest impact would occur within a half-kilometer radius of the location of an incident or home of a victim. To define the highest impacted areas, contour lines are generated from the intensity surfaces at 25 percent, 50 percent, 75 percent, and 95 percent intervals to delineate the spatial limits of the conflict. For example, the 25 percent contour line is interpreted as the smallest area within which 25 percent of the measured intensity occurs. The analysis is repeated, using the same set of parameters, for collections of incident locations for each of the five major groups of victims. The intensity contours are then used to relate the spatial distributions of incident locations to population density, as well as the spatial distribution of the homes of victims to socioeconomic factors.

Socioeconomic and Other Factors

Other studies have suggested that several socioeconomic and geographic factors might have influenced patterns of conflict-related incidents in Northern Ireland, including the degree of religious segregation within neighborhoods, proximity to physical interfaces separating predominantly Catholic from predominantly Protestant communities, population density, and levels of deprivation (Shirlow and Murtagh 2006). Here we use GIS-based techniques to explore the relationships among these four factors and both overall trends in conflict-related deaths in Belfast and spatial patterns for specific victim groups. Data were obtained on the religious makeup, population density, and levels of deprivation from the 1991 U.K. Census of Population recorded at the enumeration district (ED) level. In total there were 575 EDs in the Belfast study area in 1991 with a mean population of 486 persons per ED.

Equivalent information was not collected or geographically referenced for previous population censuses in Northern Ireland, so only 1991 data were used as surrogates for the duration of the entire conflict. Although socioeconomic characteristics likely varied somewhat during this time period, other studies suggest that spatial patterns, especially with respect to religious segregation, have remained relatively stable over much of the time period (Doherty and Poole 1997) and justify use of this data set in the absence of any viable alternative.

Once census data were accessed, the next stage involved comparing overall trends in the spatial intensity of incident locations and home addresses of victims to the spatial distribution of population density and levels of deprivation using GIS-based mapping techniques. For deprivation, the Robson index (Robson, Bradford, and Dees 1994) is used as a surrogate measure for social and economic deprivation. The Robson index incorporates nine measures of health, education, family, shelter, physical environment, income, and jobs to quantify deprivation. Greater positive Robson index values indicate increasing levels of social deprivation, whereas negative values indicate the opposite trend. Population density and Robson index values for EDs in Belfast are then both mapped to visualize their spatial distribution. Next, the intensity contours are overlain for both incident locations and homes of victims to visualize how the surfaces relate to both population density and Robson index values. At this stage, basic summary statistics are also computed to compare Robson index values for incident locations with similar values computed for the city of Belfast as a whole.

For the purposes of this study, EDs were assigned to one of three categories based on religious composition, as per Shirlow and Murtagh (2006). Predominantly Catholic EDs are defined as those with >90 percent Catholic residents and predominantly Protestant EDs are defined as those with >90 percent Protestant residents. "Intermixed'" EDs are characterized as those with a dominant religious community that comprised <90 percent of the total number of residents. After mapping the spatial distribution of EDs on this basis, the layer is used for two main purposes. First, intensity contours are overlain for each victim group over the map of religious distribution to visualize how well they correspond. Then, the number of deaths for each victim group that fell within the three categories of EDs is computed to help understand the spatial dimensions of the conflict with respect to religious segregation.

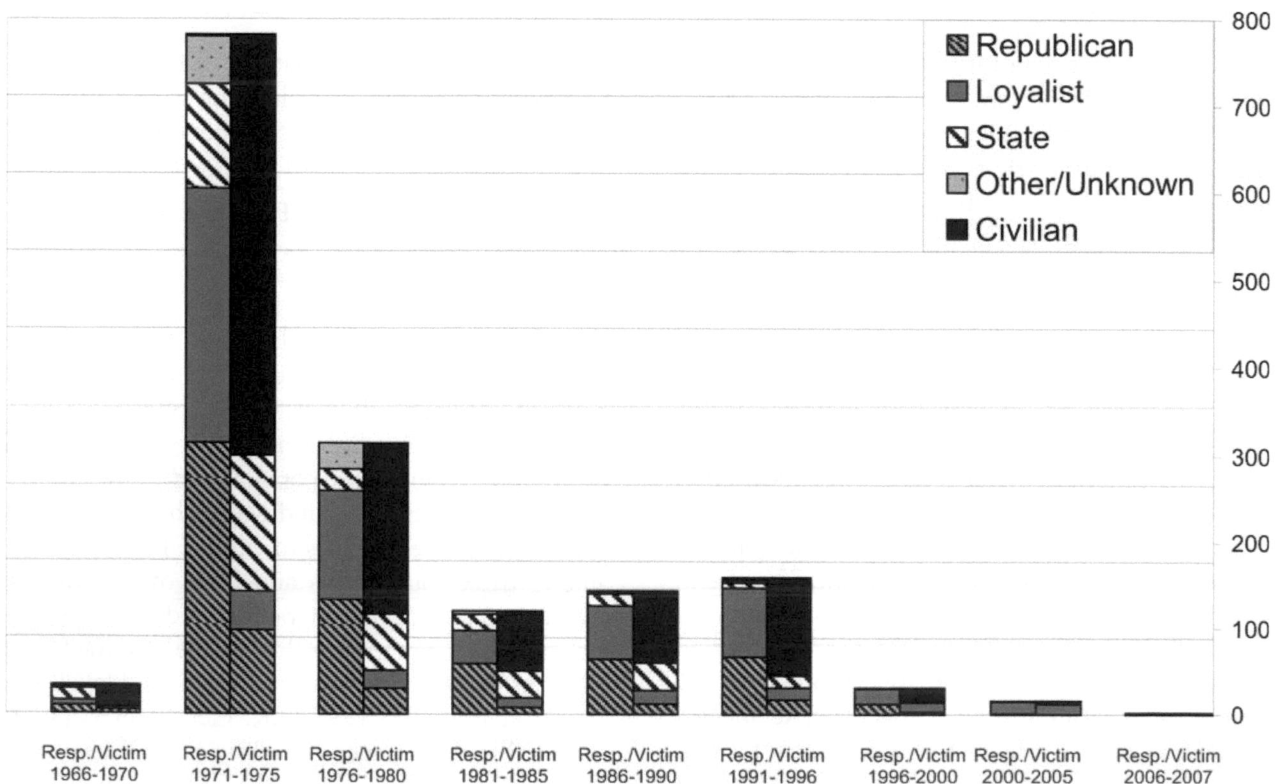

Figure 3. Deaths by victim group and responsible party, 1966–2007.

The final stage of the methodology determines whether significant numbers of conflict-related deaths are located within close spatial proximity of community interfaces and physical barriers. These barriers are euphemistically known as *peacelines*,[2] and were erected to separate predominantly Catholic and predominantly Protestant communities from one another. Again, intensity contours are compared to overall and victim-specific patterns of deaths in Belfast using mapping techniques as previously described for other factors. Once overlain, the comparison determines the spatial distance between incident locations and physical barriers with the same calculation for random points within the urban area of Belfast to measure whether deaths tended to occur closer to barriers than would be expected by chance.

Results

Victim Groups and Responsible Parties

Figure 3 shows temporal trends in conflict-related violence in Belfast, with the greatest number of deaths occurring during the 1970s, much lower levels during the 1980s and early 1990s, and very few occurring after 1995. Figure 3 also compares the distribution of responsible parties and victims for each of the five major groups over the duration of the conflict. The most striking feature is that the republican and loyalist paramilitary groups tended to be responsible for a greater number of deaths than they had victims, whereas the state and local government group sustained more deaths than those for which they were directly responsible. Civilians sustained a large proportion of casualties throughout the conflict, but as a whole, they were responsible for very few of the deaths.

Tables 1 through 3 list the numbers of deaths per victim group by more specific responsible parties and causes. Catholic civilians sustained the greatest number of victims (677) with various loyalist paramilitary groups (63 percent), the British Army (12 percent), and republican paramilitaries (17 percent) responsible for the majority of deaths. Protestant civilian causalities were 307 in total, with republican and loyalist paramilitary groups responsible for 57 percent and 32 percent of the deaths, respectively. Only twelve Protestant civilians were killed by the British Army. State and local government victims (303) largely comprised of members of the British Army, RUC, and UDR, were almost exclusively victims of republican paramilitary groups, primarily the IRA (85 percent). Loyalist deaths (130)

Table 1. Civilian deaths by responsible party

Responsible party		Civilian victim		
Category	Group	Catholic	Protestant	Other/ unknown
Other	Accidental	5	2	1
	Unknown	20	12	0
	Trauma	4	1	0
State	British Army	84	12	2
	RUC	14	1	0
	UDR	2	0	0
Republican	IRA	80	102	3
	Republican	9	53	0
	IRA(O)	8	3	0
	CIRA	2	0	0
	INLA	7	9	0
	IPLO	3	7	0
	CDL	0	1	0
	DAAD	7	0	0
	RIRA	1	0	0
Loyalist	Loyalist	22	15	1
	LVF	4	0	0
	UVF	178	39	1
	RHC	5	2	0
	RHD	2	0	0
	UDA/UFF	205	42	2
	PAF	12	0	0
	PAG	1	0	0
Civilian	Catholic	0	6	0
	Protestant	2	0	0
	Total	677	307	10

Notes: RUC = Royal Ulster Constabulary; UDR = Ulster Defence Regiment; IRA = Irish Republican Army; IRA (O) = Official Irish Republican Army; CIRA = Continuity Irish Republican Army; INLA = Irish National Liberation Army; IPLO = Irish People's Liberation Organisation; CDL = Civilian Defence League; DAAD = Direct Action Against Drugs; RIRA = Real Irish Republican Army; LVF = Loyalist Volunteer Force; UVF = Ulster Volunteer Force; RHC = Red Hand Commando; RHD = Red Hand Defenders; UDAUFF = Ulster Defence Association/Ulster Freedom Fighters; PAF = Protestant Action Force; PAG = Protestant Action Group.

were shared almost evenly between republican and loyalist paramilitaries, the state group, and accidents. Republicans (171) were largely victims of the British Army (28 percent), their own group (30 percent), and loyalist paramilitaries (15 percent), as well as a substantial proportion of accidents (23 percent); however, if these data are analyzed even more carefully, less well-known messages emerge.

Of the 171 victims of republican violence who were noncombatants, 40 were IRA members killed due to accidents (usually premature explosions). Nearly twice as many republicans (50) died due to internal republican feuding or allegations of being informants compared to the 26 loyalists killed by republicans. Among the 130

Table 2. State deaths by responsible party

Responsible party		State victim			
Category	Group	British Army	RUC	UDR	Other
Other	Accidental	0	0	1	0
	Unknown	0	2	0	1
State	British Army	5	1	0	0
	RUC	1	0	0	1
	UDR	0	2	1	0
Republican	IRA	155	57	16	29
	Republican	1	0	1	2
	IRA(O)	2	0	0	0
	INLA	3	5	2	1
	IPLO	0	1	0	1
Loyalist	Loyalist	1	0	1	0
	UVF	0	2	1	3
	UDA/UFF	0	2	2	0

Notes: RUC = Royal Ulster Constabulary; UDR = Ulster Defence Regiment; IRA = Irish Republican Army; IRA (O) = Official Irish Republican Army; INLA = Irish National Liberation Army; IPLO = Irish People's Liberation Organisation; UVF = Ulster Volunteer Force; UDA/UFF = Ulster Defence Association/Ulster Freedom Fighters.

Table 3. Loyalist deaths by responsible party

Responsible party		Loyalist victim		
Category	Group	UVF	UDA/UFF	Other
Other	Accident	5	5	0
State	British Army	4	9	0
	RUC	1	2	0
	UDR	1	1	0
Republican	IRA	8	19	0
	Republican	0	2	0
	INLA	3	1	1
	IPLO	0	2	1
Loyalist	LVF	1	1	0
	UVF	14	15	3
	RHD	0	1	0
	UDA/UFF	6	20	4
	Total	43	78	9

Notes: UVF = Ulster Volunteer Force; UDA/UFF = Ulster Defence Association/Ulster Freedom Fighters; RUC = Royal Ulster Constabulary; UDR = Ulster Defence Regiment; IRA = Irish Republican Army; INLA = Irish National Liberation Army; IPLO = Irish People's Liberation Organisation; LVF = Loyalist Volunteer Force; RHD = Red Hand Defenders.

combatant victims of loyalist violence, excluding those who died accidentally, a similar pattern emerges, with 37 deaths of republicans and 65 deaths of loyalists. In both instances there were more victims of intraloyalist and republican violence than there were between these combatant groups. As a footnote, high levels of within-community violence are rarely measured or publicized, and they are most certainly not an enduring part of the Northern Irish conflict that the rest of the world immediately recalls.

Overall Spatial Patterns

Overall trends in the spatial distribution of conflict-related deaths by incident location and home address

Figure 4. (a) Spatial intensity of incident locations (25 percent, 50 percent, 75 percent, and 95 percent contours) in Belfast compared to population density; and (b) spatial intensity of homes of victims compared to levels of social deprivation as measured by the Robson index.

Figure 5. Spatial intensities of incident locations (25 percent, 50 percent, 75 percent, and 95 percent contours) by victim group compared to religious classification of ED and presence of physical interfaces, by (a) Catholic civilians, (b) Protestant civilians, (c) state, (d) loyalist paramilitaries, and (e) republican paramilitaries.

of victims displayed almost identical patterns (Figure 4). This is consistent with the findings by Shirlow and Murtagh (2006), who documented that most victims were killed very close to their homes. Figure 4a relates the intensity of deaths to population density, illustrating how deaths tended to occur in more densely populated areas of the city, with the highest concentration found in the North and West Belfast Parliamentary

Constituencies. In terms of economic distributions, Figure 4b illustrates how victims tended to reside in more deprived areas of the city, as measured by the Robson index, whereas relatively affluent communities experienced very little impact. The burden of violence on deprived communities in comparison to more affluent places creates an uneven interpretation of violence between such places. Robson index values for the city as a whole ranged from –11.5 to 14.8, although 77 percent of deaths occurred in EDs with Robson index values greater than 0 and 44 percent occurred in areas with values greater than 5.0. Lastly, the analysis on proximity to peacelines found that deaths by location of incidents were located at a mean 517 m ($\sigma = 539$ m) from physical interfaces, compared to 1,412 m ($\sigma = 1,162$ m) for random points located across the Belfast urban area.

Victim Group-Specific Spatial Patterns

The five victim groups displayed different patterns of spatial intensity of conflict-related deaths in Belfast, particularly when compared to the religious classification of EDs and presence of physical interfaces separating religious communities (Figure 5). Incidents of Catholic civilian deaths displayed the most widespread distribution of all the victim groups, with higher intensities occurring in predominantly Catholic neighborhoods as well as near community interfaces (Figure 5a). Fifty-six percent of Catholic deaths occurred in predominantly Catholic neighborhoods, and only 24 percent and 20 percent occurred in predominantly Protestant and intermixed neighborhoods, respectively. On the other hand, higher intensities of Protestant civilian deaths were observed in predominantly Protestant neighborhoods that were located near community interfaces, with 56 percent occurring in EDs classified as such. Approximately 47 percent of Protestant civilians died in predominantly Catholic EDs, with most of these occurring near barriers separating segregated neighborhoods (Figure 5b). Intensities of deaths of state and local government victims showed a similar pattern to that observed for Catholic civilians, with 80 percent of those deaths occurring in predominantly Catholic EDs (Figure 5c). The intensity contour lines also tended to match the spatial pattern of physical interfaces, with the highest intensities observed in predominantly Catholic neighborhoods that were bordered by predominantly Protestant EDs.

Spatial patterns of paramilitary groups also showed distinct trends related to religious segregation and the presence of community interfaces. Seventy-seven percent of republican deaths occurred in predominantly Catholic EDs, with the highest intensities occurring near interfaces in the center of the city (Figure 5d). Only 13 percent occurred in intermixed EDs and 10 percent in predominantly Protestant EDs. Loyalist deaths displayed the opposite corresponding pattern, with 70 percent occurring in predominantly Protestant EDs, 20 percent in intermixed areas, and 10 percent in predominantly Catholic neighborhoods (Figure 5e). Again, the highest intensities of deaths occurred in highly segregated neighborhoods in the center of Belfast that were also proximal to physical interfaces.

Discussion and Conclusions

Northern Ireland has become synonymous with conflict and ethno-religious sectarian violence. Experts have focused debate extensively on the political, paramilitary, religious, and cultural factors driving the conflict, yet a GIS analysis also has an important role in helping to quantify and visualize spatial relationships between the location of violence and potential explanatory variables, such as segregation, deprivation, and physical divisional barriers. To that extent, the GIS results outlined in this article suggest that the vast majority of fatalities in Belfast during the conflict occurred within segregated communities composed of over 90 percent Catholics or Protestants, within areas of high deprivation as measured by the Robson index, and close to peacelines. Although these results are not wholly unexpected, the spatial analysis provides two major contributions to the established literature. First, this work demonstrates that poorer, more segregated neighborhoods in western, northern, and central Belfast suffered more fatalities and were disproportionally affected by the conflict compared to their more affluent counterparts in the southern and eastern portions of the city. Second, the GIS approach provides a structured methodology with which to help resolve locational disputes by facilitating the precise geo-referencing of violence, along with providing rapid access to a consistent database from which public policies may be formulated and tested (Imrie, Pinch, and Boyle 1996). More important, this statistical approach to mapping conflict and suffering can be used to generate a more balanced debate over the legitimacy of victims, interpretations of victimhood, and the responsibility of involved parties, all of which have implications for the resolution of peace.

Further, the database on conflict-related deaths in Belfast specifically illuminates some thoughts on victim representation, especially when challenging monocultural notions of responsibility. In general it is understood that during the Troubles, republicans fought an anticolonial struggle against state repression and loyalist intimidation. In contrast, loyalists viewed the war as conditioned by protecting the Protestant people and repelling the threat from armed republicans. Less is certain about the role of the state as a defender because there are no accurate figures concerning the extent of collusion and thus state violence can only be interpreted by known responsibility; however, when analyzing the database, some more cogent conclusions emerge regarding the nature of the conflict. The most striking finding is the comparison of violence within and between republican and loyalist paramilitary groups. Although the conflict is often viewed as a struggle between these two conflicting forces, the reality is that a greater number of fatalities were caused by members of victim's own group than the opposition. In spatial terms, the fact that republican and loyalist violence took place mostly within their own territories also undermines the notion that they often engaged in "heroic" actions behind "enemy lines." The reality of violence is that loyalists usually killed republicans within predominantly Protestant territories, and that republicans often killed loyalists within predominantly Catholic neighborhoods. Similarly, much higher levels of state fatalities occurred in predominantly Catholic communities, which helps explain why highly segregated Catholic neighborhoods in the center of Belfast experienced such a high intensity of the conflict compared to other areas of the city.

Additionally, the analysis shows that, although the conflict has been viewed largely as a struggle among republicans, loyalists, and state forces, it was the civilians, both Catholic and Protestant, that suffered by far the greatest number of casualties. At the same time, civilians were responsible for very few deaths, which perhaps legitimizes their status as victims of the conflict. Catholic and Protestant civilian victims share this legacy, but the GIS analysis of their patterns of deaths points to divergent experiences of conflict with respect to spatial location and responsible parties, because nearly one fourth of Catholic civilian deaths occurred in predominantly Protestant neighborhoods compared to 47 percent of all Protestant civilians being killed in predominantly Catholic EDs. Although many people died close to their homes throughout the course of the conflict, particularly for members of paramilitary groups, the reality is that many civilians were killed when they entered neighborhoods dominated by the "other" religion.

In terms of conflict resolution, there is very little doubt that dialogue between community representatives has improved during recent years. So too attempts by some, many of whom are former combatants, to challenge some of the many fractures within Northern Irish society (Shirlow and McEvoy 2008). Memories are still fresh, however, and the commemoration of violence still plays an important part in the minds of people in Northern Ireland. The paramilitary ceasefires of 1994 and the restoration of political devolution and a power-sharing executive have each been progressive acts that have helped reduced the volume of political violence (Aughey 2005); however, the legacies of hostility remind the world of the complex nature of mistrust that remains between spatially polarized communities. That mistrust has also been evident in the ideologically charged battleground within which the commemoration of violence and the demand for investigations has emerged in the postconflict period. The demand for public inquiries has generally emanated within a selective understanding of victimhood that exploits selective descriptions of harm to citizens[3] (Shirlow and Murtagh 2006). In this postconflict situation a common understanding of this term, *victimhood*, is that the community a person belonged to is composed of bona fide victims. This does not mean that there was no sympathy for victims within the other community but that placing those victims on an equal plane would weaken political legitimacy.

Further on peace resolution, recovery groups are working across intercommunity frameworks with most offering counseling and therapy to survivors of the conflict. The more publicly known victims' groups generally operate along a division between pro-unionist (and strongly anti-paramilitary) groups and anti-British state groups. The main unionist political party, the Democratic Unionist Party (DUP), also centers their victims debate through prioritizing the victims of paramilitaries. Again these sentiments are understood in the terms of "rights" and a hierarchical understanding of victims. In the context set by Relatives for Justice, the state is not understood as protector, as assumed by unionists, but as state "terrorists" who facilitated assassinations, armed loyalist paramilitaries, and failed to examine such collusion rigorously. Within the intercommunity vision of postconflict recovery groups such as the Victims and Survivors trust believe that "the most fitting memorial to **all** of the victims of the war on the two islands of Ireland and Britain will be a peaceful outcome brought about through dialogue, understanding and the promotion of the truth." Emphasis on the word *all* evokes a discourse of mutual engagement, the tackling of isolation and conflict-related harm, and the removal of barriers to civic participation among survivors, irrespective of their background. "Promotion of truth," too, is a key phrase, as it suggests part of the peace resolution process is acknowledging the true nature of the conflict and coming to terms with it. Finally, this article has shown how the mapping and GIS analysis of patterns of conflict-related deaths can be used to uncover new patterns in their distribution and contribute to debates on the responsibility of involved parties and legacies of victims.

Notes

1. The terms *Protestant* and *Catholic* are used, although the conflict in Northern Ireland was not driven by religious dogma but the variant constitutional beliefs of these communities. Catholics and Protestants do not necessarily view themselves as either Irish or British, respectively; however, in the main discourses that constitute the notion of identity in Northern Ireland, the terms Protestant and Catholic are used to delineate the construction of victimhood.
2. Interfaces are denoted as the zones between Catholic and Protestant communities. Peacelines are defensive physical walls that were built to reduce the capacity to cross between communities. We derived our location of peacelines from high spatial resolution imagery of Belfast taken in 1999.
3. It should be acknowledged that former loyalist and republican prisoner groups are debating the nature and consequences of their violence.

References

Amnesty International. 1994. *Political killings in Northern Ireland*. London: Amnesty International.

Aughey, A. 2005. *The politics of Northern Ireland: Beyond the Belfast Agreement*. London and New York: Routledge.

Cosgrove, D., and S. Daniels. 1988. *The iconography of landscape: Essays on the symbolic representation, design and use of past*. Cambridge, UK: Cambridge University Press.

Doherty, P. M., and A. Poole. 1997. Ethnic residential segregation in Belfast, Northern Ireland, 1971–1991. *Geographical Review* 87:520–36.

Douglas, N., and P. Shirlow. 1998. People in conflict in place: The case of Northern Ireland. *Political Geography* 17:125–28.

Downs, J. A., and M. W. Horner. 2008. Effects of point pattern shape on home range estimates. *Journal of Wildlife Management* 72:1813–18.

Fay, M. T., M. Morrissey, and M. Smyth. 1998. *Mapping troubles-related deaths in Northern Ireland 1969–1998*. Londonderry, UK: INCORE.

Fotheringham, A. S., C. Brunsdon, and M. Charlton. 2000. *Quantitative geography: Perspectives on spatial data analysis*. London: Sage.

Imrie, R., S. Pinch, and M. Boyle. 1996. Identity, citizenship and power in cities. *Urban Studies* 33:1255–61.

McDowell, S. 2008. Commemorating dead "men": Gendering the past and present in post-conflict Northern Ireland. *Gender, Place and Culture* 15 (4): 335–54.

McEvoy, K. 2007. *Truth, transition and reconciliation: Dealing with the past in Northern Ireland*. Portland, OR: Willan.

McKittrick, D., S. Kelters, B. Feeney, C. Thornton, and D. McVea. 2004. *Lost lives: The stories of the men, women and children who died as a result of the Northern Ireland Troubles*. Edinburgh, UK: Mainstream.

McPeake, J. 1998. Religion and residential search behaviour in the Belfast urban area. *Urban Studies* 13: 527–48.

Mesev, V., R. S. Courtney, J. A. Downs, P. Shirlow, and A. Binns. 2008. Measuring and mapping conflict-related deaths and segregation: lessons from the Belfast "troubles." In *Geospatial technologies and homeland security: Research frontiers and future challenges*, ed. D. Z. Sui, 83–102. New York: Springer.

Murray, R. 1982. Political violence in Northern Ireland 1969–1977. In *Integration and division: Geographical perspectives in the Northern Ireland problem*, ed. F. W. Boal and J. N. H. Douglas, 42–64. London: Academic.

Poole, M. 1983. The demography of violence. In *Northern Ireland: The background to the conflict*, ed. J. Darby, 18–37. Belfast, UK: Appletree Press.

Robson, B., M. Bradford, and I. Dees. 1994. Relative deprivation in Northern Ireland. Occasional Paper No. 28, Centre for Urban Policy Studies, Policy Planning and Research Unit, Manchester University.

Schellenberg, J. A. 1977. Area variations of violence in Northern Ireland. *Sociological Focus* 10:69–78.

Shirlow, P., and K. McEvoy. 2008. *Beyond the wire: Former prisoners and conflict transformation in Northern Ireland*. London: Pluto Press.

Shirlow, P., and B. Murtagh. 2006. *Belfast: Segregation, violence and the city*. London: Pluto Press.

Silverman, B. W. 1986. *Density estimation for statistics and data analysis*. London: Chapman Hall.

Sutton, M. 1994. *An index of deaths from the conflict in Ireland 1969–1993*. Belfast, UK: Beyond the Pale.

Verdery, K. 1999. *The political lives of dead bodies: Reburial and postsocialist change*. New York: Columbia University Press.

What Counts as the Politics and Practice of Security, and Where? Devolution and Immigrant Insecurity after 9/11

Mathew Coleman

If critical geopolitics seeks to upend practices of statecraft as well as mainstream research about it, then the danger is that it does so in terms of spatial structures of intelligibility provided by the latter. I deal with a particular aspect of this problem: how, despite broadening the security agenda, critical geopolitics has for the most part treated geopolitics and security as synonymous with foreign policy and foreign policy studies. One important consequence, as feminist political geographers argue, is that the state and statecraft are treated as abstract forces that float above the contingencies of everyday lives and spaces. To contribute to rethinking the scales of geopolitics and security, I look at the devolution of immigration enforcement in the United States after 11 September 2001 (hereinafter 9/11). So-called 287(g) and inherent authority—two chief elements of post-9/11 local-scale immigration enforcement—have come together to constitute a microgeopolitics of risk intensification for undocumented immigrants in the United States. 287(g) deputizes nonfederal officers as immigration agents; inherent authority empowers nonfederal police to enforce immigration law without cross-designation. *Key Words: 287(g), critical geopolitics, immigration enforcement, inherent authority, microgeopolitics, security, state.*

如果批判地缘政治学所寻求的是颠倒治国实践以及对此的主流研究，那么危险的是，它这样做正是基于在后者所提供的可理解性的空间结构的条件下。我对这个问题的某一特别方面进行了分析：不管安全议程的扩大化，批判地缘政治学是如何将地缘政治学和安全的大部分当作外交政策和外交政策研究的同义词。正如女权主义的政治地理学家所认为的那样，一个重要的后果就是国家和治国之道被当作一种抽象的力量漂浮于日常生活和空间的偶然性之上。为了有助于重新考虑地缘政治和安全的尺度，本文作者对 2001 年 9 月 11 日（以下简称 9 / 11）之后美国的移民执法权力下放现象进行了分析。所谓的 287（g）法案和固有的权力，是 9/11 之后地区级别移民执法的两个主要因素，两者集合在一起，对美国的无证移民构成了一种微观政治地理的风险。 287（g）法案授权非联邦官员以移民官员的权力：其固有的权力使得非联邦警察可以无需交叉任命即可执行移民法。*关键字：287 (g)，批判地缘政治学，移民执法，固有的权力，微观政治地理学，安全，州。*

Si lo que busca la geopolítica crítica es cuestionar las prácticas del manejo del estado lo mismo que la investigación centrada en este tema, queda entonces el peligro de que esto se haga en términos de las estructuras espaciales de inteligibilidad provistas por la investigación misma. Me ocupo de un aspecto muy particular de este problema: cómo, pese a la ampliación de la agenda de seguridad, por lo general la geopolítica crítica considera a la geopolítica y la seguridasd como sinónimos de políticas extranjeras y estudios de política externa. Una consecuencia importante de esto, como arguyen los geógrafos políticos feministas, es considearar al estado y la política como fuerzas abstractas que flotan por encima de las contingencias de vidas y espacios cotidianos. A título de aporte para repensar las escalas de la geopolítica y la seguridad, concentro mi atención sobre la estricta aplicación del control de inmigración de los Estados Unidos, después del 11 de septiembre de 2001 (en lo sucesivo 9/11). La así llamada norma 287(g) y el principio de autoridad inherente—los dos elementos principales que incrementan el rigor de la ley de inmigración a escala local—han llegado a constituirse juntos en una microgeopolítica de intensificación del riesgo para inmigrantes indocumentados de los Estados Unidos. La 287(g) permite habilitar a funcionarios oficiales fuera del sistema federal como agentes de inmigración; la autoridad inherente faculta a la policía local para aplicar la ley de inmigración sin que medie designación expresa. *Palabras clave: 287(g), geopolítica crítica, control estricto de inmigración, autoridad inherente, microgeopolítica, seguridad, estado.*

Immigration policing in the United States is not just about border enforcement; it is also increasingly about the surveillance of immigrant life "on the inside." This trend dates from the 1990s. Despite the large expenditures on border enforcement during that decade, "removals" (i.e., deportable and/or inadmissable aliens expelled from the United States) increased fivefold between 1990 and 2000; however, the spike in removals since 9/11—from 180,000 in 2001 to 359,000 in 2008, a full third of all immigration apprehensions made that year—signals that interior policing is now a central component of the U.S. immigration enforcement landscape. A recent Department of Homeland Security (DHS) document, *Operation Endgame*, makes this point clear. The report classifies immigration as a keystone national security issue and aims at a "100% rate of removal for all removable aliens" in the interior by 2012 (DHS 2003, 12).

Given the limited number of interior immigration agents, this strategy relies in part on enrolling nonfederal stakeholders. In this article, I examine a specific form of enrollment, the post-9/11 devolution of immigration authority to state and local police via the 287(g) program and the inherent authority doctrine. The former refers to a 1996 federal statute through which nonfederal officers (i.e., state, county, and local police) can be deputized as immigration agents; the latter is a more recent executive legal construction that, in the name of counterterrorism, allows for the local enforcement of immigration law without formal cross-designation.

287(g) and inherent authority are important because they demonstrate that immigration enforcement is not necessarily border enforcement (Coleman 2007; Winders 2007). As locally based strategies, they also suggest the limits of a strictly national lens on U.S. immigration politics (Ellis 2006; Varsanyi 2008). A third point, however, is raised by 287(g) and inherent authority concerning how geographers think about geopolitics and security. Whereas geographers have attended to both primarily as foreign policy, I argue that 287(g) and inherent authority point to the need to rescale what counts as the politics and practice of security.

I first examine critical geopolitics' accomplishments, a subfield that has been at the forefront of discussions in geography about security for at least two decades. I then briefly review some of the major criticisms levied against critical geopolitics. In a second section, I explore how post-9/11 U.S. immigration enforcement has transformed what immigration policing looks like and where it occurs. In a third section, I explore 287(g) and inherent authority as state security practices that increase everyday insecurities for undocumented migrants; however, this does not mean that insecurity is an unanticipated by-product of these practices. Rather, 287(g) and inherent authority are about the active deployment of insecurity as a security or policing tactic. I conclude with some thoughts about the focus on risk in critical geopolitics.

Displacing and Consorting with Critical Geopolitics

Heightened U.S.–Soviet tensions during the Reagan era prompted a group of geographers to examine the intersection of geographical knowledge and geopolitical power in international politics. The project, dubbed "critical geopolitics," centered on the socio-spatial structures of intelligibility underpinning foreign policy practice. As Dodds and Sidaway (1994, 518) explained, critical geopolitics seeks "to deconstruct the representational practices of conservative foreign policy elites, to reveal how they spatialize international politics." Specific attention was paid to "formal geopolitics" (i.e., foreign policy scholarship) and "practical geopolitics" (i.e., foreign policy conduct) as an ensemble of "socio-cultural resources and rules by which geographies of international politics get written" (Toal and Agnew 1992, 193). The goal was to "challenge some aspect of taken-for-granted geopolitical knowledge by looking at its social production, the parameters of its discursive economy" (Dalby and Toal 1996, 452).

Critical geopolitics' emphasis on knowledges constitutive of geopolitical practice spurred an important rethinking of security. Of particular interest were the ways in which security was hitched by mainstream foreign policy elites and experts to spatialized tropes of identity and difference. Campbell (1992), for example, unpacked security as a boundary-drawing knowledge of performative rather than pragmatic temperament—as literally a self- and other-citing act in space. Similarly, Dalby (1990) discussed the "geo-graphs"—or spatialized discourses about self and other—that inhere in foreign policy and that reduce complex global realities to simplistic, state-territorial maps about security and danger. Dalby emphasized that the self–other formulation of state security in terms of spatial exclusion and rival territories was in fact a primary constituent of human *insecurity* insofar as it transformed lived spaces into dehumanized blocs ready for military conquest.

This abbreviated account of critical geopolitics' beginnings will be familiar for many. Nonetheless, it is important to recall because it highlights critical geopolitics' parasitical relationship with mainstream geopolitics. Toal (1994, 542; 1996) hinted at this early on in the critical geopolitics experiment, noting that in displacing orthodox geopolitics "we are consorting with a philosophical tradition, an historical code, a geographical map, an order of places." Indeed, in response to the charge that critical geopolitics sometimes fails to problematize its use of mainstream geopolitical concepts (Sharp 2000; Sparke 2000), Toal (2000, 387) conceded that "[i]n seeking to engage certain discourses in order to displace them, one invariably is dependent to a certain degree upon the organizing terms of these discourses."

Toal's explanation of the cramped relationship between critical geopolitics and its mainstream interlocutors is in terms of deconstruction and its perils; however, Foucault's notion of archaeology is also relevant. In his early essays on discourse and its transformations, for example, Foucault (1989, 41) challenged the "opposition between the liveliness of innovations and the dead weight of tradition." Instead of seeing the boundary-breaking-ness of the new against the wrong-headedness of the old, Foucault emphasized how "discoursing subjects" belong to a "discursive field"—not a monolithic plane in which authors remain constants but one in which anonymous rules of expressibility, conservation, memory, reactivation, and appropriation are in play. Put in the context of Toal's comments, archaeology suggests that the new (i.e., critical geopolitics) does not break with the old (i.e., mainstream geopolitics) but is engaged in a play of dependencies with it.

I want to expand on this problem with the goal of interrogating what counts as the politics and practice of security in critical geopolitics. If, as earlier, critical geopolitics seeks to upend the socio-spatial structures of intelligibility underpinning the practice and rationalization of statecraft, the paradox is that the field as a whole tends to reactivate a mainstream account of what constitutes security in practice. Even as critical geopolitics approaches statecraft as a cultural performance of identity, and thus in essence collapses any neat differentiation between domestic and foreign policy spaces, it still has much to say about what might constitute the politics and practice of security beyond foreign policy. This is particularly the case in critical geopolitics scholarship focused on theory building.

Critical geopolitics' conjoining of security to foreign policy has not gone unnoticed. One early critique was Dodds's (1994; cf. 2001) plea that critical geopolitics move beyond representational analyses of foreign policy and deal with "domestic" geopolitics. More recent commentators point to critical geopolitics' prioritization of interstate military conflict at the expense of more "regional" research (Power 1999; Agnew 2001; Mamadouh and Dijkink 2006). Others, without noting the field's foreign policy centrism, focus on conflict in contexts frequently read as more political geographic than geopolitical per se (Herbert 1997). Likewise, Mamadouh (1999) reminds critical geopolitics scholars—mostly in the United States and United Kingdom—of non-Anglo-American "critical geopolitical" thought (i.e., Lacoste and *Hérodote*) that attempted explicitly to delink geopolitics from international relations and draw attention to the warring aspects of public policy geographies. There is also research on the urbanization of war that suggests that the politics and practice of security are not just about foreign policy in a narrow sense (Falah and Flint 2004; Graham 2006). Reading this literature alongside the most cited critical geopolitics texts suggests the latter's reluctance to engage with statecraft's more prosaic coordinates (Painter 2006; Katz 2007; Pain and Smith 2008).

Feminist political geographers in particular have remarked on critical geopolitics' dismissal of the everyday. As Staeheli (2001) summarizes, critical geopolitics fetishizes the state and statecraft, both theoretically and empirically, such that localities and lived experiences feature infrequently in the research. The result, as Sharp notes (2000, 363), is that critical geopolitics relies on a private–public mapping of "political-effectual and non-political-ineffectual spheres" with geopolitics apparently only relevant in terms of the former. Indeed, feminist scholars note that what counts as security in critical geopolitics for the most part does not stray far from "big P" politics (i.e., warfare), whereas the question of where security occurs is frequently about abstracted macro-spaces such as the "battlefield" (Kofman 1996; Dowler and Sharp 2001; Secor 2001; Smith 2001). In this spirit, Hyndman (2004, 319) asks geographers to shift "scales to include the security of the state but in relation to the security and wellbeing of people who live *in* and *across* its borders." Indeed, Hyndman (2001) argues a need to foreground a politics of (in)security at a finer and more embodied scale and broaden what counts as the politics and practice of security to include everyday violent relationships. In short, what the feminist geopolitics literature argues is that security should be approached as something that abides by no particular practices and no particular spaces.

In an attempt to contribute to rethinking geographies of security as suggested by feminist geographers, without losing critical geopolitics' problematization of security, I next explore post-9/11 immigration policing in the U.S. context. I am interested in how, in the name of national security, the localization of immigration enforcement targets immigrants' every day.

Immigration Enforcement and Devolution after 9/11

One long-standing consequence of critical geopolitics' foreign policy centrism has been muted attention to how public policy spaces and politics are geopolitical or to how public policy and foreign policy implicate increasingly indistinct practices and spaces. More recently, despite attention across the social sciences to homeland defense and how post-9/11 immigration policy in particular blurs the lines between public and foreign policy (Tirman 2004), critical geopolitics has been surprisingly quiet about how everyday spaces of immigrant regulation and incorporation (Mountz 2004), or peopled mobility more generally (Hyndman 2000; Dahlman and Toal 2005), are relevant to the politics and practice of security. This lack stands at odds with the ubiquitous motif of migrants (and their purported impact on welfare, crime, language, cultural practices, etc.) as a national security problem in mainstream geopolitics. Recent research on the spatial convergence and rescaling of internal and external security practices in the European Union, around precisely immigration enforcement, offers some important signposts (Bigo 2002; Samers 2004; Walters 2006). This literature is striking insofar as state security practice is understood not as a foreign policy problematic but as a multitude of quotidian techniques for governing immigrants across a variety of public and private sites and through means not typically thought of as geopolitical (Huysmans 2006).

Although there is not much like-minded research in the North American context, the emphasis on homeland security in the United States has brought about a comparable reorganization of internal and external security. The post-9/11 devolutionary trend in immigration enforcement is a prime example. Although states and localities played an early role in admitting and expelling noncitizens, a stand-alone (i.e., plenary) federal power over immigration was the norm for much of the twentieth century, when immigration control was conceived as a specifically national security power and as such preemptively the preserve of federal representatives and agencies (Lee 2005). Moreover, by virtue of its focus on territorial entry itself as a security risk, the plenary doctrine meant that immigration control was administered primarily at U.S. borders (Shachar 2007). This federally preemptive and outward-looking scheme has been fundamentally disrupted by the post-9/11 devolutionary trend. In making immigration enforcement a localized as well as inward-looking national security practice, devolution constitutes a significant challenge to who regulates immigration and to what practices security comprises. I limit my following discussion of devolution to the expansion of interior enforcement via state and local authorities. I look first at 287(g) and second at inherent authority.

287(g)

Questions about nonfederal authority over immigration gained ground in 1983 with *Gonzales v. City of Peoria*. The Ninth Circuit court decided against appellants who claimed that their detention by city police on immigration charges was racially motivated; however, the court also ruled that nonfederal authorities could deal only with criminal violations of the Immigration and Nationality Act (INA); that is, where an immigration violation is ancillary to a criminal offense. Civil aspects of the INA, including lack of employment authorization or "undocumentedness," were deemed off limits.

Since *Gonzales*, federal lawmakers have become increasingly interested in nonfederal authority over criminal immigration violations. For example, laws passed throughout the 1990s multiplied the grounds on which nonfederal police could make criminal arrests with immigration implications. The bulk of congressional efforts since *Gonzales*, however, have been about growing local–federal cooperation over immigration enforcement. Most important was the 1996 Illegal Immigration Reform and Immigrant Responsibility Act (IIRIRA), which delegated immigration authority to state and local police through a new addition to the INA, section 287(g). 287(g) enabled state, county, and city police to arrest and detain aliens for federal authorities as well as investigate immigration cases for prosecution in the courts—all while performing their regular duties (Seghetti, Vina, and Ester 2006).

Localities initially responded coolly to the delegation program. Only Salt Lake City sought 287(g) status post-IIRIRA. The situation changed dramatically after 9/11. Part of this about-face is explained by federal lawmakers' very public suturing of immigration enforcement to the war on terror, as well as by lawmakers' growing interest in immigration enforcement away from U.S. territorial borders. Despite federal lawmakers' emphasis on local law agencies as frontline national security and immigration enforcement assets, however, as in the 2001 Patriot Act, the first moves on 287(g) were made outside Washington. In December 2001, for example, Florida initiated an agreement to police immigration violators at seaports, airports, and energy plants. Alabama followed suit in 2003. Both agreements were couched explicitly in terms of post-9/11 domestic security issues and dovetailed neatly with the Department of Justice's (DoJ) Operation Tarmac, which targeted immigration violators working at national security-relevant sites.

Since 2003 nearly seventy additional 287(g)s have been signed (Figure 1). Many of the agreements are clustered in "new destination" migration states in the U.S. Southeast (i.e., North Carolina, Georgia, Tennessee) and are of two sorts: either jail enforcement models, where inmates are run through immigration databases, or task force agreements, where local police go out and actively look for immigration violators. Moreover, at least as many 287(g)s are awaiting approval. A survey of pending agreements suggests 150 287(g)s could be in force by the end of 2009 (Figure 2).

The growth of 287(g)s is important because it represents a substantial increase in the number of interior immigration agents, which numbered some 1,300 in 2002. Since 2002, close to 1,000 local police have been cross-deputized through the 287(g) program; the total could reach several thousand, including the pending agreements. We should add that 287(g)s are being grown

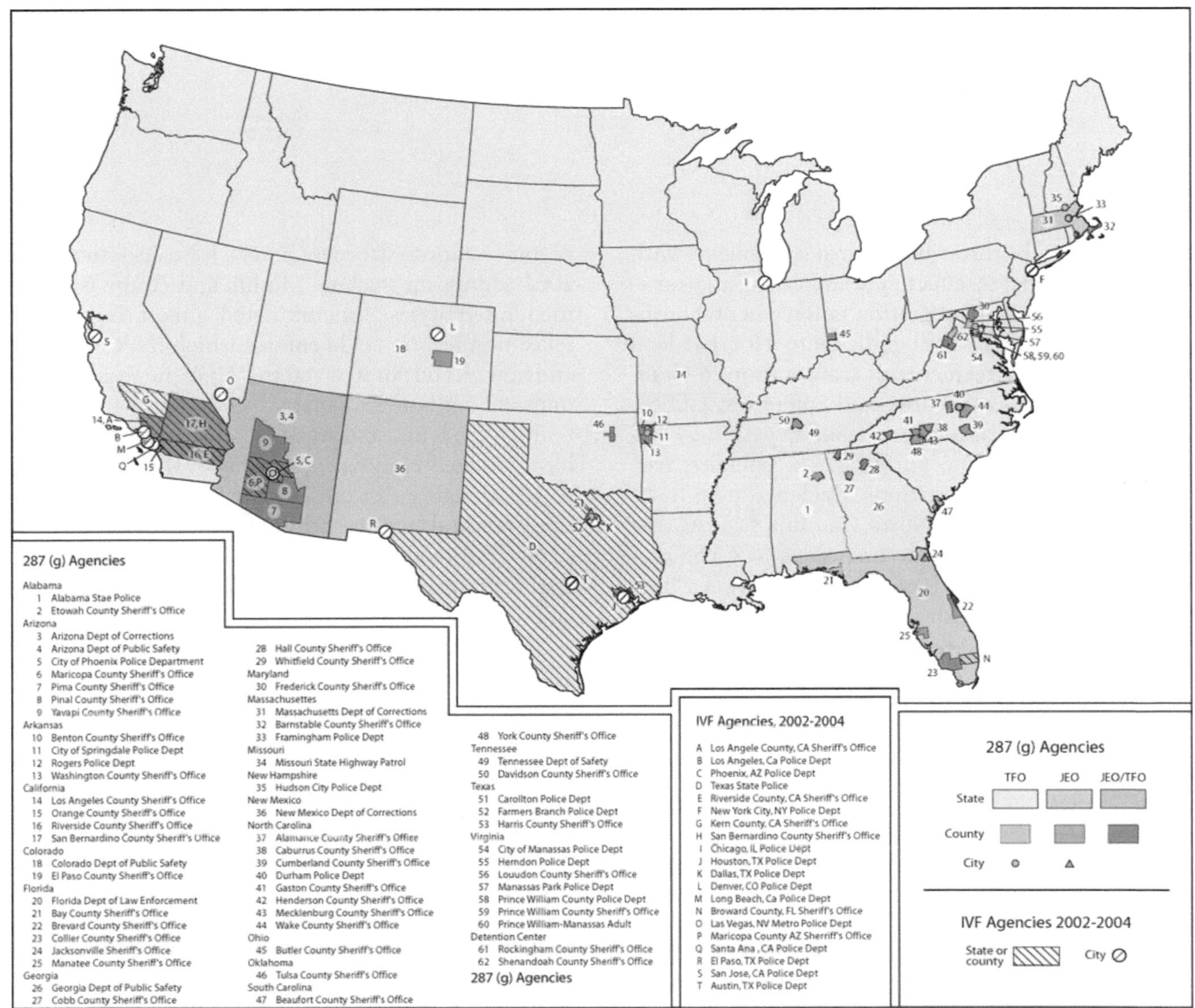

Figure 1. (g) agreements (as of August 2008) and state and local police using new immigration violator files (IVFs), 2002–2004. 287(g) data from Department of Homeland Security (2008a); National Crime Information Center data from Gladstein, Lai, and Wishnie (2005). Map produced by Jim DeGrand, Department of Geography, The Ohio State University. TFO = task force office; JEO = jail enforcement officer.

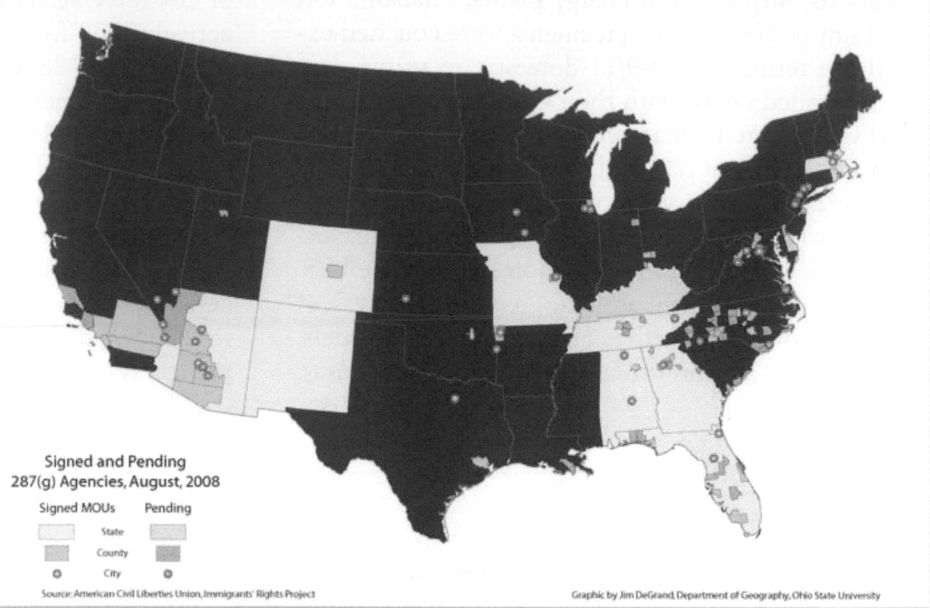

Figure 2. Projected 287(g) agreements. 287(g) data from American Civil Liberties Union (2008). Map produced by Jim DeGrand, Department of Geography, The Ohio State University. MOU = memorandum of understanding.

at the local scale through informal agreements with non-287(g) agencies, effectively increasing the numbers of police doing immigration enforcement. For instance, current 287(g)s are bundled into a local–federal initiative called Agreements of Cooperation in Communities to Enhance Security and Safety (ACCESS). ACCESS allows local police to team up with the DHS during counterterrorism, antigang, sex offender, narcotics, and other investigations. Evidence from high-density 287(g) states like North Carolina suggests that immigrants detained through the various ACCESS programs are being fed to 287(g) agencies as part of a "hub and spokes" retooling of the agreements. In other words, existing 287(g)s are extending their reach regionally through linkages with other local–federal policing operations that do not explicitly provide local authorities with the power to police INA violations.

Inherent Authority

In much the same way that once-controversial Operation Gatekeeper-style border enforcement was taken up as a cornerstone of U.S. immigration strategy, and remains so today, 287(g) is quickly becoming a routine part of the immigration enforcement landscape. Most telling is that the Obama administration is looking to shore up 287(g) enforcement (DHS 2009); however, 287(g) is not the whole story. Just as important is the inherent authority doctrine, which made a quiet appearance during Attorney General John Ashcroft's June 2002 address on tracking Muslim and Arab visitors to the United States. Ashcroft noted, almost in passing, a relaxation of who could enforce which INA violations and how. As he put it, given that "our enemy's platoons infiltrate our borders, quietly blending in with visiting tourists, students, and workers," "arresting aliens who have violated criminal provisions of the Immigration and Nationality Act or civil provisions that render an alien deportable . . . is within the inherent authority of states." Ashcroft described this authority as "limited" to counterterrorism, rather than as a general immigration enforcement tool (Ashcroft 2002).

The term *inherent authority* originated in an April 2002 DoJ Office of Legal Counsel (OLC) advisory. The advisory was not made public in 2002, but Ashcroft's elision of explicit delegation language, and his invocation of implicit nonfederal authority over civil INA violations, stood in contrast to prior OLC opinions that hewed closely to *Gonzales* (Chishti 2007). The full meaning of inherent authority became clearer in 2005, when the opinion was released. Two things stand out about it. First, it argues that no affirmative authorization by federal authorities is needed for localities to enforce immigration law—or indeed any federal law. Second, it suggests that nonfederal officers have been mistakenly prohibited from policing civil deportability. The basis of the argument is *U.S. v. Salinas-Calderon* (728 F.2d 1301), a 1984 Tenth Circuit ruling that localities have a "general arrest authority" vis-à-vis immigration violators. As one prominent proponent of inherent authority notes, the general arrest authority promised by *Salinas-Calderon* will enable a "massive force multiplier" of nearly 1 million possible immigration agents in the form of local police (Kobach 2005).

Inherent authority's dismissal of delegation has gained some support in Congress. More important, though, are extralegislative developments. Three trends deserve quick mention, all of which contradict Ashcroft's typification of inherent authority as a narrow counterterrorism tool. First, in the wake of 9/11, federal authorities have come to rely on ad hoc cooperation with state and local police outside the bounds of 287(g) authority (DHS 2007). Ohio is a case in point. The Cleveland, Columbus, and Cincinnati police departments are not cross-deputized with DHS, yet all participated in Operation Return to Sender raids in mid-July 2006, which resulted in some 150 civil immigration arrests in the state ("Ohio sweep nets 154" 2006). Second, under a new initiative called Secure Communities, non-287(g) local and county jails housing low-security inmates are now able voluntarily and without cross-designation to access federal immigration databases and initiate deportation for select inmates. Third, and most important, is the addition of immigration violator files (IVFs) to the National Crime Information Center database in mid-2003. Although deported felons have been listed there since 1996, the approximately 250,000 new files are different because they include civil violations. Indeed, the fastest growing category of IVFs accessed during routine policing relate to civil rather than criminal deportation orders (Gladstein, Lai, and Wishnie 2005). Moreover, the departments making most frequent use of the files are not 287(g) compliant (see Figure 1). As Wishnie (2004) explains, the IVF strategy is an attempt to "induce" officers not participating in 287(g) to arrest suspected immigration violators on grounds that were until very recently prohibited.

Immigrant (In)security

287(g) and inherent authority are somewhat contradictory. The former delegates federal powers to state and local authorities; in contrast, inherent authority does away with delegation, and as such seems to invalidate 287(g). Nonetheless, 287(g) and inherent authority, in concert, have spearheaded a watershed localization of immigration enforcement. As Pham (2004) argues, 287(g) and inherent authority have transformed the plenary doctrine (i.e., immigration enforcement as a preemptively federal and outward-looking power, as earlier) into a "thousand borders" immigration enforcement regime. Whereas critics typically credit 287(g) and inherent authority with an uneven array of local practices, however, I see a patterned, albeit messy, outcome: By different design, and yet in the shared name of national security, 287(g) and inherent authority are about making undocumented immigrants' everyday lives increasingly insecure. In other words, 287(g) and inherent authority deploy insecurity as a policing tactic insofar as their objective is explicitly to destabilize undocumented immigrants' lived everydays (cf. Winders 2007). The result is a sort of reversed risk society thesis in which the goal is not to manage but to *increase* radically the levels of day-to-day uncertainty for undocumented immigrants. In short, 287(g) and inherent authority constitute a post-9/11 microgeopolitics of risk intensification.

Two central aspects of this microgeopolitics warrant elaboration. The first concerns where immigration policing occurs. Interior spaces subject to immigration scrutiny have multiplied and become much more varied since 9/11. Indeed, 287(g) and inherent authority have gone hand in hand with the growth of immigration policing beyond worksite enforcement. In particular, there has been a significant growth in detentions and deportations from street corners, homes, hospitals, shopping malls, and so on. The fugitive enforcement program at the heart of DHS's Operation Endgame strategy, noted in the introduction, is a good example. The program targets immigrants who have failed to heed a formal order of removal, or who have failed to appear before a court for such an order, and involves state and local participation in primarily residential settings. Whereas in 2003 DHS fugitive teams arrested 1,900 aliens with outstanding orders of deportation, this number increased to more than 35,000 in 2008 (DHS 2008b). Moreover, funding for these operations has jumped from $9 million to more than $200 million over the past six years.

Extra-workplace raids of this sort are producing significant numbers of collateral arrests. Indeed, 22 percent of the approximately 100,000 aliens detained by the fugitive teams have been "ordinary status violator arrests" or warrantless arrests for lack of legal status (Mendelson, Strom, and Wishnie 2009). Key here is the use of racial profiling to question individuals encountered during raids about their immigration status, which critics argue will become more prevalent with increased local–federal cooperation (Romero 2002; Waslin 2003; Arnold 2007). Although the DHS has insisted that racial profiling will be limited to national security operations after 9/11, and will not implicate immigration enforcement more generally, a far more likely outcome given the nature of extra-workplace enforcement is what Olivas (2007) calls a generalized "ethnic and national origins 'tax' that will only be levied upon certain groups, certain to be Mexicans in particular, or equally likely, Mexican Americans" (55).

The second point concerns a growth in what counts as immigration enforcement. A good indication is how, given the inherent authority doctrine, a large number of local policy debates have become immigration issues since 9/11. On the back of the 2002 OLC opinion, cities, counties, and states have responded to federal inaction on immigration reform by passing hundreds of local laws. Most relevant are the illegal immigration reforms acts (IIRAs) of the sort proposed in Hazelton, Pennsylvania (Gilbert 2009). Mimicking California's unsuccessful Proposition 187, these ordinances ensure that immigrants' everydays are becoming more legible to federal and local authorities. Indeed, a key part of the IIRAs is to make local undocumented populations more visible, and hence vulnerable, to immigration authorities in the context of getting driver's licenses, paying taxes, renting, buying cars and houses, seeking hospital care, attending school, and so on. In short, 287(g) and inherent authority have not only multiplied the number of interior spaces subject to immigration oversight, but have also prompted a generalization in the forms that immigration enforcement takes.

Conclusion

Taking inspiration from Toal on how critical geopolitics works through the conceptual infrastructures of mainstream geopolitics, I have argued that critical geopolitics can be seen archaeologically as caught in a play of dependencies with mainstream geopolitics over what counts as security, and where we might find it. Instructive here is critical geopolitics' engagement with Campbell's (1992) signal text, *Writing Security*. Campbell argued that the identity and difference correlates of U.S. foreign policy have both extensive and intensive spatialities and as such that foreign policy be approached as but one aspect of an overarching process of producing and regulating foreignness (through discourses and practices of race, class, gender, and sexuality) operative "at all levels of social organization, from the level of personal relationships through to global orders" (69). Although critical geopolitics scholars vigorously debated the merits of Campbell's performative, identity-based approach to statecraft (see Dalby and Toal 1996), very little was said about his closely related problematization of internal and external security practices. I take this as more generally indicative of critical geopolitics' attention to statecraft's identity and difference discourses at the expense of its increasingly multivalent fields of practice. In other words, the focus on identity and difference has been used to recharacterize rather than deterritorialize mainstream geopolitics' domestic–foreign polarity.

My interest in these questions follows from my attempts to understand post-9/11 U.S. immigration enforcement. For example, what is important about the tactics designed to increase everyday insecurity, deployed against immigrants by various localized arms of the state "on the inside," is that they cannot be grasped through the abstract lens of foreign policy discourse or practice even as they are the public face of the so-called war on terror. Following the feminist political geographers cited earlier, 287(g) and inherent authority, even as they are rationalized as relevant to counterterrorism, are rather about composite internal–external security practices that render place-specific populations insecure through "small transformations in everyday life" (Smith 2001).

Critical geopolitics has much to offer to this discussion. For instance, the insight that state security implies coincident insecurities is an obvious aspect of both 287(g) and inherent authority. At the same time, whereas critical geopolitics tends to discuss insecurity as an unthought by-product of practices intended to otherwise secure populations and territories, 287(g) and inherent authority make insecurity a considered and intended tactic of state security. This is an important differentiation and speaks to a dominant reading of statecraft in critical geopolitics that could be rethought. For instance, using the risk society paradigm—that is, the social world as constituted by uncertainty, speed, and borderlessness—critical geopolitics scholarship tends to interpret mainstream geopolitics' insecurities as a product of its ham-fisted, anachronistic, and simple-minded territorializing practices. Indeed, statecraft's insensitivity to the flux and flow of the everyday—via inappropriate inside–outside, us–them cartographies—is held as an explanation of its violences. This more or less "classical" account of geopolitics as a state-territorial machine that produces insecurity by virtue of its geographical misimaginations does not do justice to "new" state security practices that actively constitute our contemporary vertiginousness by both actively scrambling traditional state-territorial cartographies as well as engineering widespread socioeconomic insecurity as a first-order objective. Indeed, the risk society thesis only really makes sense if one accepts security as an explicitly foreign policy strategy whose international territorializations are at odds with the spatial contingencies and social variabilities of the everyday. If one looks instead at state security as practices that, first, cut across the very state-territorial maps they propose to secure (Gregory 2004; Roberts, Secor, and Sparke 2003) and, second, promote socioeconomic turmoil on the ground, then a more worrisome geopolitics rears its head: one that effectively undoes the Cartesian, state-territorial cartographies we expect it to produce while amplifying everyday disorder and hardship on the ground.

Acknowledgments

I would like to thank Marcus Power for organizing the September 2008 State of Critical Geopolitics conference at Durham University, where this article was test-driven and fine-tuned. Thanks also to Gerry Kearns, Audrey Kobayashi, Mary Thomas, and my reviewers for their productive criticisms and comments. This article is part of a larger project on local immigration enforcement in the U.S. Southeast, funded by the Mershon Center for International Security Studies at The Ohio State University.

References

Agnew, J. 2001. Disputing the nature of the international in political geography. *Geographische Zeitschrift* 89 (1): 1–16.

American Civil Liberties Union. 2008. *Catalogue of 287(g) agreements*. Washington, DC: ACLU.

Arnold, C. 2007. Racial profiling in immigration enforcement. *Arizona Law Review* 49 (1): 113–42.

Ashcroft, J. 2002. Remarks on the NSEERS. http://www.usdoj.gov/archive/ag/speeches/2002/060502agprepared-remarks.htm (last accessed 20 February 2009).

Bigo, D. 2002. Security and immigration. *Alternatives* 27 (Suppl.): 63–92.

Campbell, D. 1992. *Writing security*. Minneapolis: University of Minnesota.

Chishti, M. 2007. Enforcing immigration rules. *NYU Journal of Legislation and Public Policy* 10 (3): 451–71.

Coleman, M. 2007. Immigration geopolitics beyond the US–Mexico border. *Antipode* 39 (1): 54–76.

Dahlman, C., and G. Toal. 2005. Broken Bosnia: The local geopolitics of displacement and return in two Bosnian places. *Annals of the Association of American Geographers* 95 (3): 644–62.

Dalby, S. 1990. *Creating the second Cold War*. London: Pinter.

Dalby, S., and G. Toal. 1996. The critical geopolitics constellation. *Political Geography* 15 (6): 451–56.

Department of Homeland Security. 2003. *Operation endgame*. Washington, DC: DHS.

———. 2007. *An assessment of ICE's fugitive operations teams*. Washington, DC: DHS.

———. 2008a. *ICE 287(g) program factsheet*. Washington, DC: DHS.

———. 2008b. *Law enforcement support center has record-breaking year*. Washington, DC: DHS.

———. 2009. Secretary Napolitano issues immigration and border security action directive. Washington, DC: DHS. http://www.bibdaily.com/pdfs/DHS%20border%20directive%201-30-09.pdf (last accessed 20 February 2009).

Dodds, K. 1994. Geopolitics and foreign policy. *Progress in Human Geography* 18 (2): 186–208.

———. 2001. Critical geopolitics after ten years. *Progress in Human Geography* 25 (3): 469–84.

Dodds, K., and J. Sidaway. 1994. Locating critical geopolitics. *Society and Space* 12 (4): 515–24.

Dowler, L., and J. Sharp. 2001. A feminist geopolitics? *Space and Polity* 5 (3): 165–76.

Ellis, M. 2006. Unsettling immigrant geographies: US immigration and the politics of scale. *Tijdschrift* 97 (1): 49–58.

Falah, G., and C. Flint. 2004. Geopolitical spaces: The dialectic of public and private space in the Palestine–Israel conflict. *Arab World Geographer* 7 (1–2): 117–34.

Foucault, M. 1989. *Foucault live*. Los Angeles: Semiotext(e).

Gilbert, L. 2009. Immigration as local politics: Re-bordering immigration and multiculturalism through deterrence and incapacitation. *International Journal of Urban and Regional Research* 33 (2): 26–42.

Gladstein, H., A. Lai, and M. Wishnie. 2005. *Blurring the lines*. Washington, DC: MPI.

Graham, S. 2006. Cities and the war on terror. *International Journal of Urban and Regional Research* 30 (2): 255–76.

Gregory, D. 2004. *The colonial present*. Oxford, UK: Blackwell.

Herbert, S. 1997. *Policing space*. Minneapolis: University of Minnesota.

Huysmans, J. 2006. *The politics of insecurity*. London and New York: Routledge.

Hyndman, J. 2000. *Managing displacement*. Minneapolis: University of Minnesota Press.

———. 2001. Towards a feminist geopolitics. *Canadian Geographer* 45 (July): 210–22.

———. 2004. Bridging feminist and political geography through geopolitics. *Political Geography* 23 (3): 307–22.

Katz, C. 2007. Banal terrorism. In *Violent geographies*, ed. D. Gregory and A. Pred, 349–61. London and New York: Routledge.

Kobach, K. 2005. The quintessential force multiplier. *Albany Law Review* 69 (1): 179–235.

Kofman, E. 1996. Feminism, gender relations and geopolitics. In *Globalization*, ed. E. Kofman and G. Youngs, 209–24. London: Pinter.

Lee, A. 2005. The unfettered executive. *Columbia Journal of Law and Social Problems* 39 (2): 223–56.

Mamadouh, V. 1999. Reclaiming geopolitics. *Geopolitics* 4 (1): 118–38.

Mamadouh, V., and G. Dijkink. 2006. Geopolitics, international relations and political geography. *Geopolitics* 11 (3): 349–66.

Mendelson, M., S. Strom, and M. Wishnie. 2009. *Collateral damage*. Washington, DC: MPI.

Mountz, A. 2004. Embodying the nation-state. *Political Geography* 23 (3): 323–45.

Office of Legal Counsel. 2002. *Non-preemption of the authority of state and local law enforcement officials to arrest aliens for immigration violations*. Washington, DC: Department of Justice.

Ohio sweep nets 154. 2006. *Columbus Dispatch* 15 July http://www.dispatch.com/live/contentbe/dispatch/2006/07/15/20060715-E1-01.html (last accessed 1 September 2009).

Olivas, M. 2007. Immigration-related state and local ordinances. *University of Chicago Legal Forum* 1 (1): 27–56.

Pain, R., and S. Smith, eds. 2008. *Fear: Critical geopolitics and everyday life*. Aldershot, UK: Ashgate.

Painter, J. 2006. Prosaic geographies of stateness. *Political Geography* 25 (7): 752–74.

Pham, H. 2004. The inherent flaws in the inherent authority position. *Florida State University Law Review* 31 (4): 965–1003.

Power, M. 1999. Provincializing geo-politics. *Space and Polity* 3 (1): 101–8.

Roberts, S., A. Secor, and M. Sparke. 2003. Neoliberal geopolitics. *Antipode* 35 (3): 886–97.

Romero, V. 2002. Devolution and discrimination. *NYU Annual Survey of American Law* 58 (3): 377–90.

Samers, M. 2004. An emerging geopolitics of illegal immigration in the European Union. *European Journal of Migration and Law* 6:27–45.

Secor, A. 2001. Toward a feminist counter-geopolitics. *Space and Polity* 5 (3): 191–211.

Seghetti, L., S. Viña, and K. Ester. 2006. *Enforcing immigration law: The role of the state and local enforcement*. Washington, DC: CRS.

Shachar, A. 2007. The shifting border of immigration enforcement. *Stanford Journal of Civil Rights and Civil Liberties* 3 (2): 165–93.

Sharp, J. 2000. Remasculinizing geo-politics? *Political Geography* 19 (4): 361–64.

Smith, F. 2001. Refiguring the geopolitical landscape. *Space & Polity* 5 (3): 213–35.

Sparke, M. 2000. Graphing the geo in geo-political. *Political Geography* 19 (4): 373–80.

Staeheli, L. 2001. Of possibilities, probabilities and political geography. *Space and Polity* 5 (3): 177–89.

Tirman, J., ed. 2004. *The maze of fear*. New York: New Press.

Toal, G. 1994. (Dis)placing geopolitics. *Society and Space* 12 (5): 525–46.

———. 1996. *Critical geopolitics*. Minneapolis: University of Minnesota Press.

———. 2000. Dis/placing the geo-politics which one cannot not want. *Political Geography* 19 (3): 385–96.

Toal, G., and J. Agnew. 1992. Geopolitics and discourse. *Political Geography* 11 (2): 155–75.

Varsanyi, M. 2008. Rescaling the alien, rescaling personhood: Neoliberalism, immigration and the state. *Annals of the Association of American Geographers* 98 (4): 877–96.

Walters, W. 2006. Border/control. *European Journal of Social Theory* 9 (2): 187–203.

Waslin, M. 2003. Immigration enforcement by local police. *NCLR Issue Brief* 9:1–20.

Winders, J. 2007. Bringing back the (b)order. *Antipode* 39 (5): 920–42.

Wishnie, M. 2004. State and local police enforcement of immigration laws. *University of Pennsylvania Journal of Constitutional Law* 6 (5): 1084–115.

Embedded Empire: Structural Violence and the Pursuit of Justice in East Timor

Joseph Nevins

East Timor faces severe limitations in its efforts to realize legal and material justice and to overcome the horrific violence associated with Indonesia's invasion and almost twenty-four-year occupation. Since the Indonesian military's withdrawal in October 1999, the now-independent country has struggled to realize legal and material justice for the war crimes and crimes against humanity committed from 1975 to 1999 and to overcome the deprivation and dispossession associated with Indonesia's invasion and occupation. In relation to these efforts, this article examines two case studies. The first is East Timor's effort to secure legal and financial restitution for damages associated with Indonesia's actions. The second involves a disagreement with Australia over oil and natural gas deposits in a shared seabed and the effort to ensure an international-law-informed resolution of the conflict. In both cases, East Timor has fallen far short of its goals. *Key Words: East Timor, empire, imperialism, justice, violence.*

东帝汶正面临着实现法律和实质正义以及克服由于印尼的入侵和近 24 年的占领所造成的恐怖暴力事件的严重困难。自从印尼军队在 1999 年 10 月撤出东帝汶，现在已经是独立国家的东帝汶一直在付诸行动，试图对 1975 年至 1999 年间所犯下的战争罪和危害人类罪实施法律和实质的正义，并克服印度尼西亚的入侵和占领所造成的贫困和侵害。关于这些付出的工作努力，本文对两个案例进行了研究。首先是针对印度尼西亚所造成的有关损失，东帝汶为确保法律和金钱方面可以提出损害赔偿的各项努力。第二个案例是关于与澳大利亚在一个共享的海床上石油和天然气资源的归属分歧，和东帝汶为确保冲突能够有一个符合国际法标准的决议所付出的努力。在上述两个案例中，东帝汶已远远落后于它的目标。*关键词：东帝汶，帝国，帝国主义，公正，暴力。*

Timor Oriental enfrenta severas limitaciones en su esfuerzo por aplicar justicia legal y material y dejar atrás la horrenda violencia asociada con la invasión de Indonesia y los cerca de veinticuatro años de ocupación. Desde la retirada militar indonesia en octubre de 1999, el ahora país independiente ha bregado duro para ejercer justicia legal y material en relación con los crímenes de guerra y crímenes de lesa humanidad que se cometieron entre 1975 y 1999, y superar la deprivación y desposeimiento derivados de la invasión y ocupación por Indonesia. Al respecto, en este artículo se examinan dos situaciones concretas. La primera es el esfuerzo de Timor Oriental por hacer la reparación jurídica y financiera por daños asociados con las acciones indonesas; la segunda, se relaciona con un desacuerdo con Australia sobre depósitos de petróleo y gas en un lecho marino compartido, y el esfuerzo por lograr una resolución del conflicto con base en normas internacionales. Muy poco es lo que Timor Oriental ha conseguido en sus pretensiones con respecto a los dos casos. *Palabras clave*: *Timor Oriental, imperio, imperialismo, justicia, violencia.*

Indonesia's 1975 invasion and subsequent occupation of East Timor—endeavors strongly backed by various Western allies (Gorjão 2002; Commission for Truth, Reception, and Reconciliation [CAVR] 2005; Dowson 2005; Fernandes 2005; Nevins 2005; Simpson 2005)—would seem to be a case of classic imperialism, the domination of a country by a superior power via territorial control. Now that the occupation has ended, one might conclude that so, too, has the associated imperialism. Yet East Timor's failure to realize justice for crimes and dispossession associated with overtly imperialist acts suggests that imperialism continues to inform the newly independent country's relations with Indonesia and powerful Western states.

The concept of imperialism remains an analytical tool for understanding international relations and the associated unjust outcomes in ways that a concept such as hegemony cannot adequately explicate (see Agnew 2005). Geographers, writing largely from a Marxist-inspired perspective, have interrogated imperialism in its contemporary guises to highlight the dynamic, complex, and sometimes contradictory linkages between nation-states, practices associated with capitalist accumulation, and various modes of global governance. These practices do not necessarily (and, indeed, rarely) involve state control over territory but rather the maintenance and enhancement of a global economy structured to privilege certain spaces over others (e.g., Harvey 2003; Smith 2003; Flint and Taylor 2007).

Although these analyses illuminate much, they tend to privilege the global economy, mechanisms of exploitation and uneven exchange, and an associated set of economic actors as imperialism's embodiment to a degree that they provide insufficient scrutiny of how the imperial order—in terms of its structural manifestations and their associated socioeconomic inequalities—plays out beyond the economic realm. As a result, various types of imperial embeddedness and diverse forms of imperial privilege and disadvantage are underexplored. As part of an effort to fill in somewhat the gaps and to facilitate a more expansive interrogation of how imperialism functions and shapes the global political economy, this article emphasizes structural violence—or institutionalized injustice—rather than uneven exchange as the key mechanism for furthering unjust relations between nation-states and their peoples.

By drawing on literature in critical theory as it relates to social difference and power, this article presents an enlarged analysis of imperialism. It shows how violence-infused power structures nation-states and the relations between them so as to maintain and produce unjust outcomes across global space. Like racism, patriarchy, and other forms of violence that enable and inhibit by (re)producing privilege and disadvantage, imperialism has no fixed essence of the form it takes (see Acker 1989; Knopp 1992). It involves a varied set of practices and relations over space–time. What all imperialisms share, however, regardless of their particular spatiality, is that their associated practices and relations unevenly shape places and peoples and affect their ability to access resources such as peace, justice, and human rights. As the case of East Timor illustrates, contemporary imperialism is, among other things, an embedded phenomenon, with a definite geography, yet it does not correspond to the model of Western empires of old, and it exhibits both overt and structural forms of violence.

Empire, Imperialism, and Violence

The two justice-related case studies examined here and their unfavorable outcomes (unfavorable from the perspective of East Timorese civil society and significant elements of the state apparatus) suggest that imperialism endures in terms of internation–state relations involving East Timor. Nonetheless, the concept of imperialism, and the critique of international relations that it embodies, is rarely raised in relation to present-day East Timor in academic analysis of the country.

This is in part a manifestation of the fact that, until recently, empire and imperialism were largely topics of historical interest among many, if not most, academics (in the West at any rate), the assumption being that we now live in a postimperial world and that the terms have declining significance. That assumption changed following 11 September 2001—at least in relation to the United States. The response of the Bush administration to that day's events was such that, as Glassman (2005) contends, it "unquestionably rescued the term 'imperialism' from the oblivion to which it seemed to be heading in most academic discourse" (1527). In the context of the Iraq war, many commentators saw evidence of a renewed U.S. commitment to imperialism. Others used the opportunity to advocate or champion (what they perceive as) a benign American imperialism (e.g., Mallaby 2002) that occasionally uses overt violence but that is not violent in any essential manner.

Such advocacy typically ignores that imperialism—like other "isms" of dispossession and accumulation that bring about, draw on, and reproduce hierarchical, unjust social relations—is inherently violent. Violence comprises not only overt acts of brutality but also structural and representational elements that contribute to fundamentally unjust, avoidable outcomes (Galtung 1969, 1990; Nevins 2005). What makes imperialism distinct from similar hierarchy-producing processes of violence is that (national) spatial difference is the principal axis around which privilege and disadvantage pivot.

The processes of violence, as Gilmore (2002a, 2002b) and Hall (1992) contend in relation to racism, are borne of the union of difference and power, which they characterize as a "fatal coupling" that profoundly informs the collective life and death chances of a particular (racialized) group, shaping its access to resources, power, and rights.

Just as racialized difference legitimates or facilitates the use of power by the relatively privileged "to treat the ethnoracial Other in ways that we would regard as cruel or unjust if applied to members of our own group" (Fredrickson 2002, 9), imperialized difference lends itself to cruel or unjust conduct toward the national spatial other. Such conduct helps to perpetuate a "violence of everyday life" (Scheper-Hughes 1993) in afflicted territories by (re)producing various forms of injury. Imperialism is present when the differences between countries are such that fundamental double standards arise as to how they are treated, when access to the global commons is highly disproportionate in relation to a particular national territory's population size, as is its ability to "call the shots," to determine who (and where) gets what and under what conditions on the global scale. A nonimperial politics is democratic; it promotes, grants, and seeks the full participation and consent of those on the receiving end of policies and practices emanating from beyond their space or territory (Muppidi 2004, 67–72). To the extent that such a politics between nation-states is lacking, imperialism—by definition—is present.

Like other violent "isms" of dispossession and accumulation, imperialism flows from and reproduces international privilege and disadvantage. It insulates the spatially and nationally privileged from carrying their fair share of the burdens and detrimental outcomes associated with the larger global order, while allowing them a certain impunity that is denied to disadvantaged "others." This order is comprised of "artifacts of past and present" that "embody generations of sociospatial relations" (Pulido 2000, 16). In other words, imperialism and other simultaneously unjustly enabling and disabling processes are embedded in space.[1] They allow the relatively privileged, regardless of their intentions (see Pulido 2000), to behave so as to maintain the overall stability of a social order of which they enjoy a disproportionate share of the benefits, thus reproducing the plight of the relatively disadvantaged. The collective interventions (collective in that they involve social relations) of the powerful are thus predicated at least in part on self-preservation (of the advantaged group).

The demise of formal empires hardly means that imperialism no longer exists, just as Jim Crow's death in the United States or apartheid's end in South Africa did not mean that racism ceased to exist in either country. Similarly, to state that the British empire is extinct is quite different from contending that British imperialism is passé. There are various imperialisms, just as there are different imperial orders. In this regard, we might think of *empire* as a generic term in relation to imperialism, just as *racial order* relates to racism. In doing so, we should keep in mind Harvey's (2003) observation that many different kinds of empire have existed over the centuries—sometimes simultaneously within the same space—providing ample options as to "how empire should be construed, administered, and actively constructed" (5). That said, to the extent that imperialism persists today, the question of its spatiality arises.

As Smith (2003) asserts, contemporary U.S. imperialism is fundamentally different from that exercised by Western European countries during the era of formal colonialism. As opposed to the old imperialist view of space as absolute or as the endowment of natural resources of a particular territory, U.S imperialists—or at least a significant slice of the imperialist class (see Glassman 2005)—perceive and treat space as malleable, the outcome (as well as constituent and constitutive) of particular political–economic processes, rather than primordial and unchanging. European colonialism facilitated the realization of this vision by helping to unite the world, integrating the "Third World" into a West-dominated world economy over which the United States would soon reign. World War II provided the opening for the United States to take advantage of the largely European-created world market. Smith (2003) characterizes this approach as one of "global economic access without colonies," paired with a geostrategic vision of "necessary military bases around the globe both to protect global economic interests and to restrain any further military belligerence" (349). How imperial power is deployed depends on perceived needs, as well as on the faction of the ruling class in power at the time (Glassman 2005).

That said, to speak of "American" imperialism risks falling into the "territorial trap" (the conflation of social relations with national territory) that Agnew and Corbridge (1995) warn us about. States do not simply project power onto the international stage from within their territories. They also do so through international institutions and mechanisms and allied actors abroad. At the same time, transnational forces and coalitions of states deploy such power (Barkawi and Laffey 1999; also Agnew 2005), the Indonesian invasion and occupation of East Timor, decisively enabled as it was by a number of Western states, being an obvious example.

As a result of their collective political, economic, and military strength, powerful states are generally far more active and influential internationally than relatively weak states. Their state power is "internationalized" (Barkawi and Laffey 1999; Glassman 1999). The most important manifestation of this power is a Western bloc centered on the states of Western Europe, Australia, New Zealand, Canada, the United States, and Japan as well (see Shaw 1997). These states—many of them largely due to their bloc "membership"—dominate international affairs in numerous ways and in a manner highly disproportionate to their population sizes. In other words, how they are situated in what we might consider an imperial order, an inherently hierarchical arrangement, reflects and shapes who (and where) gets what and who (and where) does not. Such an arrangement is not overtly violent or coercive and it need not be as the very constellation of social relations facilitates outcomes that are favorable for those at the top and detrimental as one moves down the imperial chain. To the extent that the order needs to be remade, a variety of tools—overtly coercive, indirectly or structurally violent, and persuasive—are available. In this regard, the spatiality of "empire" reflects and reproduces deeply unequal relations and unjust outcomes between nation-states (and groups within and between them). It allows for international double standards such as those that reign vis-à-vis East Timor, which both reflects and helps further the contemporary imperial order.

Injustice and Postoccupation East Timor

Indonesia's invasion and occupation of East Timor were violations of international law (Clark 1995). They were also very brutal. As documented by East Timor's CAVR (2005), the war and occupation involved myriad atrocities, including widespread torture, extrajudicial killings, "disappearances," politically created famine, and indiscriminate bombing.[2] According to the CAVR's final report, Indonesia's actions resulted in the deaths of between 102,800 and 201,900 East Timorese civilians (noncombatants), most killed by hunger and illness—out of a population that was less than 700,000 in 1975. The CAVR also estimates that Indonesian forces committed "thousands" of acts of sexual violence—acts that were "widespread and systematic," "widely accepted" within the military hierarchy, and "covered by almost total impunity."

The war and occupation involved widespread destruction of infrastructure—Indonesian forces destroyed an estimated 80 percent of the territory's buildings and infrastructure before departing in 1999—as well as the effective theft of a significant slice of the territory's resources. In the case of coffee, for example, East Timor's top agricultural export earner, an Indonesian military-backed monopoly consistently underpaid East Timorese coffee producers. In 1983 alone, East Timorese smallholders received only one sixth the price paid in neighboring (Indonesian) West Timor, resulting in an $18 million loss (CAVR 2005). Over the occupation, such underpayments likely resulted in the loss of many tens of millions of dollars in income and a significant decapitalization of East Timor's economy that stunted its development and diversification.

It is because of such atrocities, destruction, and dispossession that the commission's report calls on the international community to provide "unqualified support for strong institutions of justice"—if necessary through an international tribunal—for trying war crimes and crimes against humanity. The recommendation echoes similar calls by international nongovernmental organizations and a broad cross-section of East Timorese civil society, as well as by various investigatory United Nations (UN) commissions (Nevins 2005). Despite such calls, there has been almost no accountability. To the extent that there have been judicial proceedings for war crimes and crimes against humanity, they have been extremely limited. A hybrid international–East Timorese court (which closed in May 2005) tried low-level offenders. Established by the UN during its post-occupation administration of the territory, the court convicted eighty-four individuals, all East Timorese, typically members of militia groups created and directed by the Indonesian military, for crimes committed in 1999 (and that year only). Despite an agreement of cooperation with the UN, Indonesia did not extradite any indicted individuals under its jurisdiction. No serious pressure was forthcoming from the international community to compel Indonesia's government to do so, either. As all the key players involved in the terror (from 1999 and before) were (and are) in Indonesia, this shortcoming effectively reduced the process to irrelevance. Meanwhile, within Indonesia, a court prosecuted eighteen Indonesian citizens for crimes—again, only those from 1999. Six were convicted, but all of them were eventually cleared on appeal to a higher court.

As for an international tribunal, there has been no movement in the UN by the Secretariat or the Security Council to bring one about. This is hardly surprising given that three permanent member states—France, the United Kingdom, and the United States—(along with other Western countries such as Australia, Canada, and Japan) together provided Jakarta with billions of dollars' worth of weapons, military training, and economic aid, as well as invaluable diplomatic cover during the period between 1975 and 1999. Collectively, this assistance—the CAVR characterizes U.S. support as "fundamental"—was decisive in permitting the invasion to take place and for allowing the occupation to persist (Nevins 2005). These member-states remain steadfast allies of Indonesia.

Finally, neither apologies nor monetary reparations have been forthcoming.[3] The CAVR recommended both, not only from Indonesia but also from governments that provided military assistance during the occupation. Reparations also applied to "business corporations who benefited" from weapon sales to Jakarta. Indonesia has by and large dismissed the CAVR's recommendations, and Western countries have ignored them.[4]

In addition to enduring such impunity, East Timor also remains occupied in a sense: The Australian government continues to hold and exploit deposits of oil and natural gas contained in seabed areas in the Timor Sea, areas that, on the basis of international law, appear to belong to East Timor. Australia gained initial control of these areas through a combination of what appears to be an incorrect drawing of East Timor's east–west sea boundaries during the Portuguese colonial period and a subsequent agreement signed with Indonesia in 1989—during the occupation. Given the illegality of Indonesia's very presence in East Timor at the time, the agreement was effectively one involving stolen goods.[5] Interest in these "goods" was a significant factor (albeit only one factor among others; Dunn 1996; Way, Browne, and Johnson 2000) in informing an Australian policy of acquiescence toward and support for Indonesia's actions vis-à-vis East Timor dating back to 1975 (King 2002).

Following Indonesia's withdrawal in 1999, the embryonic East Timorese state disputed Australian control of the seabed. Although there are international legal mechanisms to adjudicate such disputes, Canberra, fearing that it would lose, changed the terms on which it accepts international dispute resolution mechanisms related to maritime boundaries. Previously, Australia had declared its acceptance of the International Court of Justice and the International Tribunal for the Law of the Sea as venues for the compulsory settlement of disputes under the United Nations Convention on the Law of the Sea (UNCLOS). In March 2002, however, Canberra announced that it was taking advantage of an UNCLOS clause that allows treaty parties to exclude certain areas from compulsory dispute resolution.

The East Timorese government could have filed a case against Canberra at the International Court of Justice following the Australian government's revocation of its commitment to compulsory jurisdiction under UNCLOS, which asserts that this commitment remains in force for a three-month period after formal notification of withdrawal is communicated to the UN. Nevertheless, Dili failed to do so within the prescribed time period. This is likely due to one of the same factors that has prevented East Timor's government from strongly advocating for some sort of international judicial process for the crimes against humanity and war crimes committed against its citizens from 1975 to 1999: fear of detrimental consequences for antagonizing a powerful neighbor (and its Western allies). Canberra was thus able to exclude sea boundaries from any would-be law-based process of adjudication, representing its move as an attempt to resolve the conflict through a less contentious means of bilateral negotiations.

Such negotiations, inevitably infused with power, greatly favored Australia due to its superior economic, military, and political strength. East Timor's economic poverty, for instance, compelled the country to reach an agreement with Australia in relatively short order to increase the revenue flow into its state coffers. The young country lacked the time—in addition to power—to compel Canberra to respect fully what it asserts to be its territory and felt compelled to sign a series of treaties that has limited its claim while allowing it to gain access to what Australia controlled de facto. Nonetheless, through its politico-diplomatic efforts, and those of allies within Australia and international solidarity activists, East Timor got Australia to agree to divide the revenues from the disputed oil and gas fields in a manner that provided more to Dili than Canberra had proposed in earlier negotiations. Still, the latest agreement (January 2006) guarantees East Timor only 60 percent of the many billions of dollars worth of revenues gained through extraction in the disputed fields, whereas, were international law to be followed, East Timor would probably receive 100 percent. In other words, the Australian state has gained billions of dollars in revenues at impoverished East Timor's expense (Nevins 2004; Cleary 2007).[6]

The loss of such revenues has the effect of reinforcing East Timor's status as a country "chained by poverty," as a 2006 report from the UN Development Program characterized it. The manifestations are multiple: 90 out of 1,000 children die before their first birthday, half the population is illiterate, 64 percent suffers from food insecurity, half lack access to access to safe drinking water, and 40 percent live below the official poverty line of 55 cents a day (UN Development Program 2006).

Conclusion: "Independence" in an Imperialized World

East Timor's experience of injustice regarding law-based accountability and restitution is neither surprising nor accidental. As Mexico's delegate to the UN's founding convention in 1945 opined, the UN Charter ensured that "the mice would be disciplined, but the lions would be free" (Eban 1995, 46). One could make a similar observation about the contemporary world order: It allows for countries of relative privilege (due to their own power or that which they have through alliances)—a privilege borne of and reproduced by various forms of violence and institutionalized injustice—a large degree of freedom (and impunity for their transgressions) on the global stage, while limiting that of the disadvantaged.

Similar factors explain Australia's ability effectively to steal much of what appears to belong to East Timor and the latter's limited capacity to remedy this dispossession. Australia's exercise of imperial privilege draws on, takes advantage of, and reinforces East Timor's political–economic marginality, which is produced in part by a war and occupation supported by Australia among others. It reproduces East Timor as an imperialized and disadvantaged place near the bottom of the global socio-spatial hierarchy, and it sustains markedly unequal power relations that profoundly inform conditions of life and death. Meanwhile, the powerful have not suffered at all for their crimes related to East Timor's suffering and dispossession. Instead, as in the case of Australia's ill-gotten oil and gas revenues, they still enjoy the benefits.

East Timor's predicament thus demonstrates how imperialism both limits and enables depending on where a particular nation-state finds itself on the international power hierarchy. Relatively privileged national actors are comparatively well positioned to ensure that their extranational–territorial interactions minimize damage to the power of their particular nation-states and the overall global order that embodies and reproduces interstate inequalities, and that is generally beneficial to them. Thus, whereas East Timor's government must restrain itself from publicly advocating for an international tribunal lest it incur the wrath of the powerful, Indonesia and its Western allies have the sociopolitical space to publicly denounce any such advocacy.

Contemporary "empire" functions in myriad ways. The U.S. occupation of Iraq is one variety, Australia's "occupation" of East Timor oil and gas deposits is another. At the same time, imperialism draws on and produces various spatialities, ranging from direct territoriality to the manipulation of internation–state relations predicated on inequality. Although having overtly violent aspects, the power of "emperors" is often deployed through political–economic relations and structures born, in significant part, of direct forms of violence, and associated inequalities, from a previous era. The weakness and unfairness of international legal mechanisms, and East Timor's inability to realize an international tribunal or to see its seabed resource case properly adjudicated, are manifestations of imperialism, as are the benefits and impunity enjoyed by those countries involved in the invasion and occupation of East Timor.

To contend that the outcomes of Indonesian and Australian imperialism examined here are demonstrations of imperialism is not to suggest that the imperially privileged always get their way. Nor is it to ignore the mosaic-like aspects of the global political economy, in which the geography of wealth and poverty is much messier than aggregate averages of individual countries (Agnew 2005).[7] Similarly, it would be incorrect to state that, in a system of racial privilege, the relatively advantaged always win and that the structurally disadvantaged are always on the receiving end of detrimental outcomes, one factor being the importance of other social categories (e.g., class and gender) in informing results. Social relations—like space—are always under construction, subject to challenge, and are thus inherently unstable (Massey 2005). The relatively weak are able at times to challenge the strong and gain concessions. In the case of contemporary imperialism, this instability is significantly due to the demise and discrediting of formal imperialism. To the extent these challenges do not alter the imperialism-informed fundamentals of the larger order, however, the inequalities—in terms of conditions of life and death, wealth and poverty, political voice and other forms of power—do not disappear. The structural violence that underlies relations between territorial states endures, as does the imperialism that it reflects and reproduces, East Timor being one of the more unfortunate examples.

Acknowledgments

I would like to thank Steve Herbert, Michael Velarde, and two anonymous reviewers for their comments on earlier drafts of this article.

Notes

1. Regarding the concept of embeddedness, see Granovetter (1985) and Ruggie (1982).
2. For the precise citations relating to information or quotes from the CAVR report contained herein, see Nevins (2007–2008).
3. There are precedents for such reparations. In 1991, the UN Security Council imposed a $52 billion reparations bill on Iraq for its invasion and occupation of

Kuwait. As of September 2008, Iraq had paid more than $24 billion—the majority to the Kuwaiti government and various corporations—despite the end of Saddam Hussein's regime.

4. East Timor's elected leaders have felt compelled to reject the report's recommendations for reparations, as well as its calls for an international tribunal for fear of offending Indonesia and its Western backers (see Kingston 2006).
5. Regarding the legality of the agreement itself, see Clark (1992) and Cleary (2007).
6. For additional background, analysis, and maps of the dispute and the various agreements, see the Web site of La'o Hamutuk (http://www.laohamutuk.org/).
7. Yet, nation-states—and the relations between them—matter profoundly in terms of livelihood-related outcomes for individuals and countries. See Flint and Taylor (2007) regarding the ongoing importance of uneven exchange. See also Agnew (2005, 181) about the importance of country of residence in accounting for individual incomes.

References

Acker, J. 1989. The problem with patriarchy. *Sociology* 23 (2): 235–40.

Agnew, J. 2005. *Hegemony: The new shape of global power*. Philadelphia: Temple University Press.

Agnew, J., and S. Corbridge. 1995. *Mastering space: Hegemony, territory and international political economy*. New York and London: Routledge.

Barkawi, T., and M. Laffey. 1999. The imperial peace: Democracy, force and globalization. *European Journal of International Relations* 5 (4): 403–34.

Clark, R. S. 1992. Timor gap: The legality of the "Treaty on the Zone of Cooperation in an Area between the Indonesian Province of East Timor and Northern Australia." *Pace Yearbook of International Law* 69 (4): 69–95.

———. 1995. The "decolonisation" of East Timor and the United Nations norms on self-determination and aggression. In *International law and the question of East Timor*, ed. Catholic Institute for International Relations and International Platform of Jurists for East Timor, 65–102. London and Leiden, The Netherlands: Catholic Institute for International Relations and International Platform of Jurists for East Timor.

Cleary, P. 2007. *Shakedown: Australia's grab for Timor oil*. Crows Nest, NSW, Australia: Allen & Unwin.

Commission for Truth, Reception, and Reconciliation. 2005. *Chega!*, Final report of the CAVR in East Timor, Dili. http://www.etan.org/news/2006/cavr.htm (last accessed 15 September 2008).

Dowson, H., ed. 2005. Declassified British documents reveal U.K. support for Indonesian invasion and occupation of East Timor, recognition of denial of self-determination, 1975–76. National Security Archive Briefing Book, 28 November. http://www.gwu.edu/~nsarchiv/NSAEBB/NSAEBB174/indexuk.htm (last accessed 27 February 2009).

Dunn, J. 1996. *Timor: A people betrayed*. Sydney: ABC Books.

Eban, A. 1995. The U.N. idea revisited. *Foreign Affairs* 74 (5): 39–55.

Fernandes, C. 2005. *Reluctant saviour: Australia, Indonesia and the independence of East Timor*. Melbourne, Australia: Scribe.

Flint, C., and P. Taylor. 2007. *Political geography: World-economy, nation–state and locality*. Harlow, UK: Prentice Hall.

Fredrickson, G. M. 2002. *Racism: A short history*. Princeton, NJ, and Oxford, UK: Princeton University Press.

Galtung, J. 1969. Violence, peace, and peace research. *Journal of Peace Research* 6 (3): 167–91.

———. 1990. Cultural violence. *Journal of Peace Research* 27 (3): 291–305.

Gilmore, R. W. 2002a. Fatal couplings of power and difference: Notes on racism and geography. *The Professional Geographer* 54 (1): 15–24.

———. 2002b. Race and globalization. In *Geographies of global change: Remapping the world*, ed. R. J. Johnston, P. J. Taylor, and M. J. Watts, 261–74. Malden, MA: Blackwell.

Glassman, J. 1999. State power beyond the "territorial trap": The internationalization of the state. *Political Geography* 18 (6): 669–96.

———. 2005. The *new* imperialism? On continuity and change in US foreign policy. *Environment and Planning A* 37 (9): 1527–44.

Gorjão, P. 2002. Japan's foreign policy and East Timor, 1975–2002. *Asian Survey* 42 (5): 754–71.

Granovetter, M. 1985. Economic action and social structure: The problem of embeddedness. *The American Journal of Sociology* 91 (3): 481–510.

Hall, S. 1992. Race, culture, and communications: Looking backward and forward at cultural studies. *Rethinking Marxism* 5 (1): 10–18.

Harvey, D. 2003. *The new imperialism*. Oxford, UK: Oxford University Press.

King, R. J. 2002. The Timor Gap, Wonosobo and the fate of Portuguese Timor. *Journal of the Royal Australian Historical Society* 88 (1): 75–103.

Kingston, J. 2006. Balancing justice and reconciliation in East Timor. *Critical Asian Studies* 38 (3): 271–302.

Knopp, L. 1992. Sexuality and the spatial dynamics of capitalism. *Environment and Planning D: Society and Space* 10 (6): 651–69.

Mallaby, S. 2002. The reluctant imperialist. *Foreign Affairs* 81 (2): 2–7.

Massey, D. 2005. *For space*. Thousand Oaks, CA: Sage.

Muppidi, H. 2004. *The politics of the global*. Minneapolis: University of Minnesota Press.

Nevins, J. 2004. Contesting the boundaries of international justice: State countermapping and offshore resource struggles between Australia and East Timor. *Economic Geography* 80 (1): 1–22.

———. 2005. *A not-so-distant horror: Mass violence in East Timor*. Ithaca, NY: Cornell University Press.

———. 2007–2008. The CAVR: Justice and reconciliation in a time of "impoverished political possibilities." *Pacific Affairs* 80 (4): 593–602.

Pulido, L. 2000. Rethinking environmental racism: White privilege and urban development in Southern California. *Annals of the Association of American Geographers* 90 (1): 12–40.

Ruggie, J. G. 1982. International regimes, transactions, and change: Embedded liberalism in the postwar economic order. *International Organization* 36 (2): 379–415.

Scheper-Hughes, N. 1993. *Death without weeping: The violence of everyday life in Brazil*. Berkeley: University of California Press.

Shaw, M. 1997. The state of globalization: Towards a theory of state transformation. *Review of International Political Economy* 4 (3): 497–513.

Simpson, B., ed. 2005. A quarter century of US support for occupation. National Security Archive Briefing Book, 28 November. http://www.gwu.edu/~nsarchiv/NSAEBB/NSAEBB174/index.htm (last accessed 27 February 2009).

Smith, N. 2003. *American empire: Roosevelt's geographer and the prelude to globalization*. Berkeley: University of California Press.

United Nations Development Program. 2006. *The path out of poverty: Timor-Leste human development report 2006*. Dili, Timor-Leste: UNDP.

Way, W., D. Browne, and V. Johnson, eds. 2000. *Australia and the Indonesian incorporation of Portuguese Timor, 1974–1976*. Carlton, Victoria, Australia: Melbourne University Press.

Armed Conflict and Resolutions in Southern Thailand

May Tan-Mullins

Since January 2004, southern Thailand has grabbed news headlines as a violent conflict zone, where the insurgency movement has taken the lives of more than 3,500 people. Stemming from various factors ranging from historical resistance to the Thai government to human rights abuses and socioeconomic marginalization, the shadowy movement has managed to infiltrate and gain sympathy among local communities; however, the majority of the local population remains resistant and resentful toward the perpetrators. This article aims to provide a comprehensive understanding of the upsurge and dynamics of the violence in southern Thailand in a two-part process. First, through focusing on the evidence obtained during my three-year stay in the Pattani province, I demonstrate how geopolitical dimensions intensified the separatist sentiments in the region. In the second section, I propose a proactive and preemptive set of resolutions as a roadmap to avoid further conflict and bloodshed. *Key Words: conflict resolution, geopolitics, southern Thailand, terrorism, transnational linkages*.

2004 年 1 月以来，泰国南部作为暴力冲突地区成为了新闻热点，那里的叛乱运动已造成了 3500 多人丧生。由于多种因素而引发，比如历史上对泰国政府侵犯人权的抵制，社会经济边缘化等，曾经隐藏的抵抗运动设法渗透并且在当地社区中获得了同情；但是，当地人口的大多数仍然反对叛乱运动，对叛乱肇事者明显不满。本文通过两部分的分析，旨在对该运动的高涨和泰国南部暴力行为提供一个全面和动态的了解。首先，通过重点分析本文作者在彭塔尼省居住三年所取得的证据，作者论证了地缘政治方面如何加强了该地区的分离主义情绪。在第二部分，作为对未来路线的建议，本文作者提出了一系列主动和先发制人的方案，以避免进一步的流血和冲突。*关键词：解决冲突，地缘政治，泰国南部，恐怖主义，跨国联系。*

Desde enero de 2004, la parte sur de Tailandia ha accedido a los titulares de las noticias como zona de conflicto violento, donde el movimiento insurgente ha cobrado las vidas de más de 3.500 personas. Alimentado por varios factores que comprenden desde la resistencia histórica al gobierno tailandés, hasta abusos de los derechos humanos y marginación socioeconómica, el oscuro movimiento se ha dado sus mañas de infiltrar y ganar simpatía de las comunidades locales; no obstante, todavía la mayoría de la población local resiste y resiente los subversivos. Este artículo busca un entendimiento más comprensivo de la aparición y dinámica de la violencia en el sur de Taiulandia, en un proceso de dos partes. Primero, concentrándome en la evidencia obtenida durante mi estada de tres años en la provincia Pattani, demuestro cómo las dimensiones geopolíticas intensificaron en la región los sentimientos separatistas. En la segunda sección, propongo un conjunto proactivo y prioritario de resoluciones a manera de hoja de ruta para frenar el conflicto y el baño de sangre. *Palabras clave: resolución de conflictos, geopolítica, sur de Tailandia, terrorismo, vínculos transnacionales*.

In the Muslim-majority provinces of Narathiwat, Pattani, and Yala of southern Thailand, more than 3,500 people have been killed since violence erupted in January 2004. Various explanations for the violence range from historical grievances to the central government's long-standing economic and political marginalization of the Muslim provinces. The human rights abuses and mismanagement in the Kru se Mosque and Tak Bai incidents of 2004 further hardened the fear and distrust of the southerners toward the central forces. Amid emerging discourses peppered with terms that capture the locals' imaginations, (such as *decentralization*, *autonomy*, and *special zones*), the central Thai government's need to maintain political control and exercise legitimate authority in the peripheral border zone is ever more pressing and evident.

Since Prime Minister Thaksin Shinawatra was ousted in September 2006, the chances of conflict resolution have diminished in Thai national politics, with a quick succession of governments, continuous clashes between the various political parties, and a lack of sustainability in policies targeted toward the south. This situation has been further complicated by the promulgation of ideas of a "network monarchy" (McCargo 2006,

2008) and possible transnational linkages to foreign radicals, reinforced by media coverage surrounding the capture of Jemaah Islamiah's (JI) operational head, Riduan Isamuddin, near Bangkok on 11 August 2003. With terrorism discourse dominating global relations since the 11 September 2001 (hereinafter "9/11") attacks, there has been speculation that the current insurgency is linked to the Southeast Asia terrorist network, or JI, and is part of greater global jihadist activities.

Investigation of the southern Thailand conflict has been mainly dominated by political science and terrorism narratives. Key concepts such as historical and religious identities have been tapped to explain the insurgency (Harish 2006; Liow 2006; Askew 2008). Other researchers prefer to examine the insurgency through global terrorism and jihadist angles (Croissant 2005; Gunaratna, Acharya, and Chua 2005; Storey 2008), however, to demonstrate the difficulties of linking the local conflict to global terrorism due to the lack of claims of responsibility for such violent activities. Reports by the Australian-based think tank International Crisis Group (2005, 2008) focus on conflict resolution and recommendations to deescalate the violence in the region, and McCargo (2006, 2008) proposes an interesting possibility of the monarchy's involvement in the southern conflict.

There also exists a wide array of geographical literature relevant to this southern Thai context. In particular, emerging research on territoriality and geopolitical boundaries infuses new insights into deconstructing this conflict. The "accidental" mapping of four Muslim provinces into the Thai geobody after the colonial powers pulled out of Southeast Asia, and subsequent separatist activities, demonstrated that the central question can no longer be "Where is the boundary?" but "How, by the way of what practices, and in the face of what resistance is the boundary imposed and ritualized?" (Dodds 1994, 193). Geographers such Newman (1999–2000) and Sidaway (2000) have also illustrated how geopolitical boundaries shape everyday life for those living on the borderlands. By focusing on "space," Routledge's (1997) concept of "terrain of resistance" further explains how much "social movements are affected by, and respond to, historical, economic, political, ecological and cultural processes and relations that are themselves place-specific. Regionalist and separatist movements have their origin in specific places or regions. Conflicts are grounded in particular places" (221).

Using southern Thailand as a case study, I first provide a comprehensive explanation for the upsurge of violence in this region, demonstrating how geopolitical dimensions to the conflict, such as regional politics and transnational religious networks, have intensified separatist sentiments in the region. National, regional, and local politics will also be assessed in relation to the conflict. In the second section of the article, I propose proactive and preemptive resolutions for how the violence could be reduced. Any investigation of armed conflicts should not be confined to the impacts of the globalization process, issues of national integration, and democratic rights but should also address the broader canvas of the sociohistorical and geopolitical factors. The next section details the research methods, field sites, and historical background briefly. The third section provides a comprehensive analysis of the violence, followed by a proposal for a peaceful resolution.

Research Methods and Sites

This article mainly draws on field research conducted between January 2002 and December 2005 that formed part of my doctoral thesis entitled *The Political Ecology of Coastal Resources Management in Fishing Villages in Southern Thailand* (Tan 2006). In 2006, I also made six trips to Kelantan, Malaysia, and two trips to Pattani (Figure 1) representing a nongovernmental organization. The aims of these trips were to establish contacts and maintain relations with self-proclaimed separatist groups for future dialogue for peace and resolutions. Out of 263 interviews, material from twenty-three in-depth informal interviews, representing various segments of the civil society and from both Buddhist (nine) and Muslim (fourteen) ethnicities, were used extensively in this article. Verbatim excerpts from *Tok Imams* (religious teachers), overseas scholars, academics, businessmen, and fishermen are presented to illustrate their opinions and frustrations about the violent situation. I have omitted the names of informants to protect their privacy.[1]

Upsurge of Violence in the South: Reasons and Theories

Sociohistorical Relations Between the South and Central Thailand

Prior to 1909, although Siam (Thailand) had claimed territorial influence over many parts of mainland Southeast Asia, political control existed mostly in a tributary system without direct central control. The Malay State of Patani (comprising today's southernmost provinces of

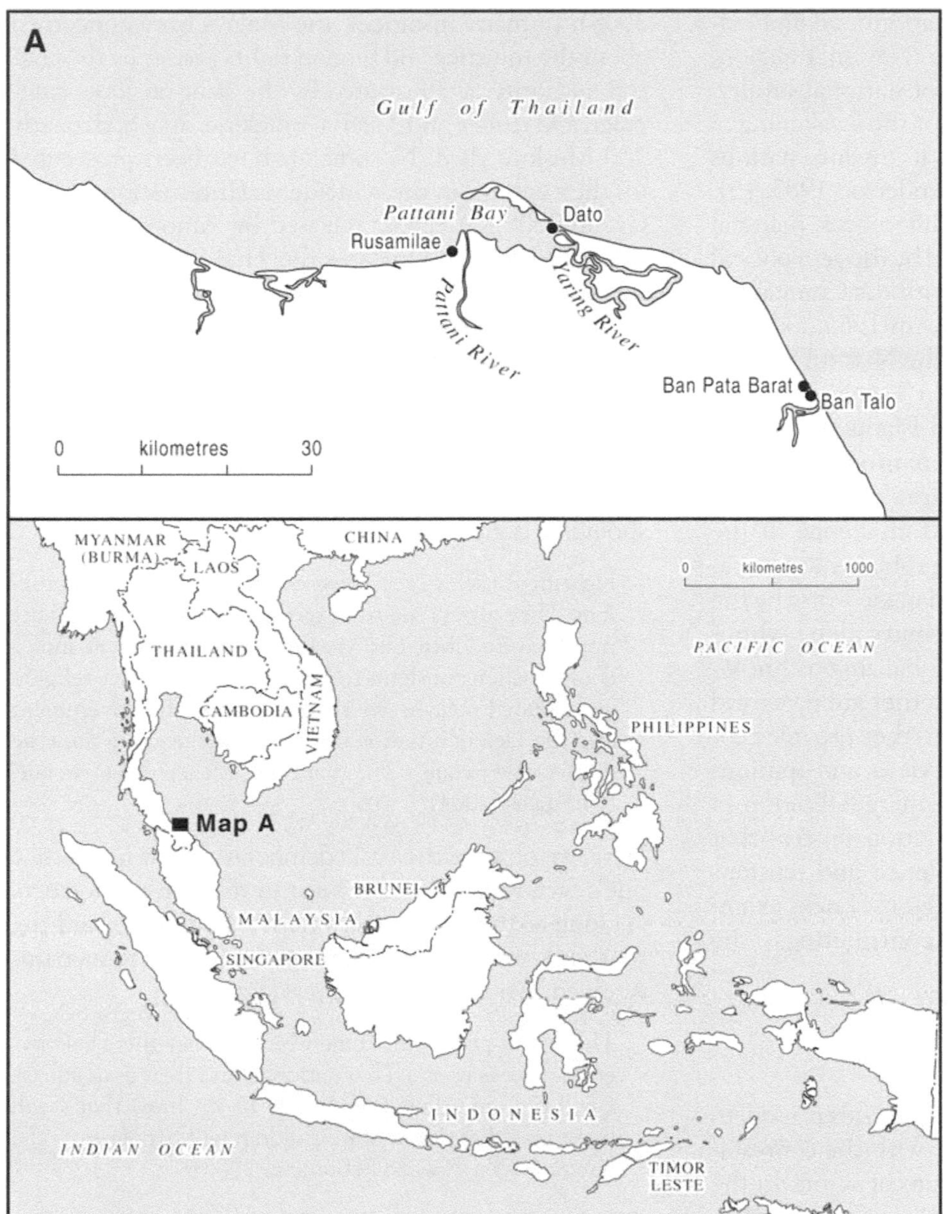

Figure 1. Map of field sites in Pattani province, southern Thailand.

Narathiwat, Pattani, and Yala) was mainly self-ruling, with a strong Islamic cultural identity. The kingdom of that time represents a golden era for the Patani Malays, regarded as a center of Islamic scholarship similar to the Sultanate of Aceh (Collins 2000, 67). This area was forcibly integrated into the Thai centralized political order, however, when the colonial powers pulled out of Southeast Asia. The political boundaries then were drawn on the basis of geographic location of a particular people rather than their ethnic spread or loyalties and without regard for the network of overlapping hierarchies and personal allegiances existing in Southeast Asia (Amitav 2000).

The behavioral distinctions between the south and the rest of Thailand are manifested in a kind of southern regionalism, a feeling of dislike for the central government and its representatives, and pride in the local dialect, culture, and history (Ruohomaki 1997, 99), further fueled by national policies. As there is no one unified, social, political Thai identity within a national boundary, the military rulers undertook a path to seek a national identity to bring its culturally diverse people from the north, the northeast, and the south together, by adopting the central region culture as the national culture and obliterating the others. Although the absolute monarchy was overthrown by a coup in 1932, it remains as a symbol and force for nationhood and cultural identity with the three pillars of "Nation, Religion, and King" constructed as the foci of national loyalty. To quote Gellner, "nationalism is not the awakening of the nations to self-consciousness; it invents nations where they do not exist" (cited in Anderson 1983, 15). The compelling need to quell any differences, political or otherwise, was further exacerbated by the geopolitical crisis in mainland Southeast Asia, with the Communist Party of Thailand (CPT) insurgency in 1965, followed by the Tet Offensive in 1968, and the Nixon Doctrine in 1969. By the early 1970s, the CPT had cultivated strongholds along every border and Islamic separatists had gained ground near Malaysia (Linantud 2008, 651).

The incorporation of most countries into the global capitalist system presents an added challenge to the region's governments, due to the problems accompanied by the economic downturn and made worse by the existing economic and political inequity along ethno-religious lines. At the same time, globalizing technologies such as mass media and the Internet are perceived to bring about more political space, as they provide people with a platform to express their views and opinions on political development. It is the marginalization of such expressions in the democratization process that has resulted in the upsurge of violence and tensions among various ethnic groups and regions. I next examine the various factors and theories contributing to the violence of the south.

Socioeconomic Grievances and Exclusions

Social and economic grievances are often used to explain the violence, linked mainly with the contestation of resources among various groups of actors. In the 1970s, new Thai-Buddhist settlers were introduced to balance the overwhelming Malay population (Satha-Anand 1987, 13), and governmental developmental projects have tended to benefit Thai (Buddhist) and Chinese residents over the Malays, widening the economic gap and further alienating the Malay community (International Crisis Group 2005). As a result, the Malays feel marginalized in many contexts, in terms of equality of economic and educational opportunities and employment in the government sector, especially in the southernmost provinces, which are among the poorest in Thailand.

At times, these grievances are peppered with bigger questions of identity, ethnicity, culture, and religion, and some argue that the suppression of such expression laid the ground for the violent rebellion (McCargo 2008). In many instances, the Malays were concerned about the injustice and human rights abuses by the central authority, as illustrated by the War on Drug campaign and Kru Se and Tak Bai incidents, in which nearly 200 Muslims died. No officials have been prosecuted for their actions in these incidents (International Crisis Group 2008). A report released by Amnesty International (2009) also indicates the Thai security forces in the southern provinces practice torture systematically. As such, the Malays feel that they are treated as second-class citizens and harbor resentment for past and continuing human rights abuses. To demonstrate the level of despair over the unjust treatment, one Malay respondent, Ba Lin, when asked what the main problem was in the recent violence in southern Thailand and what should be done to resolve it, stated:

> Fighting is useless. . . . The government is the main problem. They always see us as terrorists, just because we are not Thai-Buddhist but Muslims. How can we fit into a country when our rights to practice an alternative religion are violated? Maybe we should change the government and the idea of a national religion. What more must we do to prove we are a Thai national? Bleach ourselves fair? (07 August 2004)

Here, we observe the basic democratic right to practice one's own religion as discussed in relation to violence. As long as the Thai government's attitude toward the south remains unchanged, the situation will remain unresolved. Another Malay interviewee said:

> The day of peace will come when the day the Thai government sees us as a Thai national, and treat us as equals. I will pray to Allah (may peace be on him) that I will live to see that day and my children will benefit from the outcome. (Ba Yaw, 18 September 2005)

Another respondent indicated that social and economic grievances are indeed major concerns that intensified the willingness to fight, in particular for their future generation:

> This is not for us, but for our children. We have been tolerating this for years but we can't bear to see our children go through the same pain. For our children, we will fight if necessary to make things better. (Salim, 24 June 2006)

It is true those years of subjugation and marginalization are attributed to the willingness to participate in an armed struggle seen in the south; however, one cannot explain how the long-standing grievances, which have intensified and dissipated throughout modern times, erupted into large-scale violence again in 2004, when many Malays were also economically better off due to rising rubber prices. This explanation calls for an examination of how the greater geopolitical situation contributes to the violence in the south.

Regional Geopolitics: Pro-Thaksin and Pro-Royalist

Although economic marginalization is significant, an equally if not more important factor for the proliferation of dissent is the attitude and policies of the central government toward the south. Changes such as Thaksin's disbandment of two southern military-led security agencies (Southern Border Provinces Administration Centre [SBPAC] and Civilian Police Military Task Force [CPM 43]) in 2001 acted to reimpose central control over the opposition Democratic Party-controlled southern provinces. This move ended an important channel of communication between the southerners and Thai officialdom. Some locals further believe that the southern unrest is an expression of political resistance to Thaksin staged by various groups of elites. This distrust of the authorities—not just the Muslims toward the central authority but also by Thai-Buddhists—is summarized by Khun ML:

> No leaders, no groups, no PULO, no claims, not a person claiming responsibilities for all this violent activities! How do we know that it's the Muslims and not all these corrupted officials trying to make more money from the chaos? How bad can things be for all of us? (9 August 2005)

The SBPAC was established by Privy Council Prem Tinsulanonda to quell the unrest in a region that has been occasionally disrupted by secessionist movements, which are still active after the government's suppression and special amnesty scheme that ended most of the movements in the late 1970s. Nevertheless, various groups sharing power in the SBPAC had also used the center in seeking personal benefits. The military, the police, and local bureaucrats in the SBPAC were accused of using state power to conduct illegal trades and smuggling across the border, including staging violence in the area annually to ask for a bigger security budget from the central government. The ending of the SBPAC and the way the Thaksin government handed over the power in security affairs to the police unbalanced the military and other groups. The increased empowerment of police authorities has allowed them to use iron-fist measures to hunt down the "separatists" and their leaders. The military retaliated through the mishandling of the protesters in Narathiwat province on 25 October 2004, which resulted in more than eighty deaths.

More interestingly, the regional conflict in southern Thailand could be seen as a microcosm of the greater geopolitics being played out at the national level. Critics of Thaksin focused on his autocratic leadership style and took heart from signs of tension between him and King Bhumibol Adulyadej. McCargo (2006, 2008) viewed the series of events unraveling in both southern Thailand and Bangkok that culminated into a bloodless military coup on 19 September 2006 as not just simple military–police rivalry but a more complex triangulating power play of the pro-Thaksin (police) and pro-royalist (army) involving the monarchy. According to McCargo (2008, 9), by abolishing the military SBPAC, Thaksin had effectively staged a frontal assault on the legitimacy of the palace, unraveling the mode of virtuous rule that had been somewhat successful in curbing an earlier wave of separatist violence. He then coined the term *network monarchy* to explain the extensive network of lieutenants and supporters, mainly dominated by military figures and members of the privy council, who are supporting the political power of the monarchy through interventions such as the 2006 military coup.

At the same time, the democrats, who were politically dominant in the south but who were challenged by Thaksin since he came into power, began to exploit this confusing political space by canvassing heavily in these provinces for the 2005 election. This strategy possibly helped to explain fifty-two seats of members of parliament (out of fifty-four) captured by the Democrat party, while Thai Rak Thai could secure but one. According to a quick poll of eleven Thai-Buddhist informants who had previously voted for the Thai Rak Thai party, ten indicated that they had voted for the democrats simply because of the hard-handed policies in managing the conflict in the region. One informant also indicated that it is important to have a "check" mechanism, in this case the Democrat party, to avoid too much power vested in the hands of Thaksin (personal communications 14–15 February 2005).

In sum, the southern unrest is an expression of political resistance to Thaksin staged by various groups of elites and of reactions to greater geopolitical events. The military, politicians, and bureaucrats who have lost their personal interests under the Thaksin administration are accused of instigating the violence, and the police have been empowered in the region as the prime minister's most reliable tool to suppress the unrest. This situation inflames the old elite rivalry between the military and police at the national level, which is then manifested at the local level in the southern region. At the end of the day, Thaksin's iron-fist practices eventually isolated him, as the other elites began to unite to challenge his rule through the coup. His fall also spelled a new confusing chapter of electoral politics in Thailand, with endless mass protests and clashes between the People's Alliance for Democracy yellow (pro-royalists) and the National United Front of Democracy Against Dictatorship red (pro-Thaksin) factions in the streets of Bangkok.

Jihad? Islamic Revivalism and Transnational Linkages

Against these confusing geopolitical settings, there has also been a trend of Islamic revivalism in southern Thailand, which orients the explanation toward exogenous factors and adds an international dimension to the conflict. The concept of *ummah* (Islamic brotherhood) has enabled Islam to serve as both a powerful vehicle of criticism and a link to a larger Muslim world, thereby giving the grievances of Pattani their regional and transnational character (Kraus 1984, 423). Due to the global terrorism discourse since 9/11, however, roles of international networks such as Al Qaeda in the Middle East and JI in Southeast Asia have been questioned. In particular, the arrest of JI members in Thailand has fanned speculation that the insurgency in southern Thailand is linked to the international jihad. This was partly because there exists a minority in the region that has resorted to violent means to achieve their political and Islamic aims.

Historically, the most notable group is the Pattani United Liberation Organization (PULO). In 1988, after some leaders were arrested, the morale of the group was very low and some of its members gave themselves up to the Thai government. There are other groups that were formed over the years in the name of separatism, such as the Barisan National Pember-Basan Pattani (BIPP), Barasi Revolusi Nasional (BRN Congress), Runda Kumpulan Kecil (RKK), and BRN-Coordinate. The latter was considered by some analysts to include the main culprits of the recent spate of violence since January 2004.

A main tool used by these groups to establish a presence in this globalized world and specifically southern Thailand is media technology. For example, the PULO maintains its presence on the Internet and posts press releases on its Web site. Video compact discs (VCDs) are also brought in from Malaysia and used as propaganda for recruitment. These VCDs (Figures 2 and 3), which were sold during seminars, are Al Qaeda training videos propagating the doctrine of Osama Bin Laden and discussing conflicts in Palestine, Iraq, and Lebanon. I watched the VCDs with some youths from the villages and observed that they were quite emotional. When I asked what they could do to help the situation, many discussed the possibilities of intercultural dialogues, but some mentioned joining the "movement." These unemployed youths could be prime targets for recruitment by terrorists, possibly due to discontentment and boredom.

Other forms of transnational linkages involve sponsorships for the building of schools, wells, and other basic amenities such as public toilets in rural Muslim villages. For example, one Muslim village faced the problem of water salination, which resulted in the contamination of their drinking well. They received 600,000B (approximately US$15,000) from the Kuwaiti government to dig a 300-m-deep well and build three public toilets for the villagers (Figure 4). Such transnational linkages of assistance should not simply translate as global channels of jihadist ideas, however, due to the differences in the interpretation of Islam between Southeast Asia and the Middle East. As Mutalib (2003) states: "Southeast Asia has a different Islamic temperament and practice as compared to the heartland of Islam in Middle East. . . . Many of the extremist ideas and violent actions of Muslims in other parts of the world, such as the resort to suicide bombings are quite alien to this region" (18).

For some, transnational Islamic education was seen as a possible network base for recruitment. Thai General Kitti Rattanchaya noted that the recent spate of violence in the south indicated that long-dormant Muslim separatist insurgency has revived and spread its message through religious schools, training new young recruits to be fighters (Crispin 2004). Although there are contacts between Malays of southern Thailand and various international Muslim groups and institutions, more important, the familial linkages and ethnic similarities between southern Thai Malays and northern Malaysians present an added geopolitical dimension to the conflict. Malaysia's international policy toward southern Thailand has always been to maintain a delicate balance between national security and concerns for ethnically and culturally similar Malays across the border. Since the escalation of the violence, deteriorating relations between the two governments have at times resulted in the speculation of covert Malaysian support for the insurgent movement.

Through the interviews, however, I found that the insurgency and fighting was never viewed in connection to the greater "global holy war" managed by Al Qaeda

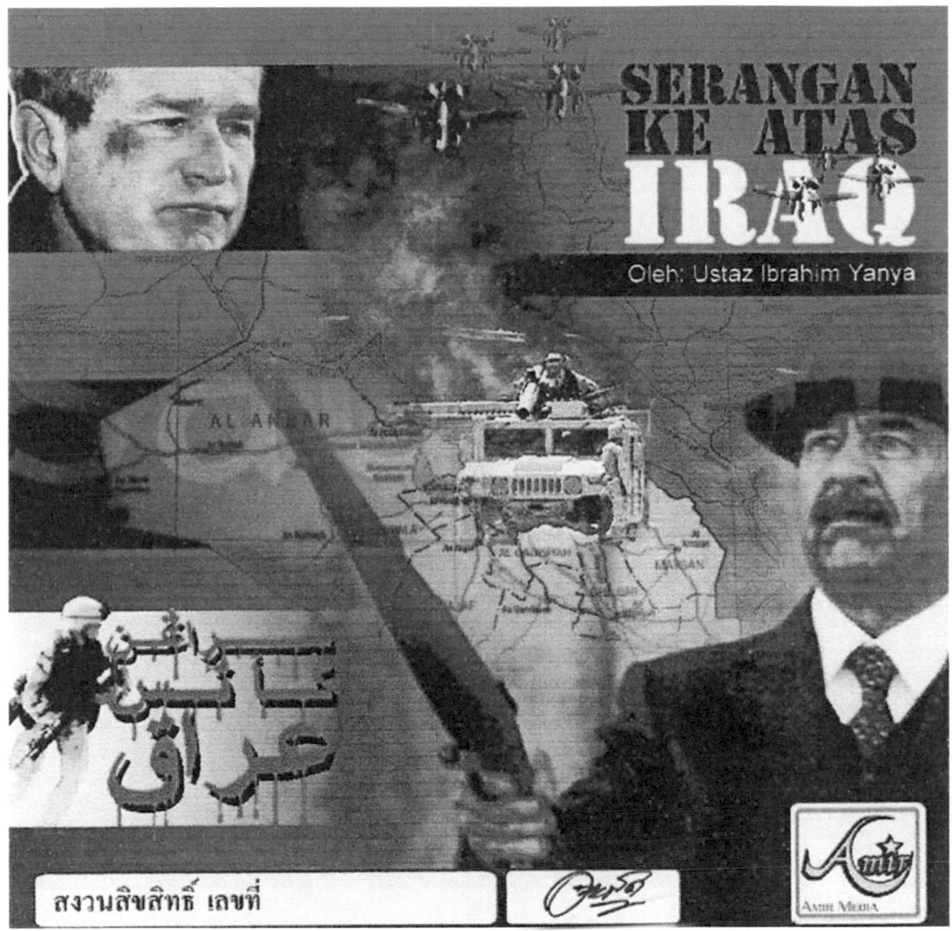

Figure 2. Promotional video compact disc (VCD) cover on the Iraq war.

and with Southeast Asia, expressed through JI's ambitions to create a pan-Islamic state, or across the border with the Malaysians. According to one informant, they will fight because of the need to defend their homeland and to fight against injustice and oppression, for the children and future generations. He stated:

> Our fight is for our children and for the legacy (land and resources) given to them by Allah (may peace be upon him). We care about the wars in Iraq and Afghanistan but this is not for them, this is for us. (Ba Ming, 18 July 2005)

Another person indicated:

> The only time we heard about JI was that the stolen arms from the camp (Narathiwat) might be sent to JI to support them in their fight and the capture of Hambali. But other than these two incidents, there wasn't much talk about JI. (Ba A, 25 November 2005)

A Malay academic based in Prince of Songkla University (PSU), who has studied in Malaysia and Singapore, questioned the transnational linkages and indicated fears concerning the influence of the insurgency in schools. He noted:

> I am very active in the religious school and as a lecturer and nobody tried to "recruit" me. But lately, the media has been reporting of pictures and evidences of militants training in the jungles of the south. I don't like them. The papers also reported that *pondoks* [schools] could be a base of recruitment for insurgents. I am worried for my *pondok* in my village and my fellow graduates, as we might be suspected as insurgents. (Ba S. H., 11 November 2004)

Of more concern to most of my informants is how the crisis has been sensationalized in the mass media, through the portrayal of killings as being ethnically and religiously instigated. This reporting further pushed the locals toward the tipping point of exasperation. Indeed, according to statistics by Deep South Watch[2] on deaths categorized by religion, from January 2004 to July 2008, more Muslims than members of any other groups were killed by insurgents. Out of 3,071 deaths, 1,665 were Muslim and 1,292 were Buddhists. The blurring of religious divisions of victims prompted many locals to

Figure 3. Promotional video compact disc (VCD) cover on Osama Bin Laden and his war.

wonder about the actual intentions of these so-called insurgents in the region. Ba-L, a Malay fisherman in Saiburi, said:

> Just who are these insurgents? Muslims do not use the name of Islam to kill our own Muslim brothers. This is not the right way of doing things. It must be the government trying to confuse us. (Ba-L, 19 January 2005)

Ba S. H. also expressed the locals' sentiments:

> It is really strange that there are lots of victims in the daily killings who are Muslims, and Muslims don't kill Muslims! But the media just don't report it. Instead, they focus on the so-called "ethnic killings." (Ba S. H., 11 November 2004)

This perhaps could explain the most important point: Despite the presence of possible avenues and factors prompting increasing sympathy or cooperation with the insurgency groups and linkages to the global jihad network, the majority of the Islamic population disagrees with their actions and resents these groups. According to a teacher in Pattani, Cik R. indicated:

> It's all Osama Bin Laden's fault! He gave our religion a bad name and America should focus on capturing him, instead of other distractions such as Iraq! Then the crisis will be over! (Cik R., 17 May 2005)

Indeed, out of the 133 Muslim interviewees, 109 were strongly against the terrorist tactics adopted by these separatist groups. They emphasized repeatedly that although they would like greater political participation, they have no sympathies for the movement and would not be part of any violent activities. The indiscriminate killings also further infuriate the locals, as it is no longer a question of disruption to the daily activities in the south but translates to real family and friends victimized by these bombings and attacks. The remaining 18 percent (twenty-four people) who have shown the slightest sympathy toward the movement either had relatives and friends victimized in the Kru se Mosque/ Tak Bai incidents or were unjustly treated by security forces. More important, the local conflict was often discussed independently from the greater Jihadist movements (of Al Qaeda and JI cells), reinforcing Routledge's (1997) point that separatist movements are often grounded in

Figure 4. Toilet and water pump donated by the Kuwaitis.

particular places but are responses to the specific historical and economic processes. It is only by addressing such processes and responses that the government will be able to win the hearts of the people and stem any support for the movement.

Resolutions

The constant change of government poses a great challenge to the sustainability of policies toward managing the violence in the south. The quick succession of governments since Thaksin saw a lack of consistency in management, which translates to insincerity in resolving the issues in the south. Perhaps one major move the Thai nation could take is to form a separate legislatively and constitutionally empowered council to manage the policies of peace building in the south. With the establishment of communication and trust, this council should then embark on various plans of development, not just economically but, more important, culturally and ethnically. An example would be to introduce the Islamic law and Yawi language as a secondary working language and to institute policies sensitive to Muslim ways of life.

The National Reconciliation Council was launched by the government in March 2005 and led by Anand Panyarachun (former prime minister of Thailand). Its main aims are to facilitate reconciliation and reduce violence in the southern provinces. Its recommendations include improvements to the government structure in the south, by establishing a Peaceful Strategic Administrative Centre for Southern Border Provinces and an unarmed peacekeeping force. These two new arms should be subsumed under the jurisdiction of the new council. There is also a need to put in place an effective justice system to address past injustice and human rights abuse, to reduce dissent, and to legitimize the peaceful coexistence of the Thais with the Malays.

Besides reforming the governance structure, the government also needs to exercise a "soft power" approach to other areas such as education. The government has barely approached education reform and schools have been considered hotbeds of conflict in the south. State-run schools are viewed by Malay nationalists as a vehicle for assimilation and indoctrination of Thainess, and the authorities suspect private Islamic schools of being breeding grounds for insurgents. It is only through such proactive and preemptive actions that the issue of injustice, misunderstanding, and marginalization will be softened and the militant movement slowly weakened.

In southern Thailand, the Muslims are currently facing a double whammy of marginalization. Throughout modern times, they have been marginalized socioeconomically due to the differences in culture and identity and later by the geopolitical crisis that began with the CPT insurgency in 1970s and then the Islamic separatist movements in the south. The global terrorism discourse also made the Thai central government unwilling to discuss the "problem" in the south. Estranged in international and national geopolitical events, the power play between various groups of actors was manifested in the (mis)management of the southern conflict. This crisis was also being relegated down the priority list by the constantly changing governments. It will be difficult to win the hearts of the Malays, but only through understanding the dynamics of violence and exercising soft power with reforms are some of the first small steps to developing a roadmap for peace in the region.

Notes

1. I have in some cases included the titles "Cik/Ba" for Malay individuals and "Khun/Mr./Ms." for Thai-Buddhist individuals.
2. See http://www.nonviolenceinternational.net/seasia/index.php?option=content&task=view&id=112 (last accessed 9 September 2008).

References

Amitav, A. 2000. *The quest for identity: International relations of Southeast Asia.* Singapore: Oxford University Press.

Amnesty International Publications. 2009. *Thailand: Torture in the southern counter-insurgency.* London: Amnesty International.

Anderson, B. 1983. *Imagined communities, reflections on the origin and spread of nationalism.* London: Verso.

Askew, M. 2008. Thailand's intractable southern war: Policy, insurgency and discourse. *Contemporary Southeast Asia* 30 (2): 186–214.

Collins, A. 2000. *The security dilemma of Southeast Asia.* New York: Macmillan.

Crispin, S. 2004. Thailand's war zone. *Far Eastern Economic Review* 11 March:12–14.

Croissant, A. 2005. *Unrest in Southern Thailand: Contours, causes and consequences since 2001.* Strategic Insights 4 (2). http://www.ccc.nps.navy.mil/si/2005/Feb/croissantfeb05.asp (last accessed 7 July 2009).

Dodds, K. J. 1994. Geopolitics and foreign policy: Recent developments in Anglo-American political geography and international relations. *Progress in Human Geography* 18:186–208.

Gunaratna, R., A. Acharya, and S. Chua. 2005. *Conflict and terrorism in Southern Thailand.* Singapore: Marshall Cavendish.

Harish, S. P. 2006. Changing conflict identities: The case of Southern Thailand discord. Working Paper 107, Institute of Defence and Strategic Studies, Singapore.

International Crisis Group. 2005. Southern Thailand; Insurgency, not jihad. http://www.crisisgroup.org/home/index.cfm?id=3436&1 =1 (last accessed 10 December 2006).

———. 2008. Thailand, political turmoil and Southern insurgency. http://www.crisisgroup.org/home/index.cfm?id=5640&1 =1 (last accessed 2 September 2008).

Kraus, W. 1984. Islam in Thailand: Notes on the history of Muslim provinces, Thai Islamic modernism and the separatist movement in the South. *Institute of Muslim Minority Affairs* 5 (2): 410–25.

Linantud, J. L. 2008. Pressure and protection: Cold war geopolitics and nation-building in South Korea, South Vietnam, Philippines, and Thailand. *Geopolitics* 13 (4): 635–56.

Liow, J. L. 2006. International jihad and Muslim radicalism in Thailand? Towards an alternative interpretation. *Asia Policy* 2:89–108.

McCargo, D. 2006. Thaksin and the resurgence of violence in the Thai South. Network monarchy strikes back? *Critical Asian Studies* 38 (1): 39–71.

McCargo, D. 2008. *Tearing apart the land: Islam and legitimacy in Southern Thailand.* Ithaca, NY: Cornell University Press.

Mutalib, H. 2003. The rise in Islamicity and the perceived threat in political Islam. In *Perspectives on doctrinal and strategic implications of global Islam*, ed. K. S. Nathan, 17–21. Trends in Southeast Asia Series: 12. Singapore: Institute of Southeast Asian Studies.

Newman, D. 1999–2000. Into the millennium: The study of international boundaries in an era of global and technological change. *Boundary and Security Bulletin* 7:63–71.

Routledge, P. 1997. Putting politics in its place. Baliapal, India, as a terrain of resistance. In *Political geography: A reader*, ed. J. Agnew, 219–55. London: Arnold.

Ruohomaki, O. 1997. *Fishermen no more: Livelihood and environment in Southern Thai maritime villages.* Bangkok, Thailand: White Lotus Press.

Satha-Anand, C. 1987. *Islam and violence: A case study of violent events in the four southern provinces, Thailand, 1976–1981.* Tampa, FL: USFG Monographs in Religion and Public Policy.

Sidaway, J. 2000. Geopolitical traditions. In *Human geography: Issues for the 21st century*, ed. P. Daniels, M. Bradshaw, D. Shaw, and J. Sidaway, 429–53. Harlow, UK: Longman.

Storey, I. 2008. Al-Qaeda takeover, a la Algeria? *The Straits Times* 1 September:A22.

Tan, P. M. 2006. The political ecology of fishing villages in Southern Thailand. PhD thesis, Department of Geography, National University of Singapore.

Crafting Liberal Peace? International Peace Promotion and the Contextual Politics of Peace in Sri Lanka

Kristian Stokke

Contemporary armed conflicts are typically intrastate conflicts in the Global South. These are often represented as global security threats, providing a justification for practical geopolitics of promoting liberal peace through elitist peace negotiations and instrumental use of humanitarian and development aid. In this context, the contemporary hegemonic discourse on peace emphasizes the synergies among liberal peace, liberal democracy, and neoliberal development. Based on recent democratic transitions, it is assumed that liberal peace can be crafted through internationally facilitated elite negotiations. This article examines this technocratic approach to peace and highlights the tension between elitist crafting of liberal peace and contextual political dynamics in conflict situations, using Sri Lanka as an exemplary case. Sri Lanka's fifth peace process is presented as a product of international and domestic power relations and stakeholder strategies, with a convergence between the government of Sri Lanka and the Liberation Tigers of Tamil Eelam around two defining characteristics: (1) crafting of peace through narrowly defined elite negotiations and (2) linking of peace and development through humanitarian and development aid. It is argued that the use of development as a precursor to peace politicized the issue of interim development administration and the combination of political exclusion of elites and social exclusion of intermediate Sinhalese classes undermined the government and its agenda for liberal peace. Sri Lanka is thus an illustrative case of international promotion of liberal peace but also the tensions between internationalized and elitist crafting of peace and contextual power relations and political dynamics in conflict situations. *Key Words: liberal peace, neoliberal development, peace process, practical geopolitics, Sri Lanka.*

当代的武装冲突通常是发生在南半球的国家内部冲突。这些武装冲突通常表现为全球的安全威胁，同时也为实用政治地理学提供了一个通过精英和平谈判以及积极使用人道主义和发展援助，从而促进自由和平地缘政治的实际理由。在这一形式下，当代霸权对和平的谈话中强调自由和平，自由民主，和新自由主义发展的协同作用。基于最近的民主转型，我们假设自由和平可以通过国际便利的精英谈判而达成。本文用斯里兰卡作为一个出色的例子，探讨达成和平的方法，并突出了在冲突局势中介于促成自由和平的精英和相关政治势力之间的紧张关系。斯里兰卡的第五阶段的和平进程是国际和国内的权力关系和利益相关者之战略的综合产品，并且是斯里兰卡政府和泰米尔伊拉姆猛虎解放组织围绕两个重要的特点的融合：（1）通过狭义定义的精英谈判而促成和平以及（2）通过人道主义和发展援助联接和平与发展。本文认为，利用发展作为和平的先锋使得临时发展管理这一问题政治化。同时，对精英阶层的政治排斥和僧伽罗人的中间阶级的社会排斥的结合破坏了政府和其自由和平的日程。斯里兰卡因此既是一个自由和平国际推广说明的范例，也是说明冲突局势中国际化的和精英制的和平以及相关权力关系和政治势力间紧张关系的范例。*关键词：自由和平，新自由主义发展，和平进程，实用地缘政治，斯里兰卡。*

En el 2005, el Programa de las Naciones Unidas para el Desarrollo (PNUD) puso en marcha en Chipre un proyecto de construcción de la paz denominado Acción para la Cooperación y la Confianza (ACT, en inglés). Este proyecto ha estado orientado a crear oportunidades para compañías bicomunales sobre protección ambiental, como estrategia para promover tolerancia intercomunitaria. En el artículo se discute críticamente la eficacia de aquel proyecto, como una contribución al debate sobre la significación de la cooperación ambiental para transformar los conflictos etno-territoriales. Nos apoyamos tanto en datos de campo como en las opiniones calificadas de chipriotas interesados en cuestiones ambientales, para mostrar que, en el caso de Chipre, el éxito de las estrategias ambientalistas de paz depende de una amplia conciencia ambiental, de la confianza en un tercer interesado (el PNUD) y del empoderamiento de la sociedad civil, lo cual, no obstante, debería complementar y no sustituir las otras intervenciones a nivel del estado. También existe evidencia de que el discurso del PNUD a partir de la idea de que "la naturaleza no reconoce límites" es más efectivo donde genera soluciones que se perciban como benéficas para todos los grupos implicados, más que cuando el medio ambiente es utilizado para construir retóricamente un "patriotismo" común por encima de identidades étnicas. *Palabras clave: resolución de conflictos, Chipre, medio ambiente, PNUD/ACT.*

The post–Cold War period has been marked by new geographies of conflict and security with associated changes in formal, practical, and popular geopolitics (Luke 2003; O'Loughlin 2005; Dalby 2008). Contemporary wars tend to be intrastate conflicts in the Global South rather than interstate in the Global North, but the representation of such conflicts has been highly contested. Whereas some observers see these "new wars" as rooted in social and economic grievances and limited opportunities for democratic political participation and peaceful social change (Kaldor 1999; Miall 2007), others argue that armed insurgencies are primarily motivated by the insurgents' greed for resources and power and that their methods amount to terrorism (Collier and Hoeffler 2001). The former representation points toward conflict resolution through political negotiations and comprehensive development and democratic transformations. The latter position lends itself to armed interventions to combat terrorism and establish political stability (Ramsbotham, Woodhouse, and Miall 2005).

Human geographic research on intrastate conflicts is somewhat uneven, containing both relative strengths and knowledge gaps. Intrastate conflicts in the Global South are a research theme that tends to fall between two traditions in a discipline that is divided by its own geography. Development geographers have paid relatively little attention to armed conflicts despite their relevance for development in the Global South; political geographers have examined northern geopolitical discourse on southern conflicts, giving little attention to contextual political and development dynamics. Although there has been a convergence between these subdisciplines around discourse analysis, there are relatively few cases of substantive integration of postcolonial studies and critical geopolitics within the research theme of armed intrastate conflicts (Gregory 2004; Kothari 2005; Korf 2006; McKinnon 2007). Scholars within critical geopolitics have especially investigated the ways in which post-Cold War intrastate conflicts have yielded new spatialized representations whereby local conflicts are construed as global security threats, especially after the attack on the World Trade Center in New York on 11 September 2001 (Cowen and Emily 2007). Some research has also been done on the manner in which these representations have paved the way for armed interventions in intrastate conflicts, but there are few political geographic studies of the nonviolent facilitation of conflict resolution and peace building (Newman 2002; Hyndman 2003; Falah 2005). I seek to address this knowledge gap at the intersection of security and development through a contextualized case study of the interplay between the practical geopolitics of key international actors and the domestic political dynamics around the recent peace process in Sri Lanka. The reason for the failure of the process is found in the disjuncture between elitist crafting of liberal peace and the domestic dynamics of political representation and mobilization. Sinhalese ethnonationalist politicization of liberal peace paved the way for a militaristic "war on terror" by the government of President Rajapakse from 2006 and a government-defined "victor's peace" after the defeat of the Liberation Tigers of Tamil Eelam (LTTE) in May 2009.

This article is based on numerous meetings and interviews with representatives of the government of Sri Lanka (GOSL), the LTTE, the Norwegian peace and development bureaucracy, the Tamil diaspora, international nongovernmental organizations (NGOs), and peace and development researchers in Sri Lanka since 2001. The contextual understanding is also derived from my continuous engagement with academic scholarship and media reports on Sri Lankan politics for the last two decades. Due to space constraints, the polarized politicization of the research topic, and the security concerns of respondents in Sri Lanka, no direct quotes or references to respondents are provided in this article.

The Interdiscursivity of Liberal Peace, Liberal Democracy, and Neoliberal Development

Contemporary human geographic research on armed conflicts is largely conducted within the framework of critical geopolitics, focusing on the production of geopolitical knowledge around international crises and interventions (Ó Tuathail 1996; Dalby 2008; Müller 2008), yielding a rich literature on formal geopolitical thinking as well as the production and circulation of geopolitical discourse in popular culture and mass media. There are relatively few geographic studies of practical geopolitics in foreign policy (Dodds 2001). Ó Tuathail (2002) argues that critical geopolitics should pay increased attention to practical geopolitics, simultaneously stressing that geopolitics is an interdiscursive field that is broader than mere geostrategic discourse. This is clearly the case in the contemporary geopolitics of crafting liberal peace, where discourses on security, democracy, and development are inextricably interwoven around the nodal point of liberal peace (Duffield 2001; Paris 2004; Richmond 2007).

Richmond (2007) observes that whereas there has been much debate among academics and policymakers about the prospects and means for negotiated transitions from intrastate conflicts to peace, the concept of peace itself is given remarkably little attention. This silence on the meaning of peace testifies to the hegemonic position of a specific discourse. Based on the thesis that liberal democratic governments are more peaceful than nondemocratic ones in both internal and international affairs, liberal peace is assumed to be both universally valid and attainable through the concerted and consistent efforts of governmental, intergovernmental, and nongovernmental actors. Based on the understanding that economic underdevelopment constitutes a fertile ground for armed insurgency, a new emphasis on neoliberal development has been added to the original liberal peace thesis, to emphasize replacing vicious cycles of underdevelopment and armed conflict with virtuous cycles of neoliberal development and liberal peace (Collier et al. 2003). Given recent experiences with internationally facilitated transitions to liberal democracy and structural adjustments to neoliberal globalization, it is assumed that liberal peace can be crafted through internationally facilitated elite negotiations supported by instrumental and conditional use of humanitarian and development aid (Anderson 1999; Carothers 2004; Paris 2004; Goodhand 2006).

There are close but underresearched links between these geopolitical discourses and practices promoting liberal peace and the elitist crafting of liberal democracy associated with the so-called third wave of democratization (Grugel 2002; Bastian and Luckham 2003). The political and academic discourse on democratic transitions since the mid-1970s has maintained a narrow focus on elitist crafting of formal rules, procedures, and institutions of liberal democracy, downplaying the importance of structural transformations and marginalizing popular forces from the democratization process (Harriss, Stokke, and Törnquist 2004). This representation lends support to the notion that liberal democracy and liberal peace can be crafted because political alliances that are conducive to democracy and peace can be encouraged or pressurized by international actors (Carothers 2004).

The discourse on democratic transitions and liberal peacemaking rests on basic claims about political representation and the links between democratic institutions and democratic politics (Törnquist, Webster, and Stokke 2009). It is assumed that elite actors who are in favor of peace and democracy carry a popular mandate that can be strengthened through the transition process, justifying the elitist focus and institutionalizing the protagonists as the principal stakeholders in negotiations. It is also assumed that the introduction of formal democratic institutions and peace deals will yield democratic politics and fulfillment of developmental rights (Bastian and Luckham 2003). In reality, these relations have proven to be much less straightforward. Recent transitions have typically produced formal rather than substantive democracies, falling short of the democratic principle of popular control over political decision making about public affairs, and many peace processes have failed to go from "negative peace" (i.e., absence of warfare) created by ceasefire agreements to lasting and substantive conflict transformation (Beetham 1999; Uyangoda 2005). In the absence of substantive popular representation and with neoliberal shrinking of the public affairs that come under democratic control, the actual fulfillment of socioeconomic rights also tends to fall short of expectations regarding developmental outcomes of transitions to peace and democracy (Jones and Stokke 2005).

These shortcomings of both democracy and development indicate that grievances that drive intrastate conflicts are altered rather than substantially transformed through elite-negotiated transitions. Consequently, elitist peace deals are often followed by resumption of warfare rather than substantive conflict transformation (Uyangoda 2005). Paris (2004) argues that transitions themselves often produce new grievances and destabilizing political mobilization. Mansfield and Snyder (2000; Snyder 2007) similarly point to the production of conflicts within the context of incomplete transitions to democracy, concluding that transitions should be sequenced so that rule of law precedes full democratic representation and popular control. Törnquist (2009) shares these concerns with flawed transitions but rejects the call for a return to politics of order, arguing that problems of flawed transitions are about the way transitions have been limited to the elites and a narrow range of public affairs. These general points shift attention from the practical geopolitics of crafting liberal peace to the politics of substantive democratization, development, and peace in transition periods, a process well illustrated by the dynamic interplay between the practical geopolitics of promoting liberal peace and the contextual politics of peace in Sri Lanka.

Contextual Politics of Liberal Peace in Sri Lanka

Sri Lanka was locked in a protracted armed conflict, from the anti-Tamil riots in July 1983 to the end of Eelam War IV in May 2009, between the government and a militant Tamil nationalist movement demanding recognition and self-determination for Tamils and their homeland (Tamil Eelam) in North-East Sri Lanka (Kleinfeld 2005). Tamil nationalism emerged as a nonviolent and democratic movement for power sharing based on federalism but was later transformed and radicalized into militant separatism (Manogaran and Pfaffenberger 1994; Stokke and Ryntveit 2000; Wilson 2000). This militant separatist movement consisted initially of five major organizations and a plethora of smaller groups but has since the late 1980s been dominated by the LTTE, claiming to be the "sole representative" of the Tamil nation but questioned by many non- and ex-militant Tamil moderates as well as most non-Tamils (Swamy 1994).

This protracted intrastate armed conflict has been interspersed by five attempts at political conflict resolution: the Thimpu Talks between the GOSL and the major separatist organizations in 1985, the Indo-Lanka Accord between the governments of Sri Lanka and India in 1987, the talks in 1989–1990 between LTTE and President Premadasa's government, the talks between LTTE and the government of President Kumaratunga in 1994–1995, and the peace negotiations between Prime Minister Ranil Wickremasinghe's government and the LTTE in 2002–2003 (Balasingham 2004; Rupesinghe 2006; Gooneratne 2007). The last peace process was characterized by active involvement of a range of international actors. India played a key role in the negotiations in the 1980s due to geopolitical interests and the links between militant Tamil groups in Sri Lanka and political leaders in Tamil Nadu. Other international actors became involved only in the late 1990s and by way of humanitarian and development aid (Bastian 2007). The peace negotiations and associated peace building were funded by the cochairs to the donor conferences (Japan, European Union, United States, and Norway) and international financial institutions such as the World Bank and the International Monetary Fund provided support for the peace process. A broad range of international NGOs and United Nations agencies have been involved in humanitarian and development programs during the "no war/no peace" period (2002–2007) and especially after the 2004 Indian Ocean tsunami disaster. Thus, the question of peace in Sri Lanka has become thoroughly internationalized, largely through securitization of aid (Goodhand and Klem 2005). Behind the convergence on technocratic concerns with development aid effectiveness are more complex and divergent geopolitical interests of defeating terrorism to ensure stability in the region and creating a sound political context for neoliberal development. As the negotiations stalled and the protagonists returned to armed hostilities, it became clear that the practical geopolitics of the United States and India were based on economic and security interests amid political instability in the region and the growing presence of China in the Indian Ocean, yielding active but conditional support for the GOSL's war against LTTE and with the other cochairs gradually becoming de facto supporters of deployment of force against a nonstate actor branded as a terrorist organization (Nadarajah and Sriskandarajah 2005; Kaplan 2009).

The practical geopolitics of peace promotion pursued by Sri Lanka's aid donors reflect post-Cold War discourses on terrorism and securitization of aid. Nevertheless, the design and dynamics of the peace process were conditioned by domestic, military, political, and economic conditions as much as by international geopolitics, especially the existence of (1) a mutually hurting stalemate between the Sri Lanka Armed Forces and LTTE, (2) crises of governance for the GOSL, and (3) crises of development for both LTTE and the GOSL (Uyangoda 2005; Bastian 2007).

Sri Lanka's fifth peace process was primarily preconditioned by the balance of power between LTTE and the GOSL. The warring parties reached a mutually damaging stalemate in the late 1990s, following a series of military victories for the LTTE that brought extensive areas under their control and created a military parity of status. Although the GOSL was put on the defensive by military setbacks, the LTTE also reached a limit for what it could achieve without unleashing negative international sanctions, especially from India (Uyangoda 2005). The Ceasefire Agreement on 22 February 2002 froze this military-territorial balance of power and segmented a de facto dual-state structure. This balance of power brought the protagonists into negotiations and kept them from resuming warfare, despite the breakdown of negotiations, until the balance was altered by changing positions among the international actors in favor of the GOSL.

The peace process was also conditioned by the existence of a crisis of governance in the sense that the government was based on a weak coalition with only a small majority in Parliament (Uyangoda and Perera 2003; Uyangoda 2005). The government was constrained by a deeply fragmented political elite, seen in contentious cohabitation between Prime Minister Ranil Wickremasinghe (United National Party) and the powerful executive President Chandrika Bandaranaike Kumaratunga from the opposition Sri Lanka Freedom Party (SLFP), as well as a demos that had been constructed and mobilized through ethnic identities and entrenched practices of ethnic outbidding (de Votta 2004). The GOSL had limited prospects for political conflict resolution and had to search for strategies of conflict management within limits set by constitutional and institutional arrangements characterized by majoritarianism and centralization of state power.

The peace process was also preconditioned by severe crises of development. Whereas the GOSL faced a growing budget crisis with soaring military expenses from the "war for peace" campaign and rising costs of living posing a serious threat to the legitimacy and electoral survival of the government, the LTTE faced a humanitarian crisis and an increasingly war-weary Tamil population who had suffered massive destruction of lives and livelihoods (Kelegama 2006; Rupesinghe 2006; Bastian 2007). This situation made both protagonists enter the peace process with a desire to address humanitarian rehabilitation and development challenges.

In this context of a frozen balance of power and crises of governance and development, the GOSL pursued a strategy of normalizing everyday life without devolution of power, assuming that this would depoliticize Tamil grievances and reduce the need for substantive power sharing. The GOSL's main objective was to further the neoliberal development process, making peace both a precondition and an instrument that could yield dividends from reduced military expenses and increased inflow of international aid (Bastian 2007; Shanmugaratnam and Stokke 2008). The LTTE followed a strategy of institutionalizing power sharing by building separate state institutions within areas under their control, producing a pretext for internal or external self-determination based on earned sovereignty. Development was instrumental: It would address immediate humanitarian needs in the war-devastated and government-neglected northeast while also furthering the LTTE state-building projects (Balasingham 2004). The strong focus on internationally funded humanitarian and development programs allowed both the GOSL and the LTTE to pursue their strategic interests, making the peace process an extension of war by other means.

This combination of international geopolitics and domestic preconditions and stakeholder strategies produced distinct features and hurdles. Sri Lanka's fifth peace process was an attempt at crafting liberal peace through internationally facilitated elite negotiations and securitization of aid, characterized by narrowly defined "track one" negotiations between the warring parties (Uyangoda and Perera 2003). Political opposition, including the left-of-center SLFP and the left Sinhalese nationalist Janatha Vimukthi Peramuna (JVP), moderate Tamils, the Muslim "second minority," and the broad diversity of civil society organizations including the Buddhist Sangha, were all excluded from the negotiations, creating numerous potential "spoilers" within the highly fragmented political elite. The negotiations were also narrowly defined to questions of power sharing between the LTTE and the GOSL in interim institutions for development administration, postponing questions of devolution of power, human rights, and substantive democratic reforms (Rainford and Satkunanathan 2008; Shanmugaratnam and Stokke 2008). Sri Lanka was at the time characterized by a de facto dual-state structure with parallel but different challenges of political transformations within two political entities. On the one hand was Sri Lankan state formation, which has a longstanding democratic tradition, but is also characterized by majoritarianism, a unitary and centralized constitution, and illiberal political practices that include ethnic constitution of demos, clientelist incorporation of people, and ethnic outbidding (Stokke 1998; de Votta 2004). On the other was the emerging state formation within LTTE-controlled areas, where the LTTE demonstrated an ability to govern but doing so by way of authoritarian centralization with few mechanisms for democratic representation (Stokke 2006). Thus, there was a strong need for comprehensive political transformations, which remained largely unaddressed as the protagonists and the international actors confined themselves to technocratic crafting of negative peace.

Development over politics characterized the peace process, with an unprecedented sequencing of the negotiation process (Sriskandarajah 2003). The negotiating teams from the GOSL and the LTTE agreed to address immediate humanitarian relief and rehabilitation needs in the war-torn areas and to establish a joint interim mechanism to plan and implement development projects in the northeast (Rainford and Satkunanathan 2008); however, the search for a mutually acceptable development administration was deeply problematic. The LTTE saw an interim administration with a fair degree of autonomy and a guaranteed position for the LTTE as an absolute necessity to ensure the fulfillment of both short-term development needs and long-term demands for internal self-determination. The GOSL and Sinhalese opposition feared that the interim administration would constitute a first step toward secession and therefore a threat to the sovereignty of the unitary Sri Lankan state. The GOSL was able to propose only minimalist institutional reforms within the framework of the existing constitution and in the face of mounting opposition from the executive president, SLFP, JVP, and Sinhalese nationalist groups. The government's proposals were rejected by the LTTE on grounds that they failed to guarantee the LTTE's position in decision making and delivery of rehabilitation and development (Balasingham 2004). These disagreements between the protagonists over an interim administration for the northeast provided an opportunity for the political opposition to bring the peace process to a halt. They also politicized the role of the international actors. The cochairs in turn imposed aid conditions, but these were an ineffective mechanism, as they were externally imposed and vaguely defined and could only be accommodated at a high political cost for the GOSL (Goodhand and Klem 2005; Kelegama 2006; Shanmugaratnam and Stokke 2008). The peace process stalled and the delivery of aid ended up having relatively weak strategic links to the peace process as it was channeled through international NGOs that at best could offer humanitarian relief and development aid in a conflict-sensitive manner but had few links to substantive political transformations.

Development administration became the main point of contention between the GOSL and the LTTE, as well as the rallying point for opposition among excluded political elite actors. At the same time, the government's development policy reinforced the social basis for oppositional mobilization in the intermediate Sinhalese classes. The convergence among the LTTE, the GOSL, and the international donors around a technocratic top-down approach to development strengthened the authority of the LTTE in the northeast but also allowed the government to further its neoliberal development agenda at the national scale (Bastian 2007; Shanmugaratnam and Stokke 2008). This deepening of neoliberal development ran counter to the strong legacy of statism and welfareism in Sri Lanka but also the legacy of clientelist distribution of material concessions that had coexisted with market-led development. The economic policies of the government under Prime Minister Ranil Wickremasinghe deepened the contradiction between the symbolic representation of poor people in political discourse and the material reality of intermediate classes by deepening socio-spatial inequalities and paying less attention to political integration through clientelist political networks (Bastian 2007). In the context of an elitist peace process, excluded political elites and socially excluded intermediate Sinhalese classes came together in vocal opposition to the peace process, the government, and the international actors, bringing down the government at the general elections in 2004. The new government of President Rajapakse combines a militant strategy vis-à-vis the LTTE in agreement with Sinhalese nationalism, a neoliberal development agenda in agreement with Sri Lanka's aid donors, and symbolic and material concessions to the intermediate Sinhalese classes, thereby replacing the elitist crafting of liberal peace with an authoritarian and militant pursuit of a victor's illiberal peace. This war on terror, which was facilitated by changing geopolitical constellations in the Indian Ocean, ended in the military defeat of the LTTE in May 2009 and a subsequent "political process" that sought to incorporate the Tamil minority by way of assimilation and handpicked local leadership rather than substantive minority rights and self-determination. The Sri Lankan experience highlights the disjuncture between the practical geopolitics of promoting elitist liberal peace and domestic power relations and political dynamics. It also demonstrates that war and peace are means rather than ends for domestic and international actors in their pursuit of state power and geopolitical order. These are lessons with important implications far beyond the contextual specificities of Sri Lanka's intrastate conflict and its failed fifth peace process.

Conclusion

Sri Lanka provides an illustrative case of intrastate conflicts, international promotion of liberal peace, and contextual political dynamics around peace and development. The peace process emerged within the context of a mutually damaging stalemate and crises of governance and development and was characterized by narrowly defined elitist peace negotiations and instrumental use of humanitarian and development aid. Using development as a precursor to peace politicized the issue of interim development administration and produced a stalemate between the LTTE and the GOSL, and political exclusion of elite actors and social exclusion of intermediate Sinhalese classes produced powerful opposition against power sharing, neoliberal development, and international intervention. The Sri Lankan case demonstrates key characteristics of international promotion of liberal peace after the Cold War but also the tensions between elitist crafting of peace and contextual dynamics of political mobilization for state power in conflict situations. This field of inquiry calls for new research across thematic and geographic divides within human geography.

Acknowledgments

I would like to thank the anonymous reviewers and the editor for valuable comments. I am also greatly indebted to my research partners Olle Törnquist and Jayadeva Uyangoda as well as our master's students and PhD candidates at the University of Oslo and University of Colombo for many fruitful and critical discussions.

References

Anderson, M. B. 1999. *Do no harm. How aid can support peace—Or war*. Boulder, CO: Lynne Rienner.

Balasingham, A. 2004. *War and peace: Armed struggle and peace efforts of Liberation Tigers*. Mitcham, UK: Fairmax.

Bastian, S. 2007. *The politics of foreign aid in Sri Lanka: Promoting markets and supporting peace*. Colombo, Sri Lanka: International Centre for Ethnic Studies.

Bastian, S., and R. Luckham, eds. 2003. *Can democracy be designed? The politics of institutional choice in conflict-torn societies*. London: Zed.

Beetham, D. 1999. *Democracy and human rights*. Oxford, UK: Polity Press.

Carothers, T. 2004. *Critical mission: Essays on democracy promotion*. Washington, DC: Carnegie Endowment for International Peace.

Collier, P., V. L. Elliott, H. Hegre, A. Hoeffler, M. Reynal-Querol, and N. Sambanis. 2003. *Breaking the conflict trap: Civil war and development policy*. Oxford, UK: Oxford University Press.

Collier, P., and A. Hoeffler 2001. Greed and grievance in civil war. World Bank Policy Research Working Paper 2355. Washington, DC: The World Bank.

Cowen, D., and G. Emily 2007. *War, citizenship, territory*. London and New York: Routledge.

Dalby, S. 2008. Imperialism, domination, culture: The continued relevance of critical geopolitics. *Geopolitics* 13:413–36.

de Votta, N. 2004. *Blowback: Linguistic nationalism, institutional decay, and ethnic conflict in Sri Lanka*. Stanford, CA: Stanford University Press.

Dodds, K. 2001. Political geography III: Critical geopolitics after ten years. *Progress in Human Geography* 25 (3): 469–84.

Duffield, M. 2001. *Global governance and the new wars: The merging of development and security*. London: Zed.

Falah, G.-W. 2005. The geopolitics of "enclavisation" and the demise of a two-state solution to the Israeli–Palestinian conflict. *Third World Quarterly* 26 (8): 1341–72.

Goodhand, J. 2006. *Aiding peace? The role of NGOs in armed conflict*. Bourton on Dunsmore, UK: ITDG.

Goodhand, J., and B. Klem 2005. *Aid, conflict and peace-building in Sri Lanka*. Colombo, Sri Lanka: The Asia Foundation.

Gooneratne, J. 2007. *Negotiating with the Tigers (LTTE)*. Pannipitiya, Sri Lanka: Stamford Lake.

Gregory, D. 2004. *The colonial present*. Oxford, UK: Blackwell.

Grugel, J. 2002. *Democratization: A critical introduction*. New York: Palgrave Macmillan.

Harriss, J., K. Stokke, and O. Törnquist, eds. 2004. *Politicising democracy: The new local politics of democratisation*. New York: Palgrave Macmillan.

Hyndman, J. 2003. Aid, conflict, and migration: The Canada–Sri Lanka connection. *The Canadian Geographer* 47 (3): 251–68.

Jones, P., and K. Stokke, eds. 2005. *Democratising development: The politics of socio-economic rights in South Africa*. Leiden, The Netherlands: Martinus Nijhoff.

Kaldor, M. 1999. *New and old wars: Organised violence in a global era*. Cambridge, UK: Polity Press.

Kaplan, R. 2009. Center stage for the 21st century: Power plays in the Indian Ocean. *Foreign Affairs* 88 (2): 16–32.

Kelegama, S. 2006. *Development under stress: Sri Lankan economy in transition*. New Delhi, India: Sage.

Kleinfeld, M. 2005. Destabilizing the identity-territory nexus: Rights-based discourse in Sri Lanka's new political geography. *GeoJournal* 64: 287–95.

Korf, B. 2006. Who is the rogue? Discourse, power and spatial politics in post-war Sri Lanka. *Political Geography* 25:279–97.

Kothari, U. 2005. Authority and expertise: The professionalisation of international development and the ordering of dissent. *Antipode* 37 (3): 425–46.

Luke, T. 2003. Postmodern geopolitics. In *A companion to political geography*, ed. J. Agnew, K. Mitchell, and G. Ó Tuathail, 219–35. Oxford, UK: Blackwell.

Manogaran, C., and B. Pfaffenberger, eds. 1994. *The Sri Lankan Tamils: Ethnicity and identity*. Boulder, CO: Westview.

Mansfield, E. D., and J. Snyder. 2000. *Electing to fight: Why emerging democracies go to war*. Cambridge, MA: MIT Press.

McKinnon, K. 2007. Postdevelopment, professionalism, and the politics of participation. *Annals of the Association of American Geographers* 97 (4): 772–85.

Miall, H. 2007. *Emergent conflict and peaceful change*. New York: Palgrave Macmillan.

Müller, M. 2008. Reconsidering the concept of discourse for the field of critical geopolitics: Towards discourse as language and practice. *Political Geography* 27:322–38.

Nadarajah, S., and D. Sriskandarajah. 2005. Liberation struggle or terrorism? The politics of naming the LTTE. *Third World Quarterly* 26 (1): 87–100.

Newman, D. 2002. The geopolitics of peacemaking in Israel-Palestine. *Political Geography* 21:629–46.

O'Loughlin, J. 2005. The political geography of conflict: Civil wars in the hegemonic shadow. In *The geography of war and peace: From death camps to diplomats*, ed. C. Flint, 85–110. Oxford, UK: Oxford University Press.

Ó Tuathail, G. 1996. *Critical geopolitics: The politics of writing global space*. Minneapolis: University of Minnesota Press.

———. 2002. Theorizing practical geopolitical reasoning: The case of the United States' response to the war in Bosnia. *Political Geography* 21:601–28.

Paris, R. 2004. *At war's end: Building peace after civil conflict*. Cambridge, UK: Cambridge University Press.

Rainford, C., and A. Satkunanathan. 2008. *Mistaking politics for governance: The politics of interim arrangements in Sri Lanka*. Colombo, Sri Lanka: International Center for Ethnic Studies.

Ramsbotham, O., T. Woodhouse, and H. Miall. 2005. *Contemporary conflict resolution*. 2nd ed. Cambridge, UK: Polity.

Richmond, O. P. 2007. *The transformation of peace*. New York: Palgrave Macmillan.

Rupesinghe, K. 2006. *Negotiating peace in Sri Lanka: Efforts, failures and lessons*. Vol. I and II. Colombo, Sri Lanka: The Foundation for Co-Existence.

Shanmugaratnam, N., and K. Stokke. 2008. Development as a precursor to conflict resolution: A critical review of the fifth peace process in Sri Lanka. In *Between war and peace in Sudan and Sri Lanka: Deprivation and livelihood revival*, ed. N. Shanmugaratnam, 93–115. Oxford, UK: James Currey.

Snyder, J. 2007. *From voting to violence: Democratization and nationalist conflict*. New York: Norton.

Sriskandarajah, D. 2003. The returns of peace in Sri Lanka: The development cart before the conflict resolution horse? *Journal of Peacebuilding and Development* 2 (1): 21–35.

Stokke, K. 1998. Sinhalese and Tamil nationalism as post-colonial political projects from "above," 1948–1983. *Political Geography* 17 (1): 83–113.

———. 2006. Building the Tamil Eelam state: Emerging state institutions and forms of governance in LTTE-controlled areas in Sri Lanka. *Third World Quarterly* 27 (6): 1021–40.

Stokke, K., and A. K. Ryntveit. 2000. The struggle for Tamil Eelam in Sri Lanka. *Growth and Change* 31 (2): 285–304.

Swamy, M. R. N. 1994. *Tigers of Lanka: From boys to guerillas*. Colombo, Sri Lanka: Vijitha Yapa.

Törnquist, O. 2009. Introduction: The problem is representation! Towards an analytical framework. In *Rethinking popular representation*, ed. O. Törnquist, N. Webster, and K. Stokke, 1–10. New York: Palgrave Macmillan.

Törnquist, O., N. Webster, and K. Stokke, eds. 2009. *Rethinking popular representation*. New York: Palgrave Macmillan.

Uyangoda, J., ed. 2005. *Conflict, conflict resolution and peace building*. Colombo, Sri Lanka: University of Colombo, Department of Political Science and Public Policy.

Uyangoda, J., and M. Perera, eds. 2003. *Sri Lanka's peace process 2002: Critical perspectives*. Colombo, Sri Lanka: Social Scientists' Association.

Wilson, A. J. 2000. *Sri Lankan Tamil nationalism: Its origins and development in the 19th and 20th centuries*. New Delhi, India: Penguin.

"Nature Knows No Boundaries": A Critical Reading of UNDP Environmental Peacemaking in Cyprus

Emel Akçalı and Marco Antonsich

In 2005, the United Nations Development Program (UNDP) set up in Cyprus a peace-building project called Action for Cooperation and Trust (ACT). This project has aimed to create opportunities for bicommunal partnerships on environmental protection as a way to promote intercommunal tolerance. This article discusses critically the efficacy of this project to contribute to the debate on the significance of environmental cooperation in transforming ethno-territorial conflicts. We rely on both survey data and the qualified opinions of Cypriot environmental stakeholders to show that, in the case of Cyprus, successful environmental peacemaking strategies are dependent on widespread environmental awareness, trust in the "third party" (UNDP), and civil society's empowerment, which, however, should complement and not substitute for intervention at a state level. There is also evidence to suggest that the UNDP discourse about "nature knows no boundaries" is most effective when it generates solutions that are perceived to be beneficial to all parties involved, rather than when it uses the environment to discursively construct a common "patriotism" beyond ethnic identities. *Key Words: conflict resolution, Cyprus, environment, UNDP-ACT.*

2005 年，联合国开发计划署（UNDP）在塞浦路斯建立了一个和平建设项目，名字叫做"合作与信任行动（ACT）"。这个项目旨在创造条件，通过两族在环境保护上的伙伴关系，以此促进两族之间的容忍度。关于环境合作能够改造种族领土冲突的重要意义，一直存在争议，本文对上述项目的成效进行了批判性的讨论，有助于我们对这一问题的理解。我们的分析依赖于调查数据和有资格的塞浦路斯环境利益相关者的意见，分析表明，在塞浦路斯，环境缔造和平这一战略的成功依赖于广泛的环保意识，对第三方（UNDP）的信任，以及民间社会的力量，然而，应当是作为补充而不是取代在国家水平上的介入。也有证据显示，联合国开发计划署对"自然无边界"这一口号的推广，当它能够生成被认为是有利于所有各方的解决方案时，而不是被用来不着边际地通过环境构建一个超越民族特性的共同的"爱国主义"的时候，是最有效的。*关键词：冲突解决，塞浦路斯，环境，联合国开发计划署的合作与信任行动。*

En el 2005, el Programa de las Naciones Unidas para el Desarrollo (PNUD) puso en marcha en Chipre un proyecto de construcción de la paz denominado Acción para la Cooperación y la Confianza (ACT, en inglés). Este proyecto ha estado orientado a crear oportunidades para compañías bicomunales sobre protección ambiental, como estrategia para promover tolerancia intercomunitaria. En el artículo se discute críticamente la eficacia de aquel proyecto, como una contribución al debate sobre la significación de la cooperación ambiental para transformar los conflictos etno-territoriales. Nos apoyamos tanto en datos de campo como en las opiniones calificadas de chipriotas interesados en cuestiones ambientales, para mostrar que, en el caso de Chipre, el éxito de las estrategias ambientalistas de paz depende de una amplia conciencia ambiental, de la confianza en un tercer interesado (el PNUD) y del empoderamiento de la sociedad civil, lo cual, no obstante, debería complementar y no sustituir las otras intervenciones a nivel del estado. También existe evidencia de que el discurso del PNUD a partir de la idea de que "la naturaleza no reconoce límites" es más efectivo donde genera soluciones que se perciban como benéficas para todos los grupos implicados, más que cuando el medio ambiente es utilizado para construir retóricamente un "patriotismo" común por encima de identidades étnicas. *Palabras clave: resolución de conflictos, Chipre, medio ambiente, PNUD/ACT.*

Competing ethnonationalisms often generate conflicts over the control of the state and the redefinition of territory (Penrose 2002). In 1963, the ethnonationalist rivalry between the Greek Cypriot (GC) majority and the Turkish Cypriot (TC) minority, competing national interests between Greece and Turkey, and the Cold War context led to intercommunal violence in Cyprus. As a result, the TCs—the main target of this upheaval—withdrew from the administration of the Republic of Cyprus (RoC), founded in 1960, and created politico-territorial enclaves. In 1974, Turkey intervened militarily as a response to a coup d'etat launched against the RoC government by GC ultranationalists in conjunction with the Greek military junta to annex Cyprus to Greece. This action led to the territorial partition of the island, with population displacement among both GCs and TCs. Since then, Cyprus has been divided into the following territorial entities: an internationally recognized RoC; a de facto Turkish Republic of Northern Cyprus (TRNC), embargoed by the international community; two sovereign British military bases, present on the island since 1960; and a buffer zone controlled by the United Nations (UN) known as the "Green Line."

Whereas TCs generally welcomed the territorial partition, which they felt could offer them a safe haven, the GCs insisted on both the right of return and the temporary nature of this territorial setting (Akçalı forthcoming). In 1977, both sides agreed on a bizonal and bicommunal federal solution as the basis for negotiations; however, despite decades of peace talks under the auspices of the UN, disagreements between the two administrations have continued and the two communities remain separated today. Assuming that the European Union (EU) membership could fix the Cyprus problem, the RoC applied and was found eligible to join the EU in the 1990s. The expectation was that the EU negotiations and the UN peace talks would go hand in hand.

Yet, on 24 April 2004, the UN plan for the creation of a United Cyprus Republic with two constituent states was rejected by 76 percent of GCs, whereas 65 percent of TCs supported it. This meant that the RoC entered the EU alone in May 2004. In summer 2008, the TC and GC leaderships opened a new negotiation process for a mutually acceptable solution to the division of Cyprus, which, in any case, will be submitted to referendum on both sides. No concrete agreement, however, has so far been reached, nor has agreement been reached in relation to Cyprus's environmental problems.

Like other countries in the eastern Mediterranean, Cyprus suffers from water scarcity, droughts, heat waves, forest fires, biodiversity losses, and soil and ecosystem degradation, which are likely to intensify due to climate change (UNDP-ACT 2008a). Despite the fact that, from an ecological point of view, the island is a series of interconnected ecosystems, environmental issues have been addressed separately by the TC and GC political bodies since 1974. Aware of this problem, in 1998, the United Nations Development Program (UNDP) launched in Cyprus a series of bicommunal environmental activities (reforestation, organic farming, waste management, etc.). Experts on both sides started working in parallel on the same environmental issues, but rarely met. With the opening in 2003 of crossing points that has been closed since 1974, the UNDP launched the second phase of its cooperative projects, whereby GC and TC experts worked independently but met regularly to coordinate their efforts. In the third and current phase, emphasis is on joint projects in which communities work together to protect and maintain environmental assets. Accordingly, in 2005 the UNDP set up a peace-building program, Action for Cooperation and Trust (ACT), to create opportunities for bicommunal partnerships that can help care for the island's common natural heritage, at the same time promoting intercommunal tolerance.

This article focuses on this latter program to analyze the role of environmental cooperation promoted by a third party in transforming ethno-territorial conflicts. The article is divided into three main sections. First, we introduce the research question, by locating it within the literature of environmental peacemaking, environmental management, and ethno-territorial conflict transformation. Then, we present the UNDP-ACT initiatives and its underlying discourse. Finally, we critically discuss the UNDP-ACT environmental peacemaking efforts by relying on both survey data and opinions of environmental nongovernmental organization (NGO) representatives based in Cyprus.

The Role of the Environment in Conflict Transformation

Conca and Dabelko (2002) suggest that environmental cooperation between conflict-ridden nations might help overcome political tensions and that civil society plays a crucial role in this process. The underlying logic of this argument is that environmental issues do not recognize any state, ethnic, or religious boundaries and have the potential to affect all dimensions of human life. Emphasis on its preservation, or dealing with its problems collectively, might thus lead the conflicting parties to construct a common identity, which can then transform the conflict communication and the interests of the parties involved (Clayton and Opotow 2003). This logic stems from new understandings of borders as social and dynamic constructions that are open to new and multiple meanings and identities through societal changes (Pace and Stetter 2003) and new studies on the transformation of ethno-territorial conflicts that have shifted from state-centered solutions to a wider "human-needs/world-society framework" (Richmond 2001). Also increasingly important is the intervention of a third party that has a genuine interest in the successful resolution of the conflict (Miall, Ramsbotham, and Woodhouse 2005, 168).

Despite having some evidence of success and promise in theory, however, environmental peacemaking as a novel strategy of conflict transformation has been unable fully to convert environmental cooperation into broader forms of political cooperation (Carius 2007, 66). Empirical studies in environmental peacemaking, environmental management, and conflict transformation suggest five important points to deal with such a shortcoming.

First, as emerged in peace park studies (Saleem 2007), to cooperate, neighboring countries do not have to share common interests but a common aversion to environmental harms. This implicitly suggests that widespread environmental awareness is a key factor. Second, environmental peacemaking should be accompanied by cultural, economic, and social development policies, giving way to a so-called integrated approach to peacebuilding (Ricigliano 2003). Third, studies on river basin management show that civil society actively participates in environmental programs when they perceive, early in the process, clear evidence of change, win–win solutions, and an equal, transparent, and respectful setting, supported by financial and technical inputs (Tippett et al. 2005). Fourth, besides a civil-society-to-civil-society dialogue, pathways to peace are also created by state-to-state interactions and, most important, the interplay between the state and civil society (Conca, Carius, and Dabelko 2005). Finally, external actors need to take into account the importance of the symbolic function of borders, for both the state and its people, when designing conflict transformation policies (Wilson and Donnan 1998).

By keeping these points in mind, in the remainder of this article we critically discuss the role of the UNDP as a third party and its environmental program in Cyprus, as a way to contribute to the debate about environmental peacemaking.

The UNDP and Environmental Peacemaking in Cyprus

The first phase of environmental peacemaking efforts of the UN started in 1998, with the Bicommunal Development Program (BDP), aimed at funding peacebuilding and cooperation initiatives benefiting both GCs and TCs. In 2005, the BDP was replaced by a similar program, called Action for Cooperation and Trust (ACT). Environment is one of the four main sectors of intervention, along with civil society, cultural heritage, and education and youth. Over the four years of the program (2005–2008), twenty-five environmental projects have been funded, for a total of US$3.5 million (13 percent of the ACT budget). Although in absolute terms this figure is lower than the amount spent during the BDP program (US$4.3 million—8 percent of the BDP budget), if one considers the percentage value and the shorter time period of the ACT program (four years vs. the seven of the BDP), it seems legitimate to affirm that the UNDP in Cyprus has given the environment an increasing role in fostering reconciliation and trust.

The ACT program states that any project that seeks UNDP financial support should be designed and implemented by Cypriots and should enhance cooperation and trust between the two main communities. The program covers various sectors of environmental intervention, from sustainable agricultural practices (e.g., organic farming through the use of *mycorrhizae*) to measures of nature conservation (e.g., the rich biodiversity of the buffer zone), restoration (e.g., abandoned quarries), and management (e.g., ecology of artificial wetlands). Moreover, in January 2007, thanks to ACT financial support, the Cyprus Environmental Stakeholder Forum was launched. The Forum, constituted around the Cyprus Technical Chamber (ETEK) and the Union of Chambers of Turkish Cypriot Engineers and Architects (KTMMOB), represents the most ambitious project to date to link organizations and people from both communities to work on a list of common environmental priorities.

ACT has also supported the creation of an online environmental directory, which offers detailed information and contact references about all Cypriot environmental NGOs.[1] By doing so, ACT hopes to offer both GCs and TCs a tool to identify common interests and generate networking opportunities for environmental stakeholders across the island.

Simply put, the philosophy behind ACT is that nature knows no boundaries. "For those who care about the environment in Cyprus, the Buffer Zone, otherwise known as the 'Green Line,' does not exist. Air pollution does not stop before continuing on its way across, nor are airborne viruses hampered by barbed wire" (UNDP-ACT 2008b, 3). This discursive rendition of the environment aims not only to construct

Cyprus as one but also to contribute to the emergence of a common "civic" identity beyond ethnic differences (UNDP-ACT 2008a, 46), which can help desecuritize the Cyprus question. The belief is that shared environmental awareness and commitment might indeed generate intercommunal solidarity and trust, which, in turn, might lead to wider cooperation on political issues.

As a way to engage critically both this discourse and the actual ACT environmental program, we discuss survey data regarding the views of both ordinary Cypriots and Cypriot environmental NGO representatives.

Cypriots and the Environment

Three recent surveys (Eurobarometer 2008; UNDP-ACT 2008c, 2008d) are used here to present Cypriots' opinions about three relevant topics: environmental awareness, knowledge of bicommunal initiatives, and intercommunal contacts and trust.[2] Eurobarometer data show that environmental awareness (as measured in terms of knowledge about climate change) is not particularly high among Cypriots. In fact, in 2008, the cumulative percentage of respondents who declared themselves to be "well informed" about the causes, consequences, and ways to fight climate change was, respectively, among GCs and TCs, slightly below (54.3 percent) and well below (39.3 percent) the EU-27 average (54.6 percent; Eurobarometer 2008, 122–24). Overall, GCs show a higher civic commitment to protect the environment, particularly in relation to young and future generations, whereas TCs are relatively less committed and more concerned about the monetary costs involved (Eurobarometer 2008, 140).

As for knowledge of bicommunal initiatives and programs, in 2008 only 12 percent of TCs and 28 percent of GCs had heard about them and only 14 percent of TCs and 12 percent of GCs had participated in bicommunal events (UNDP-ACT 2008c). These data clearly point to a problem in reaching out to the whole population, despite the efforts of UNDP-ACT in publicizing its initiatives in local mass media. It is significant that, during the same year, UNDP-ACT was known only by 12 percent of TCs and 8 percent of GCs (UNDP-ACT 2008c) and civil society organizations were known only to 15 percent of GCs and 12 percent of TCs (UNDP-ACT 2008d).

Limited environmental awareness and poor knowledge of both initiatives and actors involved in the bicommunal process add to a context of rare intercommunal contacts and trust. In 2008, 86 percent of GCs and 80 percent of TCs declared that they did not have daily contacts with a member of the other community (UNDP-ACT 2008c). In the same year, daily border crossings amounted, on average, to between 5,000 and 6,000 (equal to 0.6 percent of Cyprus's total population)—a figure, however, that also includes tourists (United Nations Peacekeeping Force in Cyprus 2009). Reciprocal mistrust characterizes both TCs (78 percent) and GCs (66 percent), as also shown by the fact that both communities equally refuse (1.6 out of 4—where 4 stands for "totally acceptable") the possibility of intermarriage for their children (UNDP-ACT 2008c).

Survey data, however, show potential for future collaboration, as well. In fact, 66 percent of GCs and 42 percent of TCs affirmed, in 2008, that "there are a lot of common elements between the two communities" (UNDP-ACT 2008c), whereas only 3 percent of GCs and 11 percent of TCs felt that "there are no common elements." Moreover, data show that those who come into contact with members of the "other" community tend to become more trusting, although even in this case, TCs (2.8 on a four-point scale) remain more skeptical than GCs (3.3). There is, therefore, some room for joint initiatives and it is significant that respondents mentioned governmental agencies, NGOs, international organizations, ordinary citizens, and the media as those that can have the greatest power in tackling environmental issues. This points to a mix of top-down and bottom-up interventions, as also suggested by the literature. Yet, neither TCs nor GCs listed environmental cooperation among the primary bicommunal programs or initiatives able to promote a solution to the Cyprus problem (UNDP-ACT 2008c), which indirectly calls for an "integrated approach" in peacemaking.

Cypriot NGOs and the Environment

As a way to investigate further the UNDP-ACT environmental peacemaking program, we designed a short questionnaire (Table 1) that, in June 2008, was sent out to all Cypriot NGOs listed in the UNDP-ACT Environmental Directory. This list includes a total of sixty-nine NGOs: forty GC, nineteen TC (nine of them, however, are no longer active or they are just an extension of a larger NGO, such as KTMMOB), and one mixed (recently, however, this latter organization has stopped working, due to financial constraints more than bicommunal problems). A few of them feature expatriates (mainly British) among their active members. Their size varies from the very large (twenty or more staff members) to the very small (only one person).

Table 1. Frequencies for Greek Cypriots' and Turkish Cypriots' answers

Questions	Yes	No	Yes and no	Don't know	Not answered	Total
Greek Cypriots						
1	9	6	7	4	—	26
2	8	6	7	4	1	26
3	5	15	1	3	2	26
4	9	10	3	1	3	26
5	14	2	—	1	9	26
Turkish Cypriots						
1	3	6	4	—	—	13
2	8	3	2	—	—	13
3	1	4	8	—	—	13
4	12	—	1	—	—	13
5	13	—	—	—	—	13

Notes: Questions were as follows: 1. Do you think the ACT environmental projects have been effective in raising people's awareness about environmental issues in Cyprus and sensibility to the existence of one ecosystem on the island? 2. Do you think this emphasis on environmental issues has been effective in promoting intercommunal tolerance and mutual understanding? 3. Do you think these projects have had an impact on all people on the island? 4. What kind of shortcomings or obstacles do you see in the idea of using the environment to promote intercommunal tolerance and mutual understanding? 5. From your experience, is there anything that could have been done differently to enhance the goal of promoting cooperation and trust through environmental initiatives?

Respondents were invited not only to answer yes or no to our questions but to motivate their answers. After the original e-mail, letter, or faxes, we again contacted the NGOs by phone and sent additional e-mails or faxes to solicit their answers. The response ratio was 64 percent for the GC NGOs and 68 percent for the TC NGOs.

Both the cover letter and the questionnaire were written in English. In general, this did not prove to be an obstacle among GCs, but it turned out to be more so among TCs. In a few instances, therefore, we sent out the questionnaire again translated into Turkish and allowed the respondents to answer in Turkish—the native language of one of the authors. The format of the questionnaire facilitated the coding process. Data were first coded according to the character (positive, negative, positive and negative) of the opinions expressed and additional codes were generated to capture recurring motives behind these opinions. Analysis relied both on frequencies and content analysis. In the next section the respondents' answers are analyzed separately for GC and TC NGOs.

Greek Cypriot NGOs

The majority of GC respondents thought that UNDP-ACT initiatives have been effective in generating both environmental awareness and intercommunal understanding and tolerance (Table 1). Yet, an even greater majority believed that the effects of these initiatives have not been felt evenly among the population, privileging those who already had some environmental awareness.

Those who did not see any obstacles to the idea of using the environment to promote intercommunal tolerance and understanding argued, in line with the UNDP discourse, that "environment is an area that has no boundaries" and that touches equally on "our quality of life." From this perspective, environmental awareness is seen to generate a *we* feeling inclusive of the whole population, beyond ethnic or religious divides, thus confirming the UNDP discursive strategy of using the environment to foster a common identity. Yet, this view should be analyzed in context. In fact, among GCs, the support for the UNDP-ACT discourse might resonate not only with genuine environmental feelings but also with a political goal, as the majority of GCs cherish the idea of a reunified island under one central state (International Crisis Group 2006, 20). The same feeling was suggested by the overwhelming majority of GCs, who in 2004 rejected the referendum on the UN plan for the creation of a bizonal and bicommunal federal Cyprus Republic. This ambiguous overlapping of environmental and political motives also surfaced in some comments of the GC respondents; for instance, in the following quote: "When there is a major fire in the North people in the South get sentimental and upset, but I think mostly within the concept that the areas that used to be ours are now destroyed."

The majority of respondents who acknowledged the shortcomings of the use of the environment in transforming the conflictual situation on the island pointed to the unsolved "political problem." This was articulated in various terms: "the possession and occupation of Cyprus by the Turkish army," "the sensitive property issue" (i.e., GCs claiming properties that they left behind in northern Cyprus); the "deep resentment (so that I will not use a harsher word), suspicion and mistrust" between the two communities; "ethnic difficulties/differences," and the fact that "people have principles, they don't want to cross on the other side." All these views can be synthesized in the following terms: "We need to solve the Cyprus problem first, then environmental cooperation can take place; not the other way round."

In all these cases, the respondents, whatever their degree of environmental awareness, put their ethnopolitical belonging first. As a consequence, the environment ceases to be articulated as one; *our* environment, instead, is cast against *their* environment, thus defusing any environmental peacemaking: "People in Cyprus worry about environmental problems when it comes to their personal and financial interest. . . . Some of them even think that if there is more pollution or destruction over there, perhaps it is for the best as that would suggest that fewer tourists will visit the North." Clearly, *our* environment and *their* environment do not matter to the same extent.

A couple of respondents openly rejected the idea that the environment can play any role in this process. For them, environmental issues concern very few people. On the contrary, always according to them, more effective results could come from daily activities (e.g., shared business, trade exchanges, and tourism), as these involve many more members of the two communities.

Almost all respondents found that things could have been done differently to enhance intercommunal cooperation and trust through environmental initiatives. The most frequent comments revolved around two related issues: more efforts to advertise environmental initiatives through mass media and to attract ordinary people, not just "the professionals of the environment." This point clearly emerged from the comments of some respondents, who indeed called for more environmental initiatives to be held in schools and among young people in general. Yet, some comments also highlighted the difficulty of "passing the information" to certain categories of people: older generations, new immigrants from Eastern Europe ("their only interest is to make a better life for themselves and to help families left behind in their native countries"), the uneducated, and rural people. Quite a few respondents also advocated more opportunities for GC and TC NGOs to meet directly and one respondent observed that the UNDP "should not use the same methods as in Britain or USA to achieve things," because "civil society in Cyprus is still very weak."

Turkish Cypriot NGOs

Also in the case of TC NGOs, the majority of the respondents maintained that the UNDP-ACT projects have been effective in promoting mutual tolerance and understanding (Table 1); however, only a few of them believed that these projects have successfully raised people's awareness about the existence of one ecosystem on the island. The following quote is illustrative:

> Any effort targeting the preservation of cultural heritage and the environment raises tolerance, respect, and love among people. Environmental awareness efforts are thus effective in raising tolerance among Cypriots in the long term. However, a very long period of common life is necessary in order to raise sensibility to the existence of one ecosystem on the island. I am not sure if Greek Cypriots know about the environmental problems in TRNC. The Turkish Cypriots do not have knowledge about the environmental problems in the South, for example.

According to quite a few of the respondents, the UNDP-ACT projects did not have an impact on all people living on the island, because they targeted mainly NGO members rather than ordinary citizens. Like their GC counterparts, TC respondents also argued that the essence of the problem was elsewhere, as the island's environmental problems simply mirror the political ones. Accordingly, almost half of the respondents mentioned that the international nonrecognition of the TRNC and the disadvantages of its de facto situation constitute a main obstacle to establishing intercommunal tolerance and understanding: "Any progress on any issue, whether it is the environment or something else always becomes part of this political problem." And again, "The Cyprus problem is a political problem. Without resolving it, any initiative of this sort will always have shortcomings. Cooperation and trust can only be achieved through the cooperation of the governments of the two parties." This suggests two points already discussed by the literature. First, a new emphasis on bottom-up approaches should not come at the expense of politico-institutional cooperation, as survey data for Cyprus have also indicated. Second, environmental cooperation alone cannot make a difference, as this should be incorporated into a more integrated approach to peacemaking. Yet, this also suggests that for some TCs, as for some GCs, environmental protection matters as a by-product of peace and reconciliation, not as a tool to foster them.

TC respondents also showed some skepticism toward bicommunal initiatives—something that also emerged in the previously discussed surveys. In their comments, this attitude at times combined with a sense of mistrust toward both the GC counterparts and the third party's involvement:

> There is no common understanding among the NGOs of both sides. The NGOs on the Greek Cypriot side are opportunists. They declare nuclear energy as an alternative energy because it's for the benefit of their state. They are hesitant to do common projects with us. The NGOs in TRNC are more courageous and seem more forceful, but

> they do not oppose the environmental degradation that the government policies are causing. The EU, UNDP, and USAID, on the other hand, run superficial projects with the Greek Cypriot NGOs which do not have any serious effect for environmental protection.

This quote exposes a contradictory understanding between TC and GC civil societies about environmental issues. Furthermore, it demonstrates some doubts about the genuine engagement of the third party (the UNDP) with environmental protection and confirms what we have already seen earlier: Any attempt to make peace through cooperation on environmental issues cannot escape the political realm or the "national interests" logic.

Finally, all TC respondents agreed that things could have been done differently. In particular, they commented on the inefficacy of some initiatives and suggested spreading environmental awareness to people as a whole, especially to young people, through educational projects. They also pointed out that NGOs are financially underfunded, which explains why they are unable to implement cooperative projects such as bicommunal youth camps.

Conclusions

This short article has critically discussed UNDP-ACT environmental peacemaking in Cyprus as a way to join the debate on the role of environmental cooperation in transforming conflictual situations. As revealed by both the UNDP surveys and the comments of the respondents to our questionnaire, there is a high level of ethno-territorial identification, mistrust of the other community, and loyalty to the state among both the public and, to a lesser extent, the environmental stakeholders in Cyprus. Often the environment is divided along ethno-territorial lines, so that *our* environment matters more than *their* environment. Alternatively, the environment tends to be perceived as an isolated matter, rather than a tool for fostering discourses of peace and reconciliation. Environmental cooperation becomes a by-product of the transformation of the conflict, rather than a factor in such transformation. This suggests that the conflict communication in Cyprus has not been contained within the ethno-territorial conflict proper but has become an omnipresent reference in societal communication.[3] This situation obviously complicates the bottom-up efforts of environmental peacemaking in Cyprus and calls for continuing politico-institutional cooperation that can complement these efforts.

This complicated scenario is, however, compensated for by a sense of shared commonality and support for bicommunal activities, which, according to the UNDP surveys, exist among TCs and, even more so, among GCs. Although not alone, but as part of an integrated approach to peacemaking (Ricigliano 2003), environmental cooperation, therefore, has the potential to help generate pathways to peace in Cyprus. Yet, this potential is somewhat conditioned by four factors: the degree of environmental awareness among the population involved, the level of trust of the civil society toward the third party, the degree of civil-society-to-civil-society dialogue, and the degree of effective interaction between civil society and state institutions.

Survey data suggest that much more needs to be done to spread environmental awareness among Cypriots and, in particular, among TCs. Similarly, the respondents to our questionnaire pointed to the fact that the UNDP-ACT initiatives have too often had an impact only on those already environmentally aware, rather than on the population as a whole. Being the commissioner of two of the surveys discussed in this article, the UNDP is well aware of this problem and in the future is planning to target younger generations more directly and to strengthen the Cypriot civil society through a new 2008 to 2011 bicommunal financial scheme.[4] This project will hopefully also tackle some of the concerns expressed by a few TCs about the credibility of the UNDP as a third party. The new financial scheme is expected to endow both TC and GC civil societies with new financial and technical means to ease the dialogue between them. Our argument, however, is that this dialogue should not remain confined within the civil society alone, as a civil-society-to-civil-society dialogue must complement rather than simply substitute for the traditional politico-institutional negotiations and new forms of dialogue between civil society and state institutions should also be supported.

Finally, the UNDP "nature knows no boundaries" discourse should be reconsidered to address TC skepticism. In fact, although this discourse can generally be favored by the GCs, in line with the official territorial integrity discourse of the RoC, the same discourse might encounter the resistance of those TCs who still hold bitter memories of the RoC before 1974. Therefore, instead of using this discourse to foster a common "civic" identity, a better strategy would be to emphasize more the positive and mutual changes emanating from bicommunal environmental cooperation. It is not a coincidence that, among the twenty-five projects, the protection of biodiversity in the buffer zone, for example, has been one of the most successful initiatives, also in terms of public visibility, as both sides have equally perceived the benefits. Similar projects that target the small scale and leave out the "big" politics also help the empowerment of the civil society and enhance trust and credibility in the third party as well.

Acknowledgments

We are extremely grateful to both the UNDP in Cyprus and the respondents to our questionnaire. Many thanks also to the two anonymous reviewers and the editor.

Notes

1. The directory is available at http://mirror.undp.org/cyprus/endir/search.asp.
2. The Eurobarometer survey (Eb 69.2) was administered during April 2008, on a sample of 504 GCs and 500 TCs. RAI Consultants carried out both UNDP-ACT surveys, the first, in September 2008, on a sample of 1,153 GCs and 1,011 TCs (UNDP-ACT 2008c), and the second, in December 2008, on a sample of 412 GCs and 248 TCs (UNDP-ACT 2008d). These surveys are available on request from UNDP-ACT.
3. This argument has been adapted from Stetter (2003).
4. Anonymous interviews with a UNDP-ACT senior representative (29 August 2008, 12 January 2009).

References

Akçalı, E. Forthcoming. *Chypre: Un enjeu géopolitique actuel* [Cyprus: A current geopolitical stake]. Paris: l'Harmattan.

Carius, A. 2007. Environmental peacebuilding. ECSP Report 12. Washington, DC: Woodrow Wilson International Center for Scholars. http://www.wilsoncenter.org/topics/pubs/CariusEP12.pdf (last accessed 14 February 2009).

Clayton, S., and S. Opotow, eds. 2003. *Identity and the natural environment*. Cambridge, MA: MIT Press.

Conca, K., A. Carius, and G. Dabelko. 2005. Building peace through environmental cooperation. In *State of the world 2005*, ed. Worldwatch Institute, 144–55. New York: Norton.

Conca, K., and G. Dabelko, eds. 2002. *Environmental peacemaking*. Washington, DC: Woodrow Wilson Center.

Eurobarometer. 2008. *Europeans' attitudes towards climate change*. http://ec.europa.eu/public_opinion/archives/ebs/ebs_300_full_en.pdf (last accessed 20 January 2009).

International Crisis Group. 2006. *The Cyprus stalemate*. http://www.crisisgroup.org/home/index.cfm?id=4003&1=1 (last accessed 16 February 2009).

Miall, H., O. Ramsbotham, and T. Woodhouse. 2005. *Contemporary conflict resolution*. Oxford, UK: Polity.

Pace, M., and S. Stetter. 2003. A literature review on the study of border conflict and their transformation in the social sciences. EUBorderConf project. http://www.euborderconf.bham.ac.uk/publications/files/stateofthea-rtreport.pdf (last accessed 14 February 2009).

Penrose, J. 2002. Nations, states and homelands. *Nations and Nationalism* 8:277–97.

Richmond, O. P. 2001. A genealogy of peacemaking. *Alternatives* 26:317–48.

Ricigliano, R. 2003. Networks of effective action: Implementing an integrated approach to peacebuilding. *Security Dialogue* 34:445–62.

Saleem, H. A., ed. 2007. *Peace parks*. Cambridge, MA: MIT Press.

Stetter, S. 2003. Methodological annex to "The EU and the transformation of border conflicts: Theorising the impact of integration and association." EUBorderConf project. http://www.euborderconf.bham.ac.uk/publications/files/methodology1.pdf (last accessed 14 February 2009).

Tippett, J., B. Searle, C. Pahl-Wostl, and T. Rees. 2005. Social learning in public participation in river basin management. *Environmental Science and Policy* 8:287–99.

United Nations Development Program Action for Cooperation and Trust. 2008a. Corporate environmental responsibility in Cyprus. http://www.undp-act.org/main/data/articles/cer_en.pdf (last accessed 16 February 2009).

———. 2008b. Environment. http://www.undp-act.org/main/default.aspx?tabid=63 (last accessed 11 February 2009).

———. 2008c. *Level of trust between the two communities in Cyprus*. Nicosia, Cyprus: RAI Consultants.

———. 2008d. *Perceptions on sustainable development and climate change in Cyprus*. Nicosia, Cyprus: RAI Consultants.

United Nations Peacekeeping Force in Cyprus. 2009. Crossing details. http://www.unficyp.org/nqcontent.cfm?a_name=crossing_details_1 (last accessed 20 January 2009).

Wilson, T. M., and H. Donnan, eds. 1998. *Border identities*. Cambridge, UK: Cambridge University Press.

Innovative Approaches to Territorial Disputes: Using Principles of Riparian Conflict Management

Shaul Cohen and David Frank

Many belligerents in ethno-territorial conflicts claim they have an absolute right to contested space, operate on a zero-sum basis, and use maximalist negotiation strategies. This article draws on ongoing fieldwork that examines ethno-territorial conflict and focuses on the transition from rights-based to needs-based negotiations over sites of worship, parading routes, and national borders. These three sites represent different scales and expressions of spatial conflict, including accommodations for Jewish and Muslim worshipers in the Cave of the Patriarchs in Hebron/al-Khalil, West Bank, the terms and conditions agreed to for sectarian parading in the town of Derry/Londonderry, Northern Ireland, and the return and subsequent lease of land on the Israeli–Jordanian border. In each case, negotiated arrangements allow for the belligerents to meet their minimal territorial needs, even as the broader dynamics of the conflict persist. Central to these arrangements is a shared recognition that space is a mutable resource and that needs-based negotiation can allow sharing of contested territory at a variety of scales. The article draws from analogous dynamics in international river treaty negotiations, specifically the riparian model developed by the authors in earlier work. We conclude by enumerating the contributions geographers can make to theories of conflict and peace. *Key Words: conflict management, riparian model, shared space, territorial disputes.*

有许多交战方在民族和地区的冲突中声称，他们对所争夺的空间拥有绝对的权利，在一个零和的基础上运作，并使用最大限度的谈判策略。本文借鉴了正在进行的实地考察来分析民族和地区冲突，重点放在了宗教礼拜地点、游行路线和国家边界的谈判中，从权利基础到需求基础的转变上。上述三种情况代表了空间冲突的不同规模和表述，包括了在约旦河西岸，希伯伦/哈利勒的始祖墓穴给予犹太人和穆斯林朝圣者的方便；在北爱尔兰关于同意在德里/伦敦德里镇举行宗派游行的条款和条件；在以色列和约旦边境返还和后续租界土地的事例。在每种情况下，谈判的和解允许交战各方满足其最低的领土需要，即使发生边界冲突的动态因素依然存在。这些和解的核心是一项共同的认识，即空间是一个可变的资源，以需要为基础的谈判可以在各种尺度上分享有争议的领土。文章参照并类比了国际河流条约谈判的情况，特别是本文作者早期开发的河岸模式。最后，我们列举了对冲突与和平的理论作出贡献的地理学家们。*关键词：冲突管理，河岸模式，共享空间，领土争端。*

Muchos de los beligerantes en conflictos etno-territoriales reclaman tener derechos absolutos sobre el espacio disputado, disputan a partir de una base de suma cero y utilizan estrategias maximalistas de negociación. Este artículo se basa en trabajo de campo actual cuyo propósito es el examen del conflicto etno-territorial y un estudio centrado en la transición desde negociaciones basadas en derechos a otras basadas en necesidades relacionadas con lugares de culto, rutas de procesiones y límites nacionales. Estos tres sitios representan diferentes escalas y expresiones de conflictos espaciales, que incluyen el alojamiento para devotos judíos y musulmanes en la Cueva de los Patriarcas de Hebrón/al-Khalil, en el Banco Occidental; los términos y condiciones acordadas para procesiones sectarias en el pueblo de Derry/Londonderry, Irlanda del Norte; y la devolución y subsiguiente arriendo de tierra en la frontera jordano-israelí. En cada caso los arreglos negociados permiten a los beligerantes satisfacer sus necesidades territoriales mínimas, incluso dentro del contexto de que las dinámicas mayores del conflicto persistan. Algo de la mayor importancia en estos arreglos es el reconocimiento compartido de que el espacio es un recurso mutable y que una negociación basada en la necesidad puede llevar a la aceptación de que se comparta territorio disputado en una variedad de escalas. El artículo también versa sobre la dinámica análoga de negociaciones para tratados sobre ríos internacionales, específicamente con base en el modelo ribereño desarrollado por los autores en un trabajo anterior. Concluimos con la enumeración de las contribuciones que pueden hacer los geógrafos a las teorías sobre conflicto y paz. *Palabras clave: manejo de conflictos, modelo ribereño, espacio compartido, disputas territoriales.*

In this article we continue the development of a novel approach to territorial disputes that offers additional tools to those seeking to resolve conflicts over territory (and space and place; Cohen and Frank 2002). Our effort involves a conceptual exercise in which we ask "what lessons from successful international water dispute resolutions can be applied to nonriparian territorial disputes?" The goal of the exercise is to gain distance from the zero-sum dynamic that characterizes many territorial conflicts and to introduce forms of sharing, often present in riparian treaties, that can allow the needs of multiple parties to be met within the context of a negotiated agreement. Although this might at first seem fanciful, we draw on the tradition of successful riparian agreements to establish their value and versatility and then provide brief illustrations of a number of ongoing accommodations that reflect some of the principles that suggest the utility of riparian-style approaches.

The first section of the article is a rehearsal of the riparian principles of conflict management. The second section is devoted to three illustrations of nonriparian conflicts that display characteristics of riparian conflict management: the coexistence of a mosque and synagogue in the same site in the town of Hebron/al-Khalil; the transformation of the parading tradition in Derry/Londonderry, Northern Ireland, from one of violence to one of relative peace; and an agreement for Israeli use of Jordanian land on their shared border. We set forth evidence in this article that principles of riparian conflict management have great potential for nonriparian conflict and that a forward-looking vision of geographical theories of peace and conflict would include riparian models of conflict management. Our illustrations draw on areas in which we have conducted fieldwork to varying degrees, but each is, by necessity, sketched out here in only bare essentials. Our effort should thus be seen as a proposition, rather than as a fully developed study of the cases. Moreover, as each of the situations we discuss is dynamic, complex, and ongoing, we deploy them to stimulate further thinking, rather than as conclusive proof of the utility of riparian approaches.

The Promise of the Riparian Model

International disputes over water resources can lead to both militarized conflicts and nonviolent conflict management (Hensel, McLaughlin Mitchell, and Sowers 2006). Wolf's Transboundary Freshwater Dispute Database (TFDD) provides summaries and full texts of nearly 450 water-related treaties (Wolf 1999, 2000; Wolf, Stahl, and Macomber 2003) and provides the most cited database on issues related to riparian conflict. The treaties in the database span four millennia, and the research suggests that the vast majority of international water disputes yield to needs-based negotiations (Elhance 2000; Wolf, Stahl, and Macomber 2003). In the last 4,000 years, Wolf notes, "there has never been a war fought over water" (Wolf 1998, 255), although it is clear that low-level conflicts have and do break out (Postel and Wolf 2001). Water is a resource essential to life, and negotiations over water conflicts show that the belligerents recognize the right of their opponents to exist. Rights and needs are intertwined in riparian negotiations. A right to exist is contingent on the need for an adequate supply of water. Denying that need would nullify the premise of negotiation, which is that of an agreed outcome. It is possible to claim that access to water is itself a right, but because the need for water is not contested, negotiations revolve around sharing modalities, or claimed needs, of the respective parties. Working through the modes of water sharing seems to hold far more potential for productive outcomes than do negotiations over rights in a zero-sum context.

If rights and needs are intertwined in the riparian model, issues of power, domination, coercion, and resistance are cast in relational terms, which take into account the consequences of one actor using its power advantage over another. Even militarily capable and powerful states may not go to war over water because

> hydropolitical belligerence necessarily imposes very heavy costs and penalties on the aggressor. For example, a militarily strong lower riparian state may decide to occupy or annex the territory of one or more of the upper riparian states in order to control the upstream water resources, as some analysts have argued that Israel has done since 1967. However, even Israel—militarily the most powerful state in the Middle East—has discovered that, in addition to the costs in human lives, the economic and political costs—both domestic and international—of occupation or annexation of territory keep rising with time. (Elhance 2000, 214)

The party with the most power in riparian negotiations will often justify a rights/needs-based approach to negotiation with a cost–benefit analysis, concluding that negotiation is the pragmatic option. Such calculations on the part of both parties, taken from short- and long-term perspectives, are at the center of the riparian model. If the essential characteristics of riparian conflict management and this shift to addressing needs can be distilled, might they provide new insights for nonriparian disputes?

In a previous article the authors focused on analogical thinking to compare riparian and nonriparian conflict management (Cohen and Frank 2002). There are, of course, differences between conflicts over water and those involving nonriparian spaces; however, there are also similarities. Use of analogical thinking allows for a comparison of similar objects and for the identification of patterns (Sunstein 1996). For the purposes of thinking about space and territorial disputes, it is useful to compare riparian and nonriparian conflict to identify patterns of difference and similarities.

The use of analogical thinking can "become an engine for the conceptual restructuring" of territorial disputes (Cohen and Frank 2002, 750), a mode of thought differing significantly from framing space with metaphors. When geographers and others use metaphor to understand space (Livingston and Harrison 1981; Buttimer 1982; Barnes 1992), the ability to make fruitful comparisons for the purposes of rethinking how territorial conflicts might be brought into a state of accommodation is reduced. Metaphoric thought tends to fuse concepts, such as sovereignty and territory, which might require the identification of careful distinctions and nuance. For example, belligerents involved in ethno-territorial conflict might declare contested space "holy," thereby transferring their claims beyond the reach of time and human reason.

The key difference between metaphoric and analogic thinking is the strength of the relationship between the things compared. When space is framed metaphorically, it conflates identity with space, and belligerents may put the metaphor beyond the realm of negotiation. Analogic thinking places concepts in juxtaposition, allowing for the possibility that concepts might be reordered and repositioned for the purposes of accommodation. By placing into juxtaposition concepts like sovereignty and territory, those interested in conflict management can acknowledge the beliefs of antagonists and still see them as ultimately plastic and mutable. The primary purpose of the authors' use of analogical thinking is to invite a comparison of riparian and nonriparian conflict management.

Drawing on the work of Wolf and others, we have identified four essential characteristics of successful riparian negotiations to consider how they might inform nonriparian conflict management (Cohen and Frank 2002). First, whereas belligerents have powerful rights claims in relation to water, riparian negotiators have often viewed water conflicts as rooted in needs rather than absolute rights. This frame leads to needs-based negotiation (Fisher, Ury, and Patton 1991). By framing water conflict as one over needs rather than absolute rights, riparian negotiators can treat water as a mutable resource that can be shared. The empirical evidence suggests that parties in water conflicts recognize that the water needs of their opponents must be met and that arrangements can be made to share the resource:

> While many international water negotiations begin with differing legal interpretations of rights, whether measured by hydrography or chronology, they often shift rather to a needs-based criteria for water allocations, as measured by some mutually agreeable parameter such as irrigable land or population. Mostly, one is struck by the creativity of the negotiators in addressing specific language to each very specific local setting and concerns. (Wolf 2002, 2)

Framing water as essential to human life and employing needs-based negotiation are necessary but not sufficient conditions for durable riparian conflict management.

The second essential characteristic of riparian negotiations is the recognition that water can be shared in and over time. Time itself is used as a resource allowing for the sharing of water. For example, many water treaties set forth arrangements allowing one party full access to a river for a discrete period of time, and once the time allotted elapses, use moves to the other party. The need to provide all parties water to drink, irrigate land, and other uses encourages quantification of the resource, producing in turn the criteria and standards necessary for effective bargaining. Although time is linked to the quantity of water being allocated, it moves the parties away from exclusive claims over the whole of the disputed resource.

A third essential characteristic of riparian negotiations is the focus on beneficial use. There is a recurring pattern in the TFDD suggesting the negotiators recognize the multiple uses of water and the fact that degrading the resource will have a detrimental impact on subsequent use. Negotiators reach agreements allowing one party to use water for power generation and the other for irrigation, for example, with a logical sequence to maximize the benefit (or at least preserve a minimum benefit). Even though the needs might be incompatible in some respects, the negotiators create agreements allowing them to be met. A final characteristic is the use of symbols and language recognizing the local realities, constraints, and settings. Water treaties are often cast in pragmatic and concrete terms, sidestepping the reach of overarching religious and nationalist ideology. The water treaties in the database reveal patterns of language with clear statements on why the needs of all parties must be met and then outline, often in explicit and quantifiable terms, how the resource will be shared.

Using riparian negotiations as an analogue (Cohen and Frank 2002), we have suggested that the use of principles characterizing riparian negotiations might help mediators and negotiators reframe the problem of Jerusalem, and we imagined the possibility of shifting the negotiations from rights to needs, focusing on time rather than absolute sovereignty, seeking to maximize the beneficial use of the city, and developing symbols and language that feature local conditions. Although the 2000–2001 negotiations between Israelis and Palestinians failed, the "deliberations produced a new concept that may prove important in future talks: 'the sacred basin'" (Lustick 2008, 294). The concept was designed to frame Jerusalem's holy sites as nested within one basin, much like the land undergirding a lake or body of water. The assumptions of the proposal are strikingly similar to those guiding riparian negotiations. First, the basic religious needs of each of the parties would be recognized. Second, use of national symbols would be minimized, including flags. Third, Israeli and Palestinian oversight of the space would be shifted from governmental to Jewish and Muslim religious bodies. Fourth, there would be freedom of movement and access, although this freedom would be restricted during particular days and weeks deemed important by the respective religious traditions (Sher 2008). Treating Jerusalem like a basin is a creative shift from the zero-sum frame that has dominated the conflict between Israelis and Palestinians. Unfortunately, the concept of the sacred basin and principles of riparian negotiations will need to wait until the next round of serious talks over Jerusalem as the broader situation remains in a stalemate, but we note that thinking of territory in a manner typically associated with water was an explicit dimension of proposals of how to consider the problem.

The three illustrations we introduce here are not full-scale case studies, but they do illuminate expressions of riparian-like conflict management in different scales and types of disputes. Unlike our conceptual exercise regarding Jerusalem, the three illustrations are drawn from negotiated arrangements that have allowed antagonists to satisfy their basic needs for more than a decade. Belligerents in these conflicts appear to see the value of needs-based negotiations and in the face of significant religious and nationalist ideologies adhere to a pragmatic view of sharing disputed resources. This view recasts visions that hold space to be nonnegotiable, revealing the possibility that even seemingly intractable disputes can move toward constructive resolution.

The Hebron/al-Khalil Accommodation

Jews and Muslims in Hebron/al-Khalil share the same structure, which hosts a mosque and synagogue, on space considered sacred by both religions. The Mosque/Synagogue of the Patriarchs is a site of tension and violence. Yet, even in a context of great anxiety, hatred, and episodes of violence, Jews and Muslims have shared and continue to share the space. This accommodation is enforced by the Israeli Defense Force (IDF), but periodically there are violent exchanges between Israeli settlers and Palestinian residents, and access to the coercive power of the state is obviously tilted toward the Jewish population. Nevertheless, the power of the state is the guarantor of the arrangement, which, although imperfect, does allow both sides to use space amid conflicting maximalist claims. But for the IDF, Jews would not be able to worship at their sacred site, but were it not for the IDF upholding the agreement, neither would Muslims.

The space is controlled by separate religious authorities and is structured to keep Muslim and Jewish worshippers apart. The Waqf ministry, the Islamic religious authority in the West Bank, provides legal oversight of the mosque, the Office of the Chief Rabbinate supervises the functioning of the synagogue, and the IDF includes the structure in its jurisdiction and mandates for mutual and separate religious control of the site. There were significant signs of riparian conflict management in the "Protocol Concerning the Redeployment in Hebron," which included this declaration:

> Both sides shall respect and protect religious rights of Jews, Christians, Muslims and Samaritans to whit: 1. Protecting the holy sites. 2. Allowing free access to the holy sites. 3. Allowing freedom of worship and practice. (Ministry of Foreign Affairs 1995)

The protocol stipulates the division of authority over the mosque and synagogue and calls for both parties to honor "the status quo" at the site. The current arrangement affords both parties their minimum need, which is the ability to worship in the structure. This baseline could provide potential for a much more significant transformation of the conflictual environment in Hebron.

With separate enforcement authorities, the space of the Mosque/Synagogue of the Patriarchs is divided. Since 1968, Jews and Muslims were able to worship side by side in the structure, and each religious party was allotted a time during the week to worship alone with their coreligionists (Emmett 2000). In the wake of the 1994 massacre of twenty-nine Muslims who were worshipping at the mosque, the building was segregated, with Muslims and Jews having separate entrances, administered by separate religious bureaucracies. The disaggregation of authority and entrances ("the separating out of difficult issues or territories into smaller parts"; Dumper 2003, 166) might eventually yield a more collaborative arrangement. Dumper (2003) argues that the "device of disaggregation" used with some effectiveness in Hebron, "can still be put to good use" (166) in thinking through the problem of Jerusalem and might be another tool used to manage conflict over contested territory.

Flowing Through Derry

In Northern Ireland, there are thousands of sectarian parades each year—some of them quite provocative and marked by violence—that bring to the surface struggles between Catholics and Protestants over contested space. As Northern Ireland has embarked on a transition from conflict to increasing accommodation in the period since the cease fires of 1994, one particular set of parades has set a new tone marking sectarian differences in public space (Graham and Nash 2006). The Apprentice Boys of Derry parades each August and December are Northern Ireland's largest and serve as icons for the Protestant experience. These parades used to be flashpoints for considerable violence and property destruction (Jarman 2003). Derry, or Londonderry as it is called by Protestants, was the site of a siege in the late seventeenth century that ultimately resulted in a defeat of Catholic forces and cast the town as emblematic of the Protestant ethos of Northern Ireland: stubborn and successful resistance to and triumph over the larger surrounding Catholic population (McBride 1997).

This victory is marked by parades composed of members the Apprentice Boys of Derry, a commemorative organization, and the sectarian bands they hire to accompany them on a circuit of the city and its historic walls. In addition to the traditional significance of marching in Derry to recall the siege, the parade is significant because, although Catholics have been a majority in the town since the mid-nineteenth century, Protestants dominated politically until the onset of the Troubles in the late 1960s and early 1970s, a period of violence that was in part sparked by events that took place there. The symbolic portions of the city are dominated by Catholics, however, who are now in firm control of the city's political apparatus. Yet there remains a long-standing rancor about the discrimination that Catholics suffered at the hands of Protestants, making the parade even more charged, as it is seen by Catholics as a display of Protestant triumphalism (Bryan 2006; Hamilton and Bryan 2006).

The critical change in relation to the Apprentice Boys' parades is that they are now subject to negotiation between the sponsoring organization and a Catholic group affiliated with both a local neighborhood and the Republican Sinn Fein party. For each parade, the two sides meet to discuss the terms under which the march can be conducted to avoid active opposition from the Catholic side. They discuss the timing of the parade, the conduct of the bands when passing by Catholic neighborhoods and churches, the timing of the different phases of the march, arrangements to avoid disruption of daily life, and the supervision of partisans on both sides of the community. The sessions delve into minute details and touch on past iterations of negotiation and what followed, each side jockeying to get more of what it wants (more parade with fewer restrictions for the Apprentice Boys, less parade with more restrictions for the Catholics) within the framework of a negotiated outcome (Cohen 2007).

The accommodations allow Protestants to flow into Catholic-dominated parts of the city that hold significance for both sides in the conflict. The terms of this "flow" are negotiated, and time is a critical factor, inasmuch as the parade occurs on specific days, in specific times, and in specific places. The goal of both sides is to preserve the integrity of the resource, that is, the city itself, and protect it from destructive violence. In the negotiations each side makes statements about its rights, but what they actually do is address various needs that are possible when the zero-sum is abandoned. Each side frames what it offers in negotiation as proactive accommodations rather than concessions, and thus there is no compromise of the rights that they claim to bolster their positions. The Protestants claim victory in that they are marching in those spaces that they hold historically significant, and the Catholics point to the fact of negotiation and the limits on the parade as a sign of their victory. Neither side is completely satisfied, but both endorse the process and its outcome, a common pattern for riparian negotiations.

Collaborative Nonsovereignty

Relations between Israel and Jordan have always been complex, with overlapping shades of belligerence and cooperation tailored to specific circumstances and shared or diverging interests. Instances of both conflict and collaboration have been tied to specific territorial issues, among them the nature of Palestinian rule, border crossings, and the sharing of water resources (Beaumont 1997; Shlaim 1990). The period between the Jordanian–Israeli Armistice Agreement of 1949 and the 1994 peace treaty was marked by a general antagonism, but modes of cooperation eased the way for final border arrangements that reflect, in some respects, elements of the riparian dynamics critical to our model. Indeed, a history of subtle riparian accommodation in relation to allocation of the water of the Jordan River shaded into creativity in addressing other territorial concerns that were part of the overall 1994 peace agreement (Haddadin 2000).

With the broad parameters of a bilateral peace treaty in place, Jordan and Israel were left with the untidy remains of a disputed boundary and issues of sovereignty and territorial control. In two small areas on the border, an arrangement was reached whereby Israel recognized Jordanian sovereignty and yet retained access to and use of the approximately 700 acres that were "returned" as part of the peace settlement. Herein lies the type of fluidity that is suggested by the riparian way of thinking. These "special regimes," contained in Annex I of the 1994 peace treaty, specify the terms that will apply to the status and use of the Naharayim/Baqura and Zofar/Al-Ghamr areas. In both cases, Israeli farmers were using the land in question, and there was a political need for the Israeli government to find a way to leave their situation intact (i.e., to postpone uprooting or otherwise displacing Israelis). At the same time, the Jordanian government needed to assert its authority throughout the territory in which it was sovereign.

To address these seemingly incompatible needs, the land in question was formally and legally returned by Israel to Jordan but simultaneously was leased to Israel, under specific terms that both honor Jordanian sovereignty and allow significant flexibility for Israeli use. The treaty explicitly recognizes Jordan's sovereign claim to the land but also the private land use rights of Israeli "land owners" (Treaty of Peace Between the State of Israel and the Hashemite Kingdom of Jordan, Annex I (c) 2). Critically, those land owners and their invitees or employees have "unimpeded freedom of entry to, exit from land usage [*sic*] and movement within the area." Within the areas in question, Israeli law "may be applied to Israelis and their activities," and Jordanian law is applied to Jordanians and any non-Israelis, and with "the minimum of formality" Israeli police can enter the area to deal with crimes involving Israelis and their employees.

In practice, this creates a situation in which the Jordanian flag flies over an area that is part of the Israeli economy and is used by Israelis who enjoy easier access than do Jordanians, who are excluded from the land by their own government. In effect, the situation is palatable to Israel if its farmers retain access to their farms without Jordanian interference, whereas for Jordan this situation is tolerable so long as there is strict control of those entering the area, and entrance includes a recognition of Jordanian sovereignty. The farmers "flow" into their fields in the morning and flow back into Israel at the end of the day. One critical dimension of the agreement is that the land is leased for a period of twenty-five years and shall be "renewed automatically for the same periods" unless formal notice is given by one of the sides, which would trigger "consultations." This element of the treaty creates a situation of impermanent stability, in which Israeli land owners can "freely dispose of their land in accordance with applicable Jordanian law," that is, the land is saleable, inheritable, divisible, and so on. At the conclusion of the lease period, however, Jordan can trigger discussions that would terminate use of Jordanian territory by Israeli farmers, thus returning the land to a more normal—albeit noncollaborative—form of sovereignty. Inasmuch as this is part of the broader Israeli–Jordanian peace treaty, the arrangement is legally binding and formally recognized in international law.

Summary

The three illustrations presented extend an initial study considering the potential benefits of applying the principles of riparian conflict management to nonriparian disputes. Hebron's Mosque/Synagogue of the Patriarchs is shared by Muslims and Jews using an approach of disaggregation. The relationship between the two parties remains tense, but the minimum religious needs of both communities are met by allowing them to share the same structure at separate times. Similarly, Catholics and Protestants in Derry/Londonderry have reached a workable accommodation on issues of contested space. The Protestant loyalists flow through space under the de facto control of Catholics and thereby satisfy their identity maintenance needs and do not seek to assert a legal claim on the space. The needs-based negotiations have allowed both parties to achieve their minimum aspirations. Finally, Israelis and Jordanians, using a system of attenuated sovereignty, are able to share land with the Jordanians holding sovereignty and Israelis securing legal rights through a long-term lease.

The illustrations we have provided here involve conflicts between parties that are split along national, ethnic, and religious lines, but the riparian approach need not be limited to such circumstances. The riparian approach helps shift belligerents from a rights-based deadlock over territorial disputes to a focus on the minimal needs of the protagonists and how they can be met. It could thus be useful at a variety of scales, and within, not just between, communities.

Despite its often critical perspectives, it is easy for the discipline of geography and its practitioners to succumb to the temptation to adopt normative approaches to the study of war and peace (Flint 2005). Even within the subdiscipline of critical geopolitics, the concept of "intractable territorial disputes" holds some appeal, due in part to the durability of disputes that are an outgrowth of decolonization and the dissolution of empires (Ottoman and British foremost but not alone among them). The Israeli–Palestinian conflict, the situation in Kashmir, and the Western Sahara dispute, like so many others, are older now than the majority of practicing geographers and diplomats. Aggregated decades of "peacemaking" have made positive contributions to many territorial disputes, but others languish and test the creative limits of the geographic imagination. It is easy to believe that conflict is an inevitable symptom of the human condition, and that some disputes defy resolution.

Yet the history of the late twentieth century reaffirms the sense that the world is full of geographic surprises and that innovative thinking about territorial conflict can produce durable accommodations. One contribution to this thinking is the riparian model of conflict management. We make no claim for the riparian approach as a panacea, but in examining various dimensions of the situation in the mosque/synagogue, in Derry/Londonderry, and along the Israeli–Jordanian border, we find support for the proposal that some claims to territory can be treated like claims to water and that uses of space and territory can be structured in such a way as to satisfy multiple needs.

References

Barnes, T. J. 1992. Postmodernism in economic geography: Metaphor and the construction of alterity. *Environment and Planning D: Society and Space* 10 (1): 57–68.

Beaumont, P. 1997. Dividing the waters of the River Jordan: An analysis of the 1994 Israel–Jordan peace treaty. *International Journal of Water Resources Development* 13 (3): 415–24.

Bryan, D. 2006. Traditional parades, conflict and change: Orange parades and other rituals in Northern Ireland, 1960–2000. In *Political rituals in Great Britain: 1700–2000*, ed. J. Neuheiser and M. Schaich, 123–28. Augsburg, Germany: Wisner-Verlag.

Buttimer, A. 1982. Musing on helicon: Root metaphors and geography. *Geografiska Annaler* 64B:89–96.

Cohen, S. 2007. Winning while losing: The Apprentice Boys of Derry walk their beat. *Political Geography* 26 (8): 949–65.

Cohen, S., and D. A. Frank. 2002. Jerusalem and the riparian simile. *Political Geography* 21 (6): 745–65.

Dumper, M. 2003. *The politics of sacred space: The old city of Jerusalem in the Middle East conflict*. Boulder, CO: Lynne Rienner.

Elhance, A. P. 2000. Hydropolitics: Grounds for despair, reasons for hope. *International Negotiation* 5 (2): 201–22.

Emmett, C. 2000. Sharing sacred space in the Holy Land. In *Cultural encounters with the environment: Enduring and evolving geographic themes*, ed. A. B. Murphy, D. L. Johnson, and V. Haarmann, 261–82. Lanham, MD: Rowman & Littlefield.

Fisher, R., W. Ury, and B. Patton. 1991. *Getting to yes: Negotiating agreement without giving in*. Boston: Houghton Mifflin.

Flint, C. 2005. The geography of war and peace. In *The geography of war and peace: From death camps to diplomats*, ed. C. Flint, 3–18. Oxford, UK: Oxford University Press.

Graham, B., and C. Nash. 2006. A shared future: Territoriality, pluralism and public policy in Northern Ireland. *Political Geography* 25 (3): 253–78.

Haddadin, M. 2000. Negotiated resolution of the Jordan–Israel water conflict. *International Negotiation* 5 (2): 263–88.

Hamilton, M., and D. Bryan. 2006. Deepening democracy-dispute system design and the mediation of contested parades in Northern Ireland. *Ohio State Journal on Dispute Resolution* 22:133–88.

Hensel, P. R., S. McLaughlin Mitchell, and T. E. Sowers, II. 2006. Conflict management of riparian disputes. *Political Geography* 25 (4): 383–411.

Jarman, N. 2003. From outrage to apathy? The disputes over parades, 1995–2003. *Ethnopolitics* 3 (1): 92–105.

Livingston, D., and R. Harrison. 1981. Meaning through metaphor: Analogy as epistomology. *Annals of the Association of American Geographers* 81 (1): 95–107.

Lustick, I. 2008. Yerushalayim, al-Quds, and the Wizard of Oz: The problem of "Jerusalem" after Camp David II and the Aqsa Intifada. In *Jerusalem: Idea and reality*, ed. T. Mayer and S. A. Mourad, 283–302. London and New York: Routledge.

McBride, I. 1997. *The seige of Derry in Ulster Protestant mythology*. Dublin, Ireland: Four Courts Press.

Ministry of Foreign Affairs, Israel. 1995. *The Israeli–Palestinian interim agreement—Main points*. http://www.mfa.gov.il/MFA/Peace+Process/Guide+to+the+Peace+Process/The+Israeli-Palestinian + Interim +Agreement+ – +Main+P.htm (last accessed August 27, 2009).

Postel, S. L., and A. T. Wolf. 2001. Dehydrating conflict. *Foreign Policy* 126:60–67.

Sher, G. 2008. Negotiating Jerusalem: Reflections of an Israel negotiator. In *Jerusalem: Idea and reality*, ed. T. Mayer and S. A. Mourad, 303–18. London and New York: Routledge.

Shlaim, A. 1990. *The politics of partition: King Abdullah, the Zionists, and Palestine, 1921–1951*. New York: Columbia University Press.

Sunstein, C. R. 1996. *Legal reasoning and political conflict*. New York: Oxford University Press.

Wolf, A. T. 1998. Conflict and cooperation along international waterways. *Water Policy* 1 (2): 251–65.

———. 1999. The Transboundary Freshwater Dispute Database Project. *Water International* 24 (2): 160–63.

———. 2000. Indigenous approaches to water conflict negotiations and implications for international waters. *International Negotiation* 5 (2): 357–73.

———. 2002. The importance of regional cooperation on water management for confidence building: Lessons learned. Background paper presented at the 10th OSCE Economic Forum on Cooperation for the Sustainable Use and the Protection of Quality of Water in the Context of the OSCE, Prague, Czech Republic.

Wolf, A. T., K. Stahl, and M. F. Macomber. 2003. Conflict and cooperation within international river basins: The importance of institutional capacity. *Water Resources Update* 125:31–40.

Walls as Technologies of Government: The Double Construction of Geographies of Peace and Conflict in Israeli Politics, 2002–Present

Samer Alatout

Since 2002, consecutive Israeli governing coalitions have been building a separation wall in the West Bank for the declared purposes of security and separation from the Palestinian population. Building on earlier phases of control, which relied on military orders, cantonment, roadblocks, and checkpoints, the wall functions as a regime of government that colonizes Palestinian life by regulating every nexus of body and space and population and territory. Rather than establishing peace, the wall's regime of government uses separation as a double construction of peace and conflict that isolates peace on the Israeli side and conflict on the Palestinian side. *Key Words: biopower, bio-territorial power, governmentality, Israel Palestine, peace and conflict.*

自 2002 年以来，历届以色列执政联盟出于其所宣称的安全和与巴勒斯坦人民隔离的目的，在西岸地区建立了一道隔离墙。隔离墙立足于实施早期阶段的控制，依靠诸如军事命令，兵营驻地，路障和检查站，隔离墙作为政府制度的一部分，对巴勒斯坦地区进行着殖民统治，控制着与此相关的每一个人与空间，人口与领土。隔离墙并非建立和平，隔离墙作为政府制度，利用隔离来建立双重的和平和冲突，将以色列一方的和平与巴勒斯坦一方的冲突相隔离。*关键词：生物能源，生物领土的权力，治理性，以色列巴勒斯坦，和平和冲突。*

Desde el 2002, las consecutivas coaliciones de gobierno de Israel han estado levantando un muro de exclusión en el Banco Occidental, con los propósitos declarados de seguridad y separación de la población palestina. Construido a partir de elementos preexistentes de control, como los relacionados con puestos militares, acantonamientos, bloques de contención de carretera y retenes, el muro funciona como un instrumento gubernamental que coloniza la vida palestina al regular todo nexo entre cuerpo y espacio, y población y territorio. En vez de traer paz, el régimen restrictivo del muro utiliza la separación como una línea de doble cara para paz y conflicto, la cual aisla a la paz en el lado israelí y al conflicto en el palestino. *Palabras clave: bio-poder, poder bio-territorial, gobernabilidad, Israel, Palestina, paz y conflicto.*

Consecutive Israeli governments have been building a separation wall (Figure 1) in the Occupied Palestinian Territories (OPT) of the West Bank since 2002. The wall extends for 723 km (450 mi) north to south and at points zigzags deep into the West Bank for more than 16 km (10 mi) to include large constellations of illegal Israeli settlements,[1] directly affecting more than a half-million Palestinians: Some, like Qlaqilia (Figure 1), are surrounded on three sides with a narrow, gated opening on the fourth; many, like Jayyus, are isolated from their agricultural lands; and some are kept on the "Israeli" side of the wall and isolated from the rest of the West Bank (Table 1).[2] The wall appropriates 11 percent of the land area of the West Bank for its immediate use. That percentage increases to a whopping 40 percent when settlements and bypass roads are included in the calculation. It is yet unclear whether the Israeli governing coalitions will build the eastern side of the wall (now about 8 percent complete), thus surrounding the West Bank from all sides and annexing the Jordan Valley. If that were to happen, the wall would expropriate an additional 25 percent of the West Bank land area, leaving the Palestinians about 13 percent of historic Palestine.

Numerous accounts have explored the wall's humanitarian, sociospatial, territorial, and political effects. Some studies focused attention on the daily effects of the wall on Palestinian lives—from the dismantling of the educational systems of many West Bank communities (Ministry of Education and Higher Education 2004) to the disruption of access to health care services (Directorate of Education-Qalqilia 2003); from denying access to farmlands (Jayyus Municipality 2003) to general effects on Palestinian livelihoods (Dolphin 2006). Others have focused on the political effects of the wall: how it shifts the presumed borders

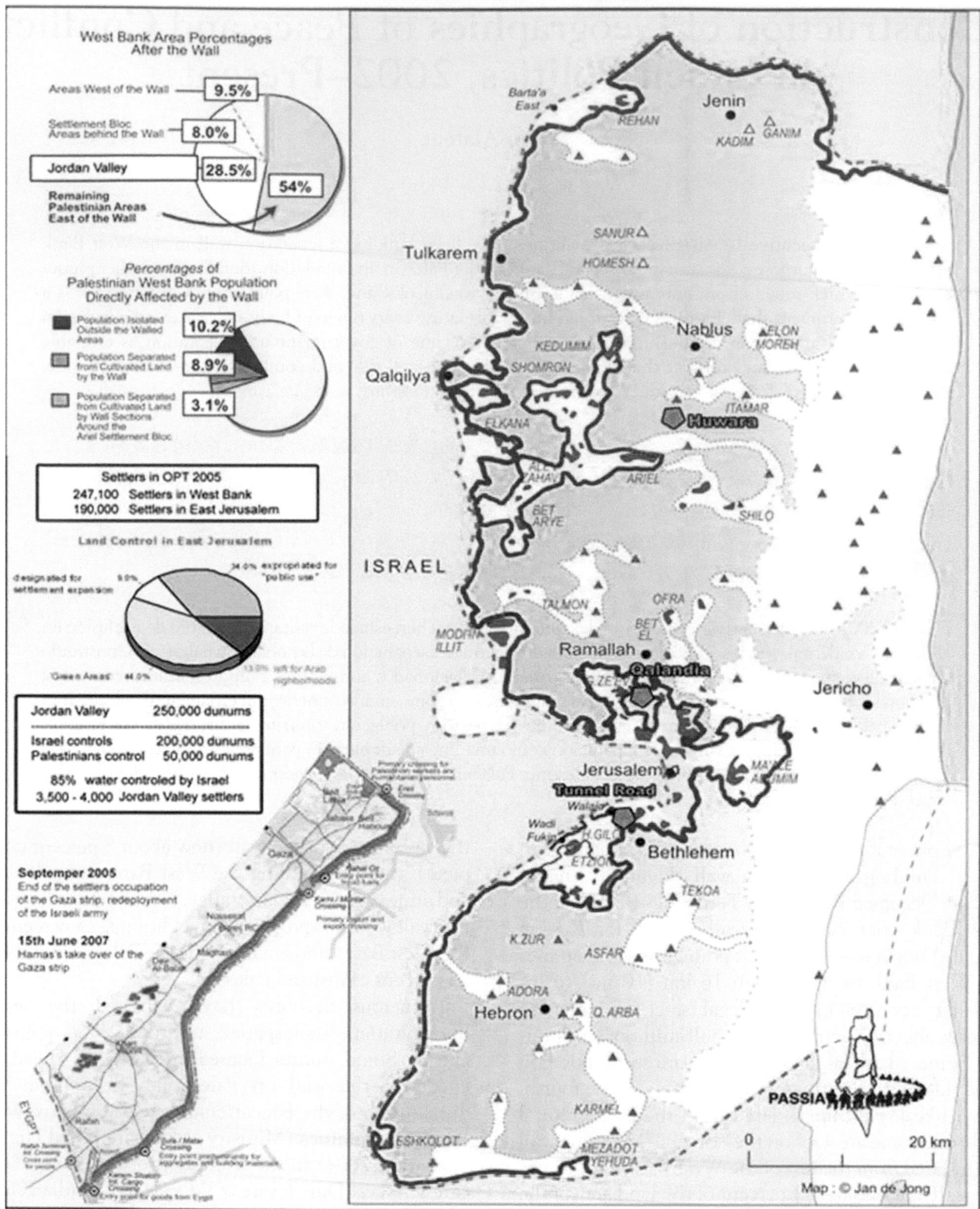

Figure 1. The Israeli Wall in the West Bank. *Source:* Palestinian Academic Society for the Study of International Affairs, Jerusalem, http://www.passia.org/index.htm (last accessed 20 September 2008).

Table 1. Palestinian population affected by the barrier's route

	Number of communities	Number of residents
Communities west of the barrier[a]	17	27,520
Communities east of the barrier that are completely or partially surrounded[b]	54	247,800
East Jerusalem	21	222,500
Total	92	497,820

[a]Residents of these towns and villages will require permits to live in their homes, and they will be able to leave their communities only via a gate in the barrier. The figure does not include three communities that are presently situated west of the barrier but lie east of the barrier according to the currently approved route.
[b]Residents of these towns and villages will not require permits or have to pass through a gate.
Source: B'Tselem's Web site (http://www.btselem.org/english/Separation_Barrier/Statistics.asp, last accessed 20 September 2008).

between Israel and Palestine from the 1967 green line deeper into Palestinian territories, creating new facts on the ground (Khalidi 2005); how the wall invests the two-state solution with new meanings and sheds doubt on both the viability of that solution and the type of a Palestinian state that might emerge under the wall regime (Usher 2005, 2006; Khalidi 2006); and how the wall underwrites both multiple forms of resistance on the national, binational, or international scales and multiple organizational strategies of grassroots disruptions of building sites, diplomatic negotiations and complaints, or legal challenges (Lynk 2005).

Most accounts treat the wall as a technology of occupation, separation, or security. As a technology of occupation, the wall is seen as temporary and subject to the strategic calculations of warring parties. Despite its heated rhetoric, probably the most important proponent of this argument is the Palestinian National Authority,[3] which rejects the wall as a fundamentally flawed framework for resolving the Israeli–Palestinian conflict but one that is not necessarily or inherently threatening. As a technology of separation, the wall is expected to undermine the very premise of a two-state solution and promote the ultimate conversion of Palestinians into isolated pockets of population centers with no political, civil, or human rights to speak of. In Khalidi's (2006, 182) words, Palestinians become "stateless in Palestine." Finally, as a technology of security, the wall is meant to separate Israeli citizens, including settlers, from the Palestinians, which implies an expanded notion of security that deems settlers and settlements within the West Bank part of the territory of the Israeli state (Lynk 2005; Sorkin 2005). The Israeli governing coalitions are the logical proponents of such an argument, but so are a number of nongovernmental organizations (Yacobi 2006, 752).

In all three frameworks for understanding the wall lurks an assumption that it can only serve one of these purposes (occupation, separation, or security)—all of which are presumed unique to the Palestinian–Israeli condition. Such singularity fails to recognize the general function of the wall as a double technology of government for spatially regulating bodies and populations, not of the Palestinians only but also of Israelis, settlers and otherwise. Thinking of the wall as a multipurpose technology of government that spatially invests individuals and populations in relations of power helps us understand its simultaneous dual construction of geographies of conflict and peace—understood here narrowly as violence and lack of violence.

In the following section I discuss the Israeli occupation, not as an undifferentiated strategy from 1967 until the present but as three successive regimes of government that differently organize individuals and populations in space (occupation, 1967–1994; cantonization, 1994–2002; and separation, 2002–present). In the section that follows, I provide four brief ethnographic accounts from fieldwork conducted in 2004 and 2006 to shed light on how the wall becomes a technology of government in practice. I then conclude by synthesizing the arguments of the article and by posing questions about the possible governmental effects of the wall on the Israeli population.

Israeli Control and Regimes of Government: From Occupation to Cantonization to Separation

Most state and international relations theorists focus on territory as the defining element of modern government and place emphasis on sovereign (territorial) power as a result. Discussing his notion of governmentality, Foucault (1977, [1978] 1990, 2007) shifted attention from territory to population as the most important element in relations of power and government, or what he called *biopower*. Recent critiques of Foucault (Alatout 2006; Crampton and Elden 2007; Elden 2007) rethink population and territory in relational terms, viewing government as a bio-territorial process by which categories of territory and population are continually *constructed* and *articulated* with one another (Alatout 2006).

With this theoretical intervention in mind, the Israeli occupation of the West Bank and Gaza since 1967 can be divided into three distinct periods, each with its own regime of government: occupation, 1967 to 1994; cantonization, 1994 to 2002; and separation, 2002 to present. Each period had a dominant technology of government that constructed various categories of people and land and related them to one another, and each regime of government that came at a later stage benefited from and assumed relations of power and government that preceded it. Against those techniques of government, the West Bank was turned into a fractured, disjointed space and the Palestinians into a Sisyphean people whose aspirations for nationhood are continually challenged and violently shattered.

Between 1967 and 1994 occupation and control over territory and population were managed through military orders (MOs) issued and enforced by the Israeli military commander in the occupied Palestinian territories to confine the Palestinian population to as little land as possible and to limit their access to resources (Benvenisti 1984). These regulations (see MOs 58, 59, 321, 364, 569, and 1091) resulted in the transfer of large tracts of land (more than 30 percent of the West Bank) to Israeli military control and to Israeli settlers who moved to the West Bank under military protection (Shehadeh 1988). The legal and practical processes of exerting control over territories and bodies sowed the seeds of a system of separation, anticipating the future wall regime: (1) increasingly, settlement expansion seemed to be a permanent activity; (2) occupation, as a regime of government, increasingly seemed the wrong description of what was taking place (it seemed more permanent than it was initially thought to be); (3) the security of the state of Israel expanded to include the security of settlements and their environment; and (4) as a result, Palestinians were prohibited from using most of the land mass of the West Bank (e.g., Palestinians were prohibited from using the bypass roads, which connected large settlements directly to Israel proper and were reserved for Israeli settlers and the military; Davis, Maks, and Richardson 1980; Benvenisti 1984; Davis 1987).

Between 1994 and 2002, a new regime of cantonization emerged with roadblocks, checkpoints, and bypass roads as its constituent elements. The Oslo agreement of 1994 divided the occupied territories into three categories of land (Figure 2) subject to different regimes of movement, surveillance, and management: Area A, heavily populated by Palestinians presumably under full Palestinian control (civil and security); Area B, under Palestinian civil control but shared security arrangements; and Area C, all territories not included in the first two (Israeli settlements, roads, and highways, even those connecting Palestinian towns and villages, uninhabited spaces, state land, etc.). Not only was the area under Palestinian control extremely small, but Israeli policies followed during this period disrupted the contiguous geography of Palestine. Military checkpoints and the massive network of bypass roads disrupted the north–south route (Nablus-Ramallah-Jerusalem-Bethlehem-Hebron) that was the main artery of West Bank life before and during the early years of occupation. At the same time, and to complete regional consolidation, east–west roads like that between Nablus, Tulkarm, and Jenin became more important for the regional economy of the north. The same can be said about the Hebron and Bethlehem areas. By

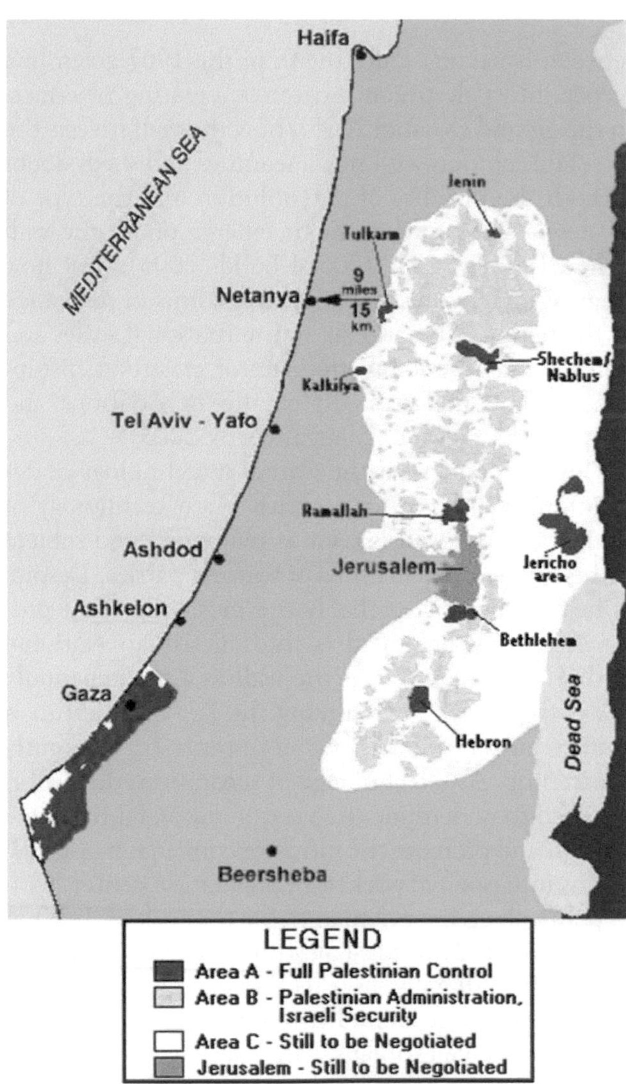

Figure 2. Land distribution according to Oslo II.

manipulating the ease or difficulty of movement throughout the West Bank, Israeli military forces were constructing different political and economic geographies that divided the West Bank into northern, central, and southern regions.[4]

Responding to the strong Palestinian resistance and to turn cantonization—what Kimmerling (2003) called *politicide* in a different context—into a permanent feature of Palestinian political life, unilateral separation has become a dominant concept in Israeli political discourse since the mid-1990s. Calls for a separation wall, initially the brainchild of Ehud Barak, Uzi Dayan, and other Labor generals, intensified during Ariel Sharon's rule in the early 2000s and the rest became history.

The assumption that it is possible to separate unilaterally from the Palestinians, while keeping intact all the miseries of occupation and cantonization, is naïve. The second *intifada* and the suicide attacks carried out in Israeli cities have proved that point. Had the separation wall been built along the green line, one would probably have had to concede that what it sought was indeed separation, but the wall does not separate Israelis from Palestinians. If at all, it dramatizes their proximity; it zigzags through Palestinian villages and towns, engulfs ever more lands, and incorporates the major settlements in the West Bank within the wall. If at all, it brings Israel closer to the occupied Palestinians, not farther from them.

If some Palestinians were at all able to escape governmental regulations under regimes of occupation and cantonization, the omnipresent wall and all the regulations surrounding its maintenance as a functioning technology of government colonize every possible nexus of body and space. Not leaving anything to chance, the wall carves space according to a calculated policy of annexation and dispossession. Whatever the wall does not achieve in its spatial-material form, it does through the unwieldy regulatory practices of permits, registrations, rules, norms, and codes.

Figure 3. Seasonal gate at the boundaries of Jayyus. © Samer Alatout 2004.

Figure 4. Daily passage gate. © Samer Alatout 2004.

Regulating Traffic: People, Lands, Agriculture, and Community Life

Lack of freedom of movement and its effect on politics, economy, education, and social life have been a constant point of tension in the occupied Palestinian territories since 1967. Since then, each regime of government constructed a regime of travel that was based, among other things, on individual biographies (how old, how involved in politics, what types of activities are practiced), communal differentiations (how collaborative is a community with occupation forces; what interests do Israeli settlers have in that community's resources; what type of a community is it, rural or urban; what type of work do most people do in that community; and how reliant is that community on work in Israel), spatial proximity to settlements and Israeli cities and towns, and the Israeli economy and its needs. Travel regimes rely on the heavy use of various documents: identity cards marked with political notations about the individual; travel permits that define where and when an individual is permitted to travel—is he or she allowed across the green line, is he or she allowed to stay in Israeli towns overnight, which ones, what times are those travels allowed and for how long;

Table 2. Gates throughout the wall, by area, July 2008

Area	Open gates	Seasonal gates	Closed gates	Total
Qalqiliya	11	3	0	14
Salfit	0	3	3	6
Tulkarm	6	4	0	10
Jenin	2	9	0	11
Jerusalem	0	0	10	10
Ramallah	1	0	7	8
Hebron	0	0	7	7
Total	20	19	27	66

Note: Data from UN Office for the Coordination of Humanitarian Affairs.
Source: B'Tselem's Web site (http://www.btselem.org/english/Separation_Barrier/Statistics.asp, last accessed 20 September 2008).

permits to travel across one of the bridges to Jordan, which bridge, when, and for how long. Even permits to travel to Jordan, for example, also indicate notations on the conditions under which this traveler should be allowed to cross the bridge and if that permit should be canceled if those conditions are not met.

The wall has sixty-six gates (Table 2) that regulate traffic on both sides. Twenty-seven are closed and used for military purposes only. Nineteen are located in places where the wall separates farmers from their lands; however, these gates are open only seasonally (Figure 3) in time for olive picking and harvest. Twenty of the gates are open daily (Figure 4) to allow, for example, farmers to tend to their crops and students to attend schools. For the most part, these gates are open for brief periods two or three times a day, usually for less than an hour in the morning, afternoon, and early evening. To cross back and forth, Palestinians have to have permits that specify the days they are allowed to cross the gate, the purposes for which they are given the permit, the duration of the permit, and the specific places they are allowed to visit.

In addition to the more prevalent situations in which the wall separates communities on the Palestinian side from their agricultural lands on the other side of the wall, a few communities were walled out (or in, as the case may be), virtually annexed to the Israeli side of the wall and thus separated from neighboring Palestinian communities with which they often constitute a social unit (sharing markets, schools, clinics, and families). Even though the wall circled them in, these communities have to acquire permits to reside where they already resided before the wall or for the purpose of crossing the gates. In any case, granting Palestinians permits to travel, across the wall in particular, depends on individual and community biographies. In the city of Jayyus, for example, a number of farmers have not been given permits to attend their olive trees and agricultural lands. In addition, only family members usually are given permission to work in the

Figure 5. Samiha walking six kilometers to cross the road. © Samer Alatout 2004.

land during harvest or olive picking—hired labor is not allowed.

The Wall and the Production of Illegal Subjects

The wall is a spatial strategy for determining who has a right to free spatial movement. Its construction disrupts the ongoing history of communities throughout the West Bank and, coupled with new legal definitions of people within and outside the wall, marks certain bodies with illegality. Figure 5 shows a portion of the wall that separates East Jerusalem from neighboring Abu-Dees. The woman shown in the picture, Samiha, lives at the northeast corner of this intersection (now in Abu-Dees) and her mother lives at the northwest corner of the intersection (now within Jerusalem). As the wall was built, she was declared a Palestinian, nonresident of Jerusalem, as her mother was being declared a legal resident of Jerusalem. The challenges posed by the wall for Samiha cannot be measured only in terms of the difficulties she undergoes daily to visit her older mother (walking six kilometers twice each day instead of a few steps across the road). Difficult as that might be, Samiha's daily experience is about government in the broadest sense, about drawing a line between legal and illegal distribution of people in territories. One of the most challenging aspects of the wall is that it marks her travel (walking, dusting, any signs of traveling from across the road) as an illegal act punishable by law. The wall defines her trip as illegal, despite the fact that her daily trip is made possible only through an opening of the wall that is yet to be completed. Neither is her trip kept secret from the state. As a matter of fact, soldiers often see her make the journey and, depending on the priorities of the day, they either let her continue with her journey or not. That is precisely what government does: Illegality is a marker allowing the state to intervene and remove certain bodies from certain spaces, but more important, it also turns mundane daily practices into political frameworks of resistance.

Figure 6. Who is the guardian of the state? © Samer Alatout 2004.

Figure 7. Selling vegetables. © Samer Alatout 2004.

The Wall as an Economy of Power: Who Protects the State?

In Figure 6, the gate is part of the wall and separates Jerusalem from Al Azariyeh, a town on the northeast border of Jerusalem. It demonstrates how the wall's status as a technology of government depends to a large degree on its ability to enroll Palestinians in performing some of its functions. This portion of the wall was delayed because of an order issued by the High Court of Israel. A Palestinian resident on the "other" side of the wall (Najeeba) objected because the wall would cause her undue harm by making her travel to work and back unnecessarily cumbersome. Najeeba is a legal resident of Jerusalem who is married to a Palestinian from Al Azariyeh with a West Bank identity card. The High Court ordered the military to stop building the wall in that portion until the dispute was resolved. Instead, the military built this particular gate to allow Najeeba access to her workplace. The key was entrusted with Najeeba. She is supposed to use it to open the gate and then lock it behind her to prevent others from crossing the wall. When I first passed by the gate (I didn't have my camera then), she had already passed through and was locking in her husband and eight-year-old son. Assignment of identity is arbitrary for sure; however, the fact that Najeeba is entrusted with the security of the state is a clear example not only of enrolling Palestinians in performing the functions of the wall; it demonstrates how the wall itself functions in an economy of power.

New Emerging Informal Economies of the Oppressed

Figures 7 and 8 show women and men crossing the wall for daily travels to set up vending locations on the streets of Jerusalem, visit family members, visit health facilities, and so on; however, these pictures draw into question the whole security argument with which the

Figure 8. New economies of separation. © Samer Alatout 2004.

Israeli state justifies building the wall. This area is not a passage point of a few desperate souls. It witnesses thousands of travels in both directions each day. It is not, in other words, a small pocket of illegality. Again in Al Azariyeh, this portion of the wall is lower than in other places. On its side, there is an edge of about one to two feet wide. People straddle the wall, stand up on the edge, pass from one side to another, and hand babies from one adult to the other. The edge drops at least thirty feet down, so there is a bodily threat to each one who passes through that portion of the wall. I saw and interviewed a number of people in that location. The stories are often similar but never the same—each surprises you in ways you have not expected before. Two women, who seemed in their sixties or seventies, each carrying a large cloth that contained assorted vegetables, were heading toward downtown Jerusalem to sit on some street corner to sell them. I asked them how far do they travel to sell the vegetables they carry on their heads and if that travel was worth it. For the most part they come from Jericho all the way to Jerusalem, the regional economic, social, and cultural center—they used to do that before the wall. There is no market for their products in Jericho or even other smaller towns like Al Azariyeh. Many other women jumped through the edge from one side to the other. Children of ten or eleven years old had the same stories of walking for an hour or two to reach the point where they can cross the wall to walk around Jerusalem selling cigarettes.

Beyond these many individual stories, there is a whole new economy of bodies, commodities, and travel emerging. The whole economy is illegal, but that is what makes it fascinating from a theory of government viewpoint. The goal of the Israeli state and the wall are not *exclusively* repressive, as it is the redistribution of people and things in spatial arrangements that marks some as illegal and others as legal. The illegal

Figure 9. Signs of resistance. © Samer Alatout 2004.

arrangements are not meant to be busted but to be monitored and surveilled, not by the army and the apparatuses of security only but by internalizing the illegality of one's body in the space in which it is traveling. It is illegal for a Palestinian from the other side of the wall to travel to Jerusalem and that illegality is marked by the difficulty of the travel but also by the presence of the wall and its transgression and by the threat of being caught at all times. Government is about the normalization of such a distribution of bodies, things, and economies.

Signs of Resistance

On the eastern Palestinian side of Abu-Dees I came across a portion of the wall that was interesting, not only because of the heavy graffiti along its length and height but because of the kaffiyeh hanging on top of the wall. I didn't have a camera then but returned the very next day with one in hand to take a picture. The kaffiyeh was not hanging on the wall anymore (as if thrown by someone from the ground) but actually affixed to the highest points of the wall (Figure 9). It is not a coincidental story; it is a story of resistance. At one moment it declared that the wall is breachable but also that its breach did not necessarily mean opening a frontal war with the Israelis. It was left there as a reminder of the possibilities and ingenuities of Palestinian resistance and that the wall can, as a matter of fact, be breached at will.

Conclusion

The separation wall in Palestine is a technology of government whose regulatory framework attempts to control all spaces and people in the West Bank. It delivers its political, economic, and socio-spatial effects through the intricate details of its material and regulatory elements: its length, height, route, gates, permits, regulation of travel, biographies of individuals and communities, surveillance, enrollments, and maintenance. These intricate details become part of the daily lives of Palestinian farmers, students, doctors, and patients. They colonize the bio-territorial framework of the population and the body and space nexus of the individual.

Although most of the discussion focused on how the wall affects the bio-territorial expression of Palestinians, the wall's effect on Israelis is ever present in the design, intent, and functioning of the wall. Not only are Israeli settlers affected, but so are Israelis living within the 1967 border. The effects, however, are different from those touching the Palestinians. The wall acts as a technology of discipline and containment on the Palestinian side, but it is meant to act as a technology of freedom (of movement, travel, economy) on the Israeli side. It is important to note, though, that further investigation is warranted of the wall's effects on Israeli daily lives and whether their perceived freedom constitutes the condition of possibility for its disciplinary effects on the Palestinians.

Finally, the different effects produced by the separation wall regime on Palestinians and Israelis should turn our attention to the fact that relations of government are indifferent to abstract questions of peace (lack of violence) and conflict (violence). In other words, there is no moral content to government and its technologies. The main concern is to manage relations of peace and conflict in ways that sustain governmental relations and not necessarily to resolve one in favor of the other. Regardless of whether it would or would not function as it is intended to in the future, the wall as a technology of government is intensifying violence on the Palestinian side while attempting to create peace (or at least absence of violence) on the Israeli side.

Acknowledgments

This research project was made possible by funding from the International Institute at the University of Wisconsin, Madison. Thanks to the reviewers of the *Annals of the Association of American Geographers* for their helpful comments. Special thanks to Audrey Kobayashi for her support and insightful comments on an earlier draft.

Notes

1. Following a number of scholars, including Said (2003) and Khalidi (2005), I decided to use the term *separation wall* to describe the structure the Israeli army has been building in the West Bank since 2002. The separation wall has been variously described as a security fence (too technical and depoliticizing of the wall), separation fence (too innocent because the structure does not only separate but makes invisible), and apartheid wall (needs a fuller understanding and accounting of the differences and similarities between the case of Israel and that of South Africa prior to 1994).
2. Many of the details about the wall, some of which are too cumbersome to recount, come from a variety of sources including personal observation and ethnographic fieldwork but also *Haaretz* newspaper between 2002 and 2008 (http://www.haaretzdaily.com), the different publications of the Applied Research Institute of Jerusalem (ARIJ; http://www.arij.org), publications of the Palestinian Hydrology Group (http://www.phr.org), publications of the Palestinian Environmental Non-Governmental Organizations Network (PENGON; http://www.pengon.org), publications of the Israeli Ministry of Defense (http://www.securityfence.mod.gov.il/Pages/ENG/default.htm), and B'tselem, the Israeli Information Center for Human Rights in the Occupied Territories (http://www.btselem.org/English/). Citations will be used when appropriate, but for a more comprehensive understanding, please visit the Web sites listed.
3. When asked about the wall and its effects on negotiations between Palestinians and Israelis, one Palestinian foreign ministry official who has been involved in negotiations replied, "Walls come down too. Look at the Berlin Wall" (interview, 22 June 2004). This was not an isolated opinion either.
4. Israeli success in constructing different, sometimes conflicting, political geographies in Gaza and the West Bank through spatial and travel manipulation seems to

influence Israeli strategic framework for dealing with the contiguous West Bank landmass.

References

Alatout, S. 2006. Towards a bio-territorial conception of power: Territory, population, and environmental narratives in Palestine and Israel. *Political Geography* 25:601–21.

Benvenisti, M. 1984. *The West Bank data base project: A survey of Israel's policies*. Washington, DC: American Enterprise Institute for Public Policy Research.

B'Tselem. 2008. *Separation barrier*. http://www.btselem.org/english/Separation_Barrier/Statistics.asp (last accessed 10 July 2008).

Crampton, J., and S. Elden, eds. 2007. *Space, knowledge, and power: Foucault and geography*. London: Ashgate.

Davis, U. 1987. *Israel: An apartheid state*. London: Zed Books.

Davis, U., A. Maks, and J. Richardson. 1980. Israel's water policy. *Journal of Palestine Studies* 9:3–31.

Directorate of Education-Qalqilia. 2003. *Behind the wall: Human instances*. Qalqilia, Palestine: Directorate of Education.

Dolphin, R. 2006. *The West Bank wall: Unmaking Palestine*. London: Pluto Press.

Elden, S. 2007. Governmentality, population, territory. *Environment and Planning D: Society and Space* 25: 562–80.

Foucault, M. 1977. *Language, counter-memory, practice: Selected essays and interviews by Michel Foucault*. Ithaca, NY: Cornell University Press.

———. [1978] 1990. *The history of sexuality: An introduction*. Vol. I. New York: Vintage Books.

———. 2007. *Security, territory, population (lectures at the College De France, 1977–1978)*. New York: Palgrave Macmillan.

Jayyus Municipality. 2003. *Information on the separation wall in Jayyus*. Jayyus, West Bank: Jayyus Municipality.

Khalidi, R. 2005. From the editor. *Journal of Palestine Studies* 35:5.

———. 2006. *The iron cage: The story of the Palestinian struggle for statehood*. Boston: Beacon Press.

Kimmerling, B. 2003. *Politicide: Ariel Sharon's war against the Palestinians*. New York: Verso.

Lynk, M. 2005. Down by law: The High Court of Israel, international law, and the separation wall. *Journal of Palestine Studies* 35:6–24.

Ministry of Education and Higher Education. 2004. *Effects of the wall of annexation and expansion on the educational process*. Ramallah, West Bank: Educational Information Department.

Said, E. 2003. A road map to where? *London Review of Books*. 25(12). http://www.lrb.co.uk/v25/n12/said01_.html (last accessed 25 September 2009).

Shehadeh, R. 1988. *Occupier's law: Israel and the West Bank*. Washington, DC: Institute for Palestine Studies.

Sorkin, M., ed. 2005. *Against the wall*. New York: The New Press.

Usher, G. 2005. Unmaking Palestine: On Israel, the Palestinians, and the wall. *Journal of Palestine Studies* 35:25–43.

———. 2006. Introduction. In *The West Bank wall: Unmaking Palestine*, ed. R. Dolphin, 1–34. London: Pluto Press.

Yacobi, H. 2006. The NGOization of space: Dilemmas of social change, planning policy, and the Israeli public sphere. *Environment and Planning D: Society and Space* 25:745–58.

Citizenship in the Line of Fire: Protective Accompaniment, Proxy Citizenship, and Pathways for Transnational Solidarity in Guatemala

Victoria L. Henderson

In the battle for political space in postconflict Guatemala, protective accompaniment presents prospects for actualizing citizenship rights. I use the term *proxy citizenship* to refer to the potential for limited rights transfer between citizen bodies, privileged and bare. Foreign accompaniment volunteers can influence the dynamics of conflict by being present and bearing witness, therein supporting the (re)placement of local human rights defenders in the politico-juridical sphere as rights-bearing citizens empowered to testify against impunity. I situate my own experiences in Guatemala within the broader literature on radical democratic theory and geographies of citizenship, examining the possibilities and problems associated with exploiting the differential "worth" of citizen bodies. Taking up the claim that "presence signals the possibility of a politics," I argue that putting citizenship in the line of fire may open modest but strategic pathways in transnational solidarity. *Key Words: accompaniment, citizenship, Guatemala, racism, radical democratic theory.*

在危地马拉冲突之后的政治空间斗争中，保护性的援助营给出了实现公民权利的前景。我使用了代理公民这一术语以表述公民，特权和无权阶层之间转移有限权利的一种可能性。援助营的外国义工们可以影响到冲突的动态发展，作为在政治和司法领域享有权利的公民，有能力可以不受惩罚地做见证，通过在场并作证，以此支持了要求（重新）实现当地人民人权的捍卫者。本文作者将自己在危地马拉的经历放到了一个更广泛的关于激进民主理论和公民地理学的文献背景内，调查了探索公民主体的不同“价值”所面临的可能性和问题。对于“存在标志着一个政治可能性”这一声明，我认为，将公民权置于火线上的作法有可能开启一条谨慎但却是战略性的途径以实现跨国团结的目标。*关键词：援助营，公民，危地马拉，种族主义，激进民主理论。*

En la pugna por espacios políticos en la Guatemala postconflicto, el acompañamiento protector se ofrece como opción para actualizar los derechos ciudadanos. Utilizo la expresión *ciudadanía proxy* para referirme al potencial para la transferencia restringida de derechos entre los cuerpos ciudadanos, los privilegiados y rasos. Los voluntarios extranjeros de acompañamiento pueden influir sobre las dinámicas del conflicto al hacerse presentes y fungir como testigos, apoyando en consecuencia el (re)emplazamiento de los defensores locales de los derechos humanos en la esfera político-jurídica, como ciudadanos abanderados de los derechos con el poder de atestiguar contra la impunidad. Sitúo mis propias experiencias en Guatemala dentro de la más amplia literatura sobre teoría democrática radical y las geografías de la ciudadanía, examinando las posibilidades y problemas asociados con la explotación de la "valía" diferencial de los cuerpos ciudadanos. Adoptando el dicho de que "la presencia es señal de la posibilidad de una política," arguyo que al poner la ciudadanía en la línea de fuego se pueden abrir rutas modestas pero estratégicas de solidaridad transnacional. *Palabras clave: acompañamiento, ciudadanía, Guatemala, racismo, teoría democrática radical.*

Leave or you're dead. . . . If you keep talking to human rights [defenders] about the massacre, things can return to the way they were in the '80s, sons of bitches.

—Death threat received in 2004 by a witness to the 1982 massacre at Plan de Sánchez, Guatemala; Coordinación del Acompañamiento Internacional en Guatemala (CAIG; 2006, 4)

In December 1996, United Nations (UN)-sponsored peace accords brought an official, if largely symbolic, end to thirty-six years of armed conflict between leftist insurgents and the state in Guatemala. Promising "a firm and lasting peace" based on "full respect for human rights" and the "(t)he genuine participation of citizens—both men and women—from all sectors of society" (MINUGUA 2003, 221–23), the accords codified and popularized the type of egalitarian discourse considered fundamental to radical democratic politics (Laclau and Mouffe 1985; Smith 1998). If the accords "opened up political space" (Jonas 2001, 49), however, it is also the case that claims to this space, particularly by citizens and civil society groups seeking justice for wartime atrocities, have been repeatedly and violently repressed. The *Unidad de Protección de Defensoras y Defensores de Derechos Humanos* (UPDDH, Human Rights Defenders Protection Unit) registered 733 attacks against human rights defenders in Guatemala between 2000 and 2005, an average of 122 attacks per year (Samayoa 2006). In 2006, the UPDDH registered 277 attacks; in 2007, 195 (United Nations High Commissioner for Refugees 2008). According to Hina Jilani, Special Representative of the UN Secretary-General, 98 percent impunity for attacks against human rights defenders "makes justice an empty word in Guatemala" (UN 2008, 2).

Focusing on the provision of nonviolent third-party intervention through protective accompaniment, I consider the potential for extending proxy citizenship to human rights defenders by examining how the voluntary movement of privileged foreign bodies into a (post)conflict zone can problematize the foreclosure of domestic political space. By being present and bearing witness, intervening bodies can assume certain security and surveillance functions from the state and, in so doing, can actualize the fundamental civil, political, and social rights of those who have been banned—or, more properly, abandoned—by sovereign power (Agamben 1998). Having conducted fieldwork in Guatemala since 2006, I have received numerous requests to provide informal protective accompaniment, leading me to question whether my short-term, practical worth to racially, culturally, economically, and geographically distinct others might derive more from the color of my skin and the coat of arms on my passport than from any research I might publish. Given that Guatemala has a long and relatively well-documented history of nonviolent third-party intervention, I am able to place my own experiences and concerns within a broader analytical framework. The synergies between protective accompaniment and radical democratic theory are of particular interest, insomuch as the latter is concerned with a concept of citizenship rights that politicizes suprainstitutional social relations and everyday forms of power (Rasmussen and Brown 2003; see also Laclau and Mouffe 1985).

Defining the Terms of Engagement

Nonviolent third-party intervention encompasses four, often corollary, strategies: protective accompaniment, observing and monitoring, interposition, and presence (Boothe and Smithey 2007). Although accompaniment volunteers might self-identify as "bodyguards" (Coy 2001, 577), this should not be taken to mean that the sole purpose and outcome of protective accompaniment is the preservation of "bare life" (Agamben 1998).[1] Bodies, both privileged and bare, are inscribed with powerful political meanings (Perera 2006, 638). The interposition of privileged foreign bodies in a highly volatile, (post)conflict landscape is intended to discharge a communicative process that sends a "message to the aggressor that his or her actions will have certain costs" and a message to the oppressed that they are accompanied in both "a personal and emotional sense" (Mata 2006, 28). Intervention, thus understood, is as much performative and transformative as it is protective or preservative. Protective accompaniment is intended to create "a broader political space for those accompanied, permitting individuals and groups to exercise their rights" (Mata 2006, 58; see also Coy 2001; Boothe and Smithey 2007, 47). It is important to understand the accompaniment role as facilitative: "We are most certainly not the main players in the struggle for justice," says a former protective accompaniment volunteer who recently completed an assignment in Guatemala. "In solidarity, we accompany the struggle of the witnesses" (Boido 2007; see also Anderson 2003).

Historically, protective accompaniment has relied on volunteers being "visibly foreign—and, most often, visibly white" (Boothe and Smithey 2007, 47), the premise being that "Westerners" are less likely to be attacked "because of the reaction that such an action would provoke on the part of the international community" (Boothe and Smithey 2007, 47; see also Coy 1997a, 2001; Hunter and Lakey 2003; Mata 2006). International accompaniment "may not have racism at its core," but it nevertheless "engage(s) the preferential dynamics of racism, and it flirts with colonialism" (Coy 1997b, 244, cited in Boothe and Smithey 2007, 47). To speak of "privileged foreign bodies," therefore, is to speak of the differential "worth" of citizen bodies, (re)produced through centuries of racism and classism and evidenced in the privileges that accrue to accompaniment volunteers in situ. Media and public thresholds in rights-based intervention are informed by a scalar logic of power so that the attention granted to foreign nationals by the media, embassies, and consulates abroad can exceed the level of care these same citizens might reasonably expect to receive within their own national borders:[2] "What can be done to a *local person* without the world hearing about it (media threshold) or doing something about it (public threshold) is more than that for an international. The same is true at an organisational level" (Abu-Zahra 2005, 45, italics added).

I recognize criticisms associated with premising intervention on the assumption that "'we' and 'they' are already distinct" (Edkins 2003, 255); however, I suggest that protective accompaniment—to the extent it takes up the challenge to construct a "we" based on the principle of democratic equivalence (Mouffe [1993] 2005)—can strategically, if incrementally, contest institutional and geopolitical boundaries and exclusions. I employ the concept of "privilege" on the understanding that, "although third party intervention seeks to shift power from global elites to local activists" it does so only by virtue of accompaniment volunteers' hierarchical placement "in relation to sources of institutional power" (Boothe and Smithey 2007, 47; see also Levitt 1999). Recognizing that "the good helper role is part of the imperialism that we carry within" (Koopman 2008, 300), this approach remains committed to critical solidarity on the grounds that "[t]here can be no pure opposition to power, only a recrafting of its terms from resources invariably impure" (Butler 1993, cited in Koopman 2008, 299).

The scope of this article does not permit full theoretical and ethical engagement with the ways in which notions of privileged and bare life intersect with issues of racism, classism, and (neo)colonialism. I offer, therefore, a modest contribution to the small but growing literature on geographies of citizenship and transnational solidarity (e.g., Ong 1999, 2006; Barnett and Scott 2007). Against the traditional "container model" of citizenship, which restricts citizenship claims to national borders, "flexible citizenship" (Ong 1999) considers "transnational solidarity premised on shared interests in issue-specific grievances against geographically dispersed objects of contention" (Barnett and Scott 2007, 307).

I use the term *proxy citizenship* to refer to the potential for limited rights transfer between citizen bodies in the exercise of protective accompaniment. Proxy citizenship so defined is insurgent rather than institutional, distinct from "shadow state citizenship" (Lake and Newman 2002) and arguably distanced from ancillary critiques of "democracy by proxy" (Hudock 1999). Whereas shadow state citizenship emphasizes the role of nongovernmental organizations (NGOs) in securing social rights for those abandoned by the state, proxy citizenship hinges on the interposition of *corpus*, an individual (subinstitutional) body through which claims to basic civil, political, and social rights can be actualized. Proxy citizenship, so defined, is based on the notion that "(t)oday's citizenship practices have to do with the production of 'presence' by those without power" (Sassen 2003, 58). In the realm of protective accompaniment, presence obtains in two forms: the interposition of privileged foreign bodies with the potential to "influenc[e] the dynamics of the conflict itself" (Hunter and Lakey 2003, 34–38, cited in Boothe and Smithey 2007, 43), and the (re)placement of local human rights defenders in the politico-juridical sphere as rights-bearing citizens empowered to testify against impunity.

The Politics of Habeas Corpus

In his discussion of habeas corpus, Agamben (1998) states: "If it is true that law needs a body in order to be in force, and if one can speak, in this sense, of 'law's desire to have a body,' [then] democracy responds to this desire by compelling law to assume the care of this body" (124). Premised on extreme inequities in the politico-judicial "care" of corpus, protective accompaniment recognizes, and at the same time exploits, the fact that "some lives are barer than others" (Pratt 2005, 1074). In Guatemala, the barest lives—those attacked with impunity, most often and in the greatest number—belong to indigenous Maya, who today account for as much as 60 percent of the population (Lovell and Lutz 1994). Of the more than 200,000 civilians killed or "disappeared" during Guatemala's armed conflict (1960–1996), 83 percent were Maya (CEH 1999). Accused by the state of harboring communist sympathies, Mayan communities faced a battery of wartime assaults, including genocide, forced relocation, rape, and other forms of individual and collective torture. International and domestic truth commissions assembled to investigate the Guatemalan conflict registered 626 massacres and declared the Guatemalan army, police, and paramilitary groups responsible for 90 percent of abuse (REMHI 1998; CEH 1999).

It is no coincidence that Guatemala's two principal accompaniment coordinators, Peace Brigades International (PBI) and the *Coordinación del Acompañamiento Internacional en Guatemala* (CAIG, Coordination of International Accompaniment in Guatemala), recruit volunteers principally (although not exclusively) to escort individuals under threat because of their decision to provide testimony, counsel, or public support for the ongoing prosecution of those responsible for human rights abuses committed during the armed conflict. Figures released in 2007 by Guatemala's Human Rights Defenders Protection Unit reveal that the vast majority of attacks are committed against human rights defenders working in the areas of truth and justice, followed by those working in areas of environmental, development, *campesino*, and labor rights (UPDDH 2007). The emphasis on truth and justice reflects an ongoing struggle for historic memory in Guatemala, where threats and attacks against human rights defenders "can be interpreted as a systematic strategy to guarantee impunity" (CAIG 2006, 2). Scholars have underscored the bid to "internationalize" Guatemala's human rights struggle as a means of putting pressure on the Guatemalan state "to recognize its international commitments through covenant to respect human rights and thereby build a safety shield for political activists" (Blacklock and Macdonald 1998, 136). Protective accompaniment is a key component of this strategy.

The Rise of Protective Accompaniment in Guatemala[3]

PBI entered Guatemala in 1983, at the height of the armed conflict, and encouraged the founding of the *Grupo de Apoyo Mutuo* (GAM), a support network for families of the disappeared, which operated from PBI's Guatemala headquarters during its first three years. By the end of 1984, two of GAM's leaders had been assassinated; one year later, military dictator General Oscar Humberto Mejía Victores threatened to expel PBI if it did not persuade GAM to refrain from "disruptive activity" (Mahony and Eguren 1997, 30). PBI refused Mejía Victores's demand and several accompaniment volunteers were subsequently expelled. PBI itself, however, remained in operation, flying in new volunteers to accompany GAM leaders. Gradually, PBI's accompaniment program was expanded to include other groups requesting assistance, always with the objective of "supporting and protecting the social movement" and "distribut(ing) information about Guatemala to the exterior" (http://www.pbi-guatemala.org, last accessed 30 May 2009).

Following the signing of the peace accords in 1996, requests for protective accompaniment decreased dramatically. In 1999, the same year that Rigoberta Menchú Tum, a K'iche' Maya activist, filed a formal complaint with the Spanish National Court charging high-ranking Guatemalan officials (including Mejía Victores) with acts of terrorism, genocide, and torture carried out during the armed conflict,[4] PBI opted to end its Guatemala mission, acting on the belief that "political space for human rights organizations" had been secured (http://www.pbi-guatemala.org, last accessed 30 May 2009). But Menchú's legal action, followed by a similar initiative launched in 2001 by the *Asociación para la Justicia y Reconciliación* (AJR, Association for Justice and Reconciliation), sparked a new round of violence and intimidation, leading PBI to reopen its Guatemala file and restore its protective accompaniment program in 2002. Today, PBI provides protective accompaniment to members of more than ten NGOs in Guatemala, including the *Centro para Acción Legal en Derechos Humanos* (CALDH, Center for Human Rights and Legal Action), the *Coordinadora Nacional de Viudas de Guatemala* (CONAVIGUA, National Coordination of Widows of Guatemala), and the *Movimiento de Trabajadores Campesinos* (MTC, Movement of Campesino Workers).

In response to renewed demands for protective accompaniment, CAIG was founded in 2000. A collaboration of ten NGOs from nine First World countries,[5] CAIG has placed approximately 300 foreign volunteers in Guatemala since its inception. Multiparty legal initiatives supported by CAIG accompaniment volunteers include the Menchú case, as well as a case filed in 2002 with the Inter-American Commission on Human Rights (IACHR) by survivors of the 18 July 1982 massacre at Plan de Sánchez, which claimed the lives of more than 250 Achí Maya. The IACHR referred the Plan de Sánchez case to the Inter-American Court, which issued a verdict in 2004 condemning the violence and ordering the state to indemnify survivors and to launch a criminal investigation with the objective of identifying, judging, and sentencing the intellectual authors of the massacre (IACHR 2004). The Court's verdict triggered a spate of death threats and attacks against witnesses and accompaniment volunteers in Plan de Sánchez, as well as human rights lawyers from CALDH who provided legal counsel in the case. From September 2003 to September 2004, CAIG (2006, Appendix 1) documented thirteen threats directly related to the Plan de Sánchez case, including the following:

> Son a bitch. We've been following you since the eighties. Stop fucking with us or we'll beat the shit out of you. (Threat received 22 September 2003 by a member of the CALDH legal team)[6]

> You will soon be dead. For being a witness of the massacre at Plan de Sánchez and for being responsible for sentencing the state of Guatemala, this is your coveted prize: you will soon be assassinated. Happy 18 of July anniversary. (Personally addressed threat to a witness in the Plan de Sánchez case; received 17 July 2004 in a box containing tear gas)

On 1 August 2004, the office of CALDH in the municipality of Rabinal (approximately ten kilometers from Plan de Sánchez) received a half-page list of death threats directed to a member of the AJR of Plan de Sánchez and personnel of CALDH, including the organization's "security," referring to CAIG's international accompaniment volunteers (CAIG 2006, 8).

Being Present and Bearing Witness

The provision of protective accompaniment for human rights defenders and participants in cases such as Plan de Sánchez is more than an attempt to protect "some cadaveric 'bare life'" (Heins 2005, 854). It is an attempt to safeguard a witness and to actualize, in whole or in part, the civil, political, and social rights fundamental to citizenship. At the most basic level, this is made possible because physical accompaniment "reduces the chance of assassination" (Hunter and Lakey 2003, 21), thereby holding open the potential for testimony to be heard and, in so doing, challenging attempts to limit who "properly" belongs in the politico-juridical sphere. Survivors lend their voices to the dead "by proxy" (Agamben 1999, 34); in turn, accompaniment volunteers—"(a)rmed with cameras, cell phones, notebooks, and their foreign citizenship" (Coy 2001, 577)—share in the production and circulation of testimony. Emergency response networks link insurgent and institutional forms of transnational solidarity. Working directly with local human rights defenders, accompaniment volunteers are well placed to draft detailed reports for rapid dissemination through globally networked groups like Amnesty International (Smith, Pagnucco, and Romeril 1994; Anderson 2003).

If we accept Agamben's (1998) argument that the decisive conflict of our time is between definitions of the biological and the political—between ascriptions of that which is common to all and that which is the realm of human potential—then the communicative process discharged through intervention is critical. The "ethical force" of witnessing "lies in resisting biopolitical efforts to separate those abandoned in bare life from speaking, fully human beings" (Pratt 2005, 1072; see also Sylvester 2006). Accounts of an emergent solidarity between accompaniers, the accompanied, and communities-at-large in Guatemala are telling.

> They want to break our momentum of providing testimony of what we saw and what we lived. . . . If I die, I die. But I have to think of my children. I feel demoralized and nervous. I can only hope that you don't abandon me; that you don't leave me alone. (Statement from a witness to the Plan de Sánchez massacre in response to the wave of threats sparked by the IACHR case: CAIG 2006, 4)

> During my training, one of the coordinators shared what someone she had been accompanying told her, which struck me for some reason. "If they really want to kill me, they are going to do it whether you are here or not. But the fact that you are here means that if they kill me the whole world will know what I died for." (Accompaniment report from a CAIG volunteer in Guatemala, Dwyer 2006)

Differentiated Citizenship as a Development Problematic

Geographers have suggested that "(w)hat we have in humanitarian spaces of exception is an intervention from beyond, where the international community takes the role of the state away from it, while preserving the localization" (Elden 2006, 483). This idea is most fully developed with respect to large-scale, multiparty interventions, such as those undertaken in Haiti, Bosnia, and Kosovo. There is reason to consider insurgent, in addition to institutional, forms of intervention, however, especially given that the latter, when tied to aid-for-peace programs, have been criticized for "discouraging grassroots political activity and claiming to be apolitical, while generating political outcomes" (Abu-Zahra 2005, 1).

Scholars wary of the ways in which intervention can buttress power suggest that "as the disarticulation of subjects from the law becomes manifest in the suspension of citizen rights by governments, activists are able to appropriate the vacated space. But they often fail to consider the state of exception as the basis of their power in the first place" (Trivedi 2005, 14; see also Agamben 1998). While recognizing as problematic a lack of reflexivity in intervention, I want to suggest that accompaniment volunteers are not simply "appropriating the vacated space" of the state. Rather, accompaniment can actualize citizen rights within domestic political space to create not only short-term benefits for human rights defenders but also long-term benefits for the institutionalization of democracy (Levitt 1999). The task of accompaniment volunteers is not to determine the behavior of local human rights defenders but to "expand the range of choices available to them" (Coy 1997a, 99).

My own experience with protective accompaniment in Guatemala is never far from mind, not least because I remain uneasy about having walked away from the first accompaniment request I received. It was the summer of 2006 and I was visiting a community of Q'eqchi' Maya in the company of an academic delegation from the University of Northern British Columbia. Forcibly and violently evicted from the *finca* (large farm or plantation) where they had lived and worked for generations, the community had set up makeshift homes along the roadside as a form of protest for wages and land reparations long overdue. Coffee prices were down; tensions were up. The *campesinos* had been attacked on several occasions by private security forces in the employ of the *finca* owner, as well as by military and police forces (Unión Verapacense de Organizaciones Campesinas 2006). One by one, community members stepped forward to recount their ordeals, asking us to document their testimony: a young boy with bullet wounds that ran the length of his calf; an elderly woman with a machete wound to her head; a widow whose 75-year-old husband had been shot in the chest and then bludgeoned to death with a rock.

As one of the few Spanish speakers in the delegation, I helped relay the words of the Q'eqchi'-to-Spanish translator to the English-speaking members of our group. I suspect this is the reason I was singled out by a community elder, who requested, through the Q'eqchi' interpreter, that I stay with the community. The request had nothing to do with me per se; any (white, foreign) body in the community's midst could equally be expected to decrease the risk of another attack. In retrospect, I have come to regard that moment as perhaps the fullest expression of my citizenship privilege: I was inscribed, drawing on the example of the Greek city-state, in a class of equals "who in principle would be fully exchangeable with one another" (Painter and Philo 1995, 110). I chose not to stay with the community for reasons that take me beyond the scope of this article. I have, however, provided informal,[7] short-term accompaniment to human rights defenders during subsequent periods of fieldwork in Guatemala, always with the knowledge that, in the words of a Guatemalan activist and friend, "there are no guarantees."

Conclusion

Radical democratic theory places citizenship at the threshold of political theory and geography, offering "a conception of democracy as a way of life, a continual commitment not to a community or state but to the political conceived as a constant challenge to the limits of politics" (Rasmussen and Brown 2003, 175). Using the term *proxy citizenship* to refer to the potential for limited rights transfer between citizen bodies, privileged and bare, I have considered the prospects for actualizing citizenship entitlements through protective accompaniment. I have argued that accompaniment volunteers can influence the dynamics of conflict by being present and bearing witness, thereby supporting the (re)placement of local human rights defenders in the politico-juridical sphere. Much research remains to be done in this area, not only to untangle the messiness of adjudicating who is deemed worthiest of protection in the face of accompaniment demands that far outstrip volunteer supply (Stoll 1994; see also Mahony and Eguren 1997) but also, and critically, to understand how questions of race and class are (re)negotiated, organizationally and interpersonally, as the differential "worth" of citizen bodies is exploited. It is argued that presence "signals the possibility of a politics" but that "(w)hat this politics will be will depend on the specific projects and practices of various communities" (Sassen 2003, 62; see also Mahony 2007). If this claim is true, then putting citizenship in the line of fire might open modest but strategic pathways in transnational solidarity.

Acknowledgments

I would like to thank the human rights defenders and accompaniment volunteers who contributed to this article by sharing their experiences, both in person and through the documentation and public dissemination of their work. My thanks are also extended to Audrey Kobayashi, W. George Lovell, Joyce Davidson, Mick Smith, and three anonymous reviewers, all of whom helped me to refine the argument presented here; to Catherine Nolin, who allowed me to join her 2006 Guatemala delegation; and to my family for their ongoing support. Funding from the Social Sciences and Humanities Research Council of Canada and Canadian taxpayers is gratefully acknowledged.

Notes

1. Following Agamben (1998), I understand "bare life" to manifest in the indistinction between biological life (common to all living beings) and political life (the realm of human potentiality). Bare life is the condition of *homo sacer*, or sacred man, a juridical category designating the individual who, deemed unworthy of protection by the state, may be killed by anyone with impunity.
2. I thank an anonymous reviewer for this observation.
3. For a review of protective accompaniment in the context of the return process for Guatemalan refugees exiled during the armed conflict, see Levitt (1999).

4. As presented in her book, *I, Rigoberta Menchú*, Menchú's (1984) testimony has drawn criticism from Stoll (1999), whose fieldwork suggests that although the violence described by Menchú did occur, it occurred, in many cases, to others and not to Menchú herself.
5. CAIG's foreign volunteers are recruited through NGOs based in Austria (*Acompañamiento de Austria*), Canada (Maritimes-Guatemala Breaking the Silence Network and *Projet Accompagnement Québec-Guatemala*), Denmark (*Mellemamerika Komiteen*), France (*Collectif Guatemala*), Germany (*Cadena para un Retorno Acompañado*), Great Britain (Guatemala Solidarity Network), Sweden (*Movimiento Sueco por la Reconciliación*), Switzerland (Peace Watch Switzerland), and the United States (Network in Solidarity with the People of Guatemala).
6. All Spanish-to-English translations are mine.
7. I use the term *informal* to denote accompaniment offered in response to a direct, personal request (i.e., not coordinated through PBI or CAIG). Accompaniment by *sueltos* (unaffiliated volunteers) has been criticized by some scholars (Levitt 1999).

References

Abu-Zahra, N. 2005. No advocacy, no protection, no "politics": Why aid-for-peace does not bring peace. *Borderlands* 4 (1): 1–47.

Agamben, G. 1998. *Homo sacer: Sovereign power and bare life*. Trans. D. Heller-Roazen. Stanford, CA: Stanford University Press.

———. 1999. *Remnants of Auschwitz: The witness and the archive*. Trans. D. Heller-Roazen. New York: Zone Books.

Anderson, K. 2003. *Weaving relationships: Canada-Guatemala solidarity*. Waterloo, ON, Canada: Canadian Corporation for Studies in Religion.

Barnett, C., and D. Scott. 2007. The reach of citizenship: Locating the politics of industrial air pollution in Durban and beyond. *Urban Forum* 18 (4): 289–309.

Blacklock, C., and L. Macdonald. 1998. Human rights and citizenship in Guatemala and Mexico: From "strategic" to "new" universalism? *Social Politics: International Studies in Gender, State & Society* 5 (2): 132–57.

Boido, O. 2007. Accompaniment report, Maritimes-Guatemala Breaking the Silence Network. http://breakingthesilencenet.blogspot.com/2008/06/human-right-accompaniment-olimbia-boida.html (last accessed 1 March 2009).

Boothe, I., and L. A. Smithey. 2007. Privilege, empowerment, and nonviolent intervention. *Peace & Change* 32 (1): 39–61.

Butler, J. 1993. Extracts from gender as performance: An interview with Judith Butler. Interview by Peter Osborne and Lynne Segal, London. *Radical Philosophy* 67 (Summer 1994). http://www.theory.org.uk/but-int1.htm (last accessed 5 February 2009).

Commission for Historical Clarification. 1999. Memory of silence (Tz'inil na'tab'al). Report of the Commission for Historical Clarification. http://shr.aaas.org/guatemala/ceh/report/english/toc.html (last accessed 10 January 2009).

Coordinación del Acompañamiento Internacional en Guatemala. 2006. Informe de observación. http://gsn.civiblog.org/_attachments/1985300/CAIG%20Informe%20Obs%20impunidad_2006.05.15_Final.pdf (last accessed 31 May 2009).

Coy, P. G. 1997a. Cooperative accompaniment and Peace Brigades International in Sri Lanka. In *Transnational social movements and global politics: Solidarity beyond the state*, ed. J. Smith, C. Chatfield, and R. Pagnucco, 81–100. Syracuse, NY: Syracuse University Press.

———. 1997b. Protecting human rights: The dynamics of international nonviolent accompaniment by Peace Brigades International in Sri Lanka. PhD dissertation, Syracuse University, New York.

———. 2001. Shared risks and research dilemmas on a Peace Brigades International team in Sri Lanka. *Journal of Contemporary Ethnography* 30 (5): 575–606.

Dwyer, E. 2006. Accompaniment report, Maritimes-Guatemala Breaking the Silence Network. http://breakingthesilencenet.blogspot.com/2008/06/human-rights-accompaniment-report-2006.html (last accessed 2 February 2009).

Edkins, J. 2003. Humanitarianism, humanity, human. *Journal of Human Rights* 2 (2): 253–58.

Elden, S. 2006. Review article: Spaces of humanitarian exception. *Geografiska Annaler: Series B, Human Geography* 88 (4): 477–85.

Heins, V. 2005. Giorgio Agamben and the current state of affairs in humanitarian law and human rights policy. *German Law Journal* 5:845–60. http://www.germanlawjournal.com/article.php?id=598 (last accessed 3 April 2007).

Hudock, A. 1999. *NGOs and civil society: Democracy by proxy?* Cambridge, MA: Polity Press.

Hunter, D., and G. Lakey. 2003. *Opening space for democracy: Third-party nonviolent intervention, training curriculum*. Philadelphia: Training for Change. http://trainingforchange.org/content/view/111/33/ (last accessed 10 March 2007).

Inter-American Commission on Human Rights (IACHR). 2004. Inter-American Court of Human Rights, Caso Masacre Plan de Sánchez vs. Guatemala. http://gaceta.tc.gob.pe/cidh-caso.shtml?x=2013 (last accessed 2 January 2009).

Jonas, S. 2001. Democratization through peace. In *Globalization on the ground: Postbellum Guatemala*, ed. C. K. Chase-Dunn, S. Jonas, and N. Amaro, 49–82. Lanham, MD: Rowman & Littlefield.

Koopman, S. 2008. Imperialism within: Can the master's tools bring down empire? *ACME: An International E-Journal for Critical Geographies* 7 (2): 283–307.

Laclau, E., and C. Mouffe. 1985. *Hegemony and socialist strategy: Towards a radical democratic politics*. London: Verso.

Lake, R. W., and K. Newman. 2002. Differential citizenship in the shadow state. *GeoJournal* 58:109–20.

Levitt, B. 1999. Theorizing accompaniment. In *Journeys of fear: Refugee return and national transformation in Guatemala*, ed. L. L. North and A. B. Simmons, 237–54. Montreal and Kingston, ON, Canada: McGill-Queen's University Press.

Lovell, W. G., and C. H. Lutz. 1994. Conquest and population: Maya demography in historical perspective. *Latin American Research Review* 29 (2): 133–40.

Mahony, L. 2007. Protection: A non-governmental organisation experience, Peace Brigades International. In *The human rights field operation*, ed. M. O'Flaherty, 243–64. Aldershot, UK: Ashgate.

Mahony, L., and L. E. Eguren. 1997. International accompaniment for the protection of human rights: Scenarios, objectives, and strategies. Institute for Conflict Analysis and Resolution, Working Paper No. 11. http://icar.gmu.edu/wp_11_ mahonyeguren.pdf (last accessed 30 April 2007).

Mata, J. 2006. International accompaniment in violent scenarios: A performative reading of Peace Brigades International in Colombia. Master's thesis, Centre for Peace Studies, University of Tromsø, Norway. http://www.ub.uit.no/theses/available/etd-06072006–132543/unrestricted/oppgave.pdf (last accessed 27 March 2007).

Menchú, R. 1984. *I, Ribogberta Menchú: An Indian woman in Guatemala*. Trans. A. Wright, ed. E. Burgos-Debray. London: Verso.

Misión de Verificación de las Naciones Unidas en Guatemala. 2003. *Acuerdos de Paz*. Guatemala City: MINUGUA.

Mouffe, C. [1993] 2005. *The return of the political*. London: Verso.

Ong, A. 1999. *Flexible citizenship*. Durham, NC: Duke University Press.

———. 2006. Mutations in citizenship. *Theory, Culture & Society* 23:499–505.

Painter, J. and C. Philo. 1995. Spaces of citizenship: An introduction. *Political Geography* 14 (2): 107–20.

Perera, S. 2006. "They give evidence": Bodies, borders and the disappeared. *Social Identities* 12 (6): 637–56.

Pratt, G. 2005. Abandoned women and spaces of the exception. *Antipode* 37 (5): 1052–78.

Rasmussen, C., and M. Brown. 2003. Radical democratic citizenship: Amidst political theory and geography. In *Handbook of citizenship studies*, ed. E. Isin and B. Turner, 175–88. London: Sage.

Recuperación de la Memoria Histórica. 1998. *Guatemala: Nunca Más. Informe del Proyecto Interdiocesano de Recuperación de la Memoria Histórica [Guatemala: Never again. Recovery of Historic Memory Project]*. Guatemala City: Oficina de Derechos Humanos del Arzobispado de Guatemala. http://www.odhag.org.gt/03publicns.htm (last accessed 10 January 2009).

Samayoa, C. V. 2006. *Front line Guatemala: Attacks against human rights defenders, 2000–2005*. Dublin, Ireland: Front Line and the Human Rights Defenders Protection Unit of the National Movement for Human Rights. http://www.frontlinedefenders.org/files/en/Front%20Line%20Guatemala%20Attacks%20against%20human%20rights%20defenders%202000–2005.pdf (last accessed 30 May 2009).

Sassen, S. 2003. The repositioning of citizenship: Emergent subjects and spaces for politics. *CR: The New Centennial Review* 3 (2): 41–66.

Smith, A. M. 1998. *Laclau and Mouffe: The radical democratic imaginary*. London and New York: Routledge.

Smith, J., R. Pagnucco, and W. Romeril. 1994. Transnational social movement organizations in the global political arena. *Voluntas* 5:121–54.

Stoll, D. 1994. Guatemala: Solidarity activists head for trouble. *The Christian Century* 112 (1): 17–21.

———. 1999. *Rigoberta Menchú and the story of all poor Guatemalans*. Boulder, CO: Westview.

Sylvester, C. 2006. Bare life as a development/postcolonial problematic. *The Geographical Journal* 172 (1): 66–77.

Trivedi, N. 2005. Biopolitical convergences: Narmada Bachao Andolan and *Homo sacer*. *Borderlands* 5 (3). http://www.borderlandsejournal.adelaide.edu.au/v015n03_2006/trivedi_biopolitical.htm (last accessed 22 April 2007).

Unidad de Protección de Defensoras y Defensores de Derechos Humanos. 2007. Movimiento nacional por los derechos humanos: Situación de defensores y defensoras de derechos humanos, Informe preliminar, enero-junio 2007 [National movement for human rights: The situation of human rights defenders. Preliminary report, January–June 2007]. http://www.protectionline.org/IMG/pdf/UPD_Informe_Enero_-_Junio_2007.pdf (last accessed 1 February 2009).

Unión Verapacense de Organizaciones Campesinas. 2006. La injusticia en Guatemala provoca otro derramiento de sangre en las comunidades de Cabañas y Moca, municipio de Senahu AV [Injustice in Guatemala provokes more bloodshed in the communities of Cabañas and Moca, Municipality of Senahu AV]. Press release. Distributed by the Guatemalan Human Rights Commission–USA. http://www.nisgua.org/themes_campaigns/land_rights/Reports/Organizations%20Denounce%20Evictions%20021506.pdf (last accessed 30 May 2009).

United Nations. 2008. Comunicado de prensa de la Representante Especial del Secretario General de Naciones Unidas sobre la situación de los defensores de los derechos humanos [Press release of the Special Representative of the Secretary-General of the United Nations concerning the situation of human rights defenders]. http://www.oacnudh.org.gt/documentos/comunicados/20093181811320.20082201213420.Press%20statement%20Guatemala%20Spanish.pdf (last accessed 30 May 2009).

United Nations High Commissioner for Refugees. 2008 (19 June). Observatory for the Protection of Human Rights Defenders, Annual report 2007–Guatemala. http://www.unhcr.org/refworld/docid/4864667ec.html (last accessed 1 March 2009).

Staging Peace Through a Gendered Demonstration: Women in Black in Haifa, Israel

Orna Blumen and Sharon Halevi

Israeli Women in Black was founded twenty years ago to demonstrate for peace and against the occupation of the Palestinian territories. Their political activism involves three spatial processes: *outing* their protest by taking to the streets, *locating* their protest through strategic siting, and *performing* their politics, thus redefining the places they occupy. The Haifa chapter has relocated its demonstrations several times in response to political opposition. Observations and interviews support our analysis of the geographical implications of their unique method of demonstration, their locational choices, and the tension between their femininity and activism played out against the city dynamic. *Key Words: activism, demonstration, gender, Haifa, Israel, Women in Black.*

以色列的黑人妇女于 20 年前就开始和平示威，抗议对巴勒斯坦领土的占领。她们的政治行动涉及到三个空间过程：在街头表示她们的抗议，通过策略性的静坐以定位她们的抗议，通过参与政治事物，从而重新定义她们所占据的地方。海法分会曾经多次改变其示威活动的地点以对应政治反对派。我们的观察和访谈结果支持我们对她们独特的示威方法、她们对地点的选择，她们的女性身份和反城市动态的政治活动之间的紧张关系之研究的地理影响。*关键词：运动，示威，性别，以色列，海法，黑人妇女。*

El grupo israelita Mujeres de Negro se fundó hace veinte años para promover manifestaciones públicas por la paz y contra la ocupación de territorios palestinos. Su activismo político es hecho a través de tres procesos espaciales: *mostrando* sus protestas al tomarse las calles, *localizando* su protesta al sentarse estratégicamente en el piso y *actuando* su política, para así redefinir los lugares que se ocupen. En respuesta a la oposición encontrada, el capítulo de Haifa ha cambiado varias veces el lugar donde hace sus manifestaciones. Las observaciones y entrevistas realizadas apoyan nuestro análisis de las implicaciones geográficas que tiene este inusual método de demostraciones, sus opciones locacionales y las tensiones resultantes al enfrentar su carácter femenino y activismo contra la dinámica de la ciudad. *Palabras clave: activismo, manifestaciones públicas, género, Haifa, Israel, Mujeres de Negro.*

Women in Black (WiB) is a twenty-year-old network founded by Israeli women to protest against the occupation of the Palestinian territories through weekly demonstrations. Activist researchers have identified the women as political actors (Helman and Rapoport 1997; Shadmi 2000; Sasson-Levy and Rapoport 2003; Benski 2005). Yet, the fact that WiB expresses itself almost exclusively through geography has been largely neglected. We consider the urban geography of demonstration and how it is practiced by women, illuminating the gendered spatiality of a common practice of democratic citizenship.

"Where" a demonstration takes place bears a significant communicative value, derived mostly from the quality of places to entwine political and cultural histories and disseminate a message through a long-standing built symbolism (Routledge 1996). The geography of a demonstration is threefold: "outing" protest (e.g., Staeheli and Mitchell 2004), locating it (e.g., Cresswell 1996), and performing it (e.g., Johnston 2007). "Outing" protest through demonstrations is a spatial strategy of communication, whereby situating opposition in the public sphere generates drama. In the democratic world, the increasing number of demonstrations has often routinized this drama, yet demonstrations remain a key means, easily accessible to activists when making themselves known (Garber 1993; Smith et al. 2001).

Locating demonstrations in material public space includes two aspects. Place symbolism refers to the power of the state to naturalize its authority by defining, building, and controlling places. The capital city is one such prime symbolic resource of legitimacy, unity, and order, accommodating compelling symbols, such as houses of Parliament.[1] Demonstrations in such places appropriate their hegemonic reputation, ascribing it with extra, transgressive meaning (Routledge 1996).Yet, areas specially designed for demonstrations make place for opposition, reflecting the hegemonic power to incorporate dissent into the political structure (e.g., Washington's Mall; Mitchell and Staeheli 2005). In ordinary, indistinctive streets, demonstrations use residual spaces that provoke a less challenging impression (Pile 1997). As is often typical of places of power, political symbolism cohabits with historical and cultural symbolism, which highlights their iconic aspect (e.g., Routledge 1996; Missingham 2002). The second, less studied, aspect of locating a demonstration is centrality. Usually its significance is taken for granted and only briefly mentioned (Oliver and Myers 1999; Mitchell and Staeheli 2005). Centrality underscores places as a practical means of access to many publics and divulges their communicative value as symbols of functional vitality.

Demonstrations are more likely to be communicative when including provocative practices such as carnivals, masquerades, and street parties, where gender can play a conspicuous performative part (e.g., Johnston 2007). By disrupting the cultural scheme that closely associates gender with the private–public divide, some women demonstrators are able to intensify the drama embedded in their protest. In Greenham Common (U.K.), women staged their "home away from home": By performing their traditional domestic roles in the open, outside an army base, they dramatized the tension embedded in the gendered dichotomy of peace and war (Cresswell 1996). The women of Plaza de Mayo form another, perhaps the most famous (nondemocratic) example of mothers searching for their missing children. Mobility, as a bodily practice, was imposed on the mothers as standing still in front of the police headquarters was outlawed; the resulting spectacle of mature women marching persistently around the Plaza intensified the drama (Bosco 2001). Talking, calling, shouting, and singing are additional familiar practices (Marston 1990; Staeheli and Mitchell 2004). In another (nondemocratic) case, Santiago de Chile, where such practices were banned, a semistructured banging on kitchen pots and pans was scheduled at lunchtime, and these feminine, domestic, and private practices enabled activists to orchestrate a drama in the city soundscape (Scarpaci and Frazier 1993).

Although some demonstrators have succeeded in recasting renowned places, their power to construct their own symbolic places is limited, and they usually facilitate only temporary interruptions of the public routine. Even when they have transformed the built environment—destroying monuments, digging trenches, building barricades, setting up camps—demonstrators have usually generated short-lasting changes, whereas the long-lasting built symbolism of the state remains in place long after a demonstration is over. This study points to how the communicative disparity between effective performative practices and inadequate locational choices impedes demonstrators' potential to carve a symbolic place in the city's spatial structure of opportunities. The next sections explore the twenty-year geography of weekly WiB demonstrations in Haifa and the application of geographical thinking as they relocated.

Demonstrating Against War

Our seven-year study of this group is drawn from five complementary procedures: close observations of some of the weekly demonstrations; numerous conversations with participants, mostly after the demonstrations; six additional in-depth "narrative" interviews (Jovchelovitch and Bauer 2000) with some of the most veteran activists who willingly shared their memories and memorabilia with us; reading the minutes regarding the last decision to relocate; and finally, verifying our understandings in discussions with some of the interviewees.[2] The respondents' friendly cooperation enhanced our insights.

Today WiB is a worldwide network of women opposing war through demonstrations. WiB was founded by Israeli women who have challenged both political and gender conventions. Conventionally, Israeli men assume the political and military preservation of the nation, and women care for its biological and social reproduction. Yet Israeli-Jewish society continues to nurture a false myth of women's equality (Blumen 2002). Another divide separates the political left, which favors returning the occupied territories in exchange for peace and recognition of Israel's right to exist within fixed and secure boundaries, from the political right, which stresses the historical right to the Promised Land. This divide has severely fractured Israeli society, which as a whole has adopted a conformist stance (in practice rather than in opinion). WiB is but one of a handful of small groups with a Jewish majority that act outside the national consensus (Bar-On 1999). WiB lacks a formal hierarchy and administration, and the participants commit only to the demonstrations they attend. At its height, Israeli WiB staged some thirty weekly demonstrations, whereas today it stages seven.

Protesting in public as women has greatly motivated WiB to express its political views, to "educate, inform and influence public opinion."[3] Haifa's activists affirmed this motive, stressing that their most important goal is "to speak out in public and declare their political vision." As in other democracies, Israeli WiB has not encountered particular formal difficulties in "outing" their political criticism through demonstrations. Moreover, in Haifa (and some other places) the small number of protestors (under fifty) has legally exempted WiB from applying for a permit, freeing it to demonstrate almost anywhere across the metropolitan public space; however, none of the local respondents reported contemplating demonstrations where political power is in situ, either in famous historical-cultural places or outside the central city. Thus, local activists effectively adopted centrality as their main locational principle, intuitively rejecting the options embedded in the multiplicity of representational places. Another locational principle is the need to deal with their opponents' hostility. Both principles, which reflect their group individuality, are the focus of our analysis. First, however, we turn to the geographical implications embedded in their more general performative choices, which are shared with other groups.

Demonstrating the Consequences of War

In its attempts "to make war an unthinkable option" Israeli WiB developed a unique set of performative guidelines for a women-only demonstration, "wearing black, standing in a public place in silent, non-violent vigils at regular times and intervals, carrying placards," which are stop-shaped signs reading "Stop the Occupation."[4] These practices show women *how*, by challenging the traditional gender and private–public distinctions, to represent their involvement in power relations within the local, Israeli-Jewish context. By placing the term *women* in their name and displaying only female bodies, WiB articulates its ability both to exclude men and to reject maternal or parental civic resistance (one shared with men). It asserts women's right to speak out on two traditionally male issues, the military and national defense (Halevi 1999). Overall, WiB asserts its demand for peace by revealing, rather than concealing, the deadly consequences of war; however, by naming its demonstration a vigil, it nevertheless adheres to women's traditional role as mourners. Death is also emphasized by their black dress, which "repeats" the movement's name while distinguishing the demonstrators on the street; being still and silent implies mourning and remembrance, while demanding public respect of the mourners. Silence both counters the popular association of femininity with speech, and reinforces it as the women realign themselves with traditional femininity (silent in public) by holding signs that speak for them. Within Israeli culture that privileges speech, silence also represents WiB as ineffectual political activists.

WiB's demonstrations are timed for early Friday afternoons before the commencement of Jewish Sabbath (at sunset). Although this enables many to attend the demonstrations, it also disregards the representational consequences of this hour of transition from secular to sacred time, after a short workday when people rush home. More important, it is when a devoted Jewish woman should conclude the Sabbath preparations at home, not display her body and views on the street. This is also when mourners are required by Jewish law to cease mourning in honor of the Sabbath; the public display of grief further escalates the challenge posed. At a broader level, Judaism, which instructs the community to comfort mourners, also obliges them to move on and "choose life"; but death in "national" circumstances is exceptional. Inspired by modern European nationalism, Israel commemorates national death and reinscribes national identity through a system of rites. This is a recurrent Israeli experience: National death, which looms over civilians and soldiers, in Israel, the occupied territories, and abroad, structures the daily reality of Israelis as an unbroken chain fusing past, present, and future (Bar-On 1999). WiB claims to be mourning both Israelis and Palestinians, past and future victims of violence;[5] its choice to call for peace and "choose life" by commemorating death as a universal experience that transcends national belonging and is shared with enemies, is often perceived as unpatriotic and disrespectful of their own nation.

These choices dramatize femininity as a political-cultural challenge that, by becoming a means to construct a new meaning to a place, counterchallenges the activists.[6] WiB's restrained demonstrations have often provoked hostility, which has magnified the challenge, intensifying its communicative value. The value of this drama is reinforced by WiB's guidelines to its chapters to stage predictable demonstrations that set meeting time and place for its local network at low cost.[7] Although unnoticed, this time-geography is also greatly advantageous for external communication: Coupled with the "cultural" drama, spatio-temporal predictability ritualizes the place of demonstration and can help WiB to carve its own place of meaning and naturalize itself and its message through the built environment. How the Haifa activists have met this challenge is discussed next.

Placing the Call for Peace

Predictability reinforces WiB's focus on place centrality as a functional, communicative value derived from the local setting. Haifa, the third largest urban area in Israel, lies on Mount Carmel, and its range of influence includes the Northern District. Haifa's population is about a quarter of a million, and the metropolitan area exceeds half a million, nearly 10 percent of the national population. Haifa is one of five Israeli cities where Arab and Jewish citizens live together, and it is known for its comfortable coexistence. Planned mainly by the British as its Middle East strategic oil and military port, Haifa has developed in a unique pattern: The city has two adjacent but separated central business districts (CBDs): Downtown, the major one, and Hadar, the secondary one, which over the last two decades has experienced a decline typical of inner-city areas. Israeli-Arabs, about 10 percent of the population, mostly reside in the older areas around Downtown.[8] The city's residential areas consist of some mountain-top neighborhoods of mostly apartment buildings with a suburban lifestyle. The outer suburbs offer a typical lifestyle of upper-middle-class families. Haifa is also known for its feminist activity (Blumen and Halevi 2005); several members of the most influential group Woman to Woman (*Isha Le'isha*) were among the first WiB activists. The demographic and political profile of the Haifa activists resembles that of other WiB groups: They are mostly in their fifties and older, secular, of European American origin, and highly educated, ranging from anarchists and anti-Zionists to left-wing Zionists. Only some define themselves as feminists (Helman and Rapoport 1997; Benski 2005).[9] In recent years, the number of Jewish women at each demonstration ranges from twenty to forty; occasionally some Arab women and a few men (three to five) join them.[10]

Locating the demand to end the occupation in Israel's public spaces and in Haifa in particular is not inconsequential. Within Israel itself the occupation and its military practices, although only minutes away, are edged out of sight; they occur "elsewhere," in the territories—regions hardly visited by civilians and nonsettlers. The most common representation of the occupation is economic—the sight of many noncitizen Palestinians (often clearly distinct in appearance) seeking employment. On Friday afternoons, when many return to the territories, WiB reintroduces the occupation as a political decision. In a mixed, Jewish-Arab city like Haifa, this representation challenges the hegemony of the Jewish majority and reminds Israeli Arabs of their Palestinian brethren in the territories, evoking their shared past including the defeat in Israel's War of Independence (the *Nakba*—catastrophe in Arabic). *Where* the occupation is challenged affects how the challenge is read, an essential point for understanding WiB in Haifa. Before analyzing its locational history, it is important to remember that the Haifa women relocated because they were persistently harassed by right-wing activists. The lack of requirement of a police permit to demonstrate resulted in police inattention at this busy hour: WiB has often asked for police protection but it was insufficiently provided, so WiB decided to relocate, an effect that is discussed later.

The Locational History of WiB in Haifa

The locational history of WiB in Haifa encompasses three main places. In 1988, the first demonstration was staged at the central bus station, the centrality of which was magnified by the adjacent main train station. It faced the station's main gate, on the sidewalk of the main highway entering the city and its main CBD from the south. The women, however, were less aware of what lay behind them; their backs were pressed against the very high iron mesh fence of the city's soccer stadium where the famous, premier-league city team plays. Here WiB was noticed by the passengers of all the buses entering and leaving the station, as well as metropolitan residents and many visitors. The communicative effect was heightened, as was the demonstrators' gender. Their standing still and silent dramatically contrasted with the busy traffic, rushing commuters, cars, and buses, and the physical activity of soccer (coded as hegemonically masculine). This locational choice shows the early activists' intuitive understanding of the geography of a demonstration as centrality, naturally implementing local knowledge; it also reflects their blindness to the "maleness" embedded in this place, which impeded WiB's attempts to recast it with a new, more feminine and political meaning. In Haifa and elsewhere, WiB's demonstrations attracted right-wing and public hostility, but the Haifa group was the only one who chose to avoid it by relocating.

The new location was inspired by the women's determination to continue to express their political views in public and find a more "comfortable" place, defined as a "central location which is not distant from Isha [the feminist center]" and was easily spotted in "less than ten minutes walk down [the mountain]" at Beit Ha'Kranot, at the heart of Hadar, the secondary CBD. Beit Ha'Kranot is a building complex close to city hall and the district court house (located today elsewhere), situated at a major crossroads that is unavoidable when traveling to this CBD from many metropolitan neighborhoods. Shortly after WiB restaged their demonstrations here, right-wing opponents started to harass the women and later staged their own corresponding demonstration, which featured loud singing, shouting, and waving huge yellow flags, overshadowing WiB's "passivity." After two years the women felt that they had had enough and decided to stage their weekly demonstration in different places. The respondents were unable to retrace their exact six-month path across the city space and could name only three places, all neighborhood centers. Hostility decreased, but so did the number of demonstrators; to maintain their network and despite right-wing hostility, WiB returned to Hadar for three more years. In 1994 after the signing of the Oslo Peace Accords, Israeli WiB felt it had succeeded in altering the political situation and ceased activity.

In September 2000, after growing disappointment with the implementation of the Peace Agreement and the outbreak of the second Palestinian uprising, WiB renewed its activity. The Haifa group returned to Hadar—"our old familiar place"—to face the resumed harassment and parallel demonstrations of right-wing activists. In the spring of 2003, after a sharp decline in the number of women demonstrating, WiB activists met to discuss their options. Many admitted openly that they were worn out by their opponents' viciousness and even some of the most dedicated professed their intention to stop attending. WiB decided to relocate. It is, however, noteworthy that a recently implemented transportation plan diverted much of the public transportation away from the streets of Hadar, downplaying its centrality. Although unmentioned, the consequences of this plan were most likely felt and might even have intuitively prompted the activists' decision to relocate.

Altogether Hadar had served WiB for nearly seven years. Here WiB achieved spatio-temporal predictability, not only as a meeting point for its network but, more important, as a focus of civic attention, a means of communication with the public. Each week, at this hour before the Sabbath serenity descended, WiB appropriated this well-known urban corner. Although it had to be shared with opponents, the fact that WiB was targeted by its opponents' violence was well known and became crucial in identifying this place with WiB; the decision to relocate discarded the high communicative value ascribed to it.

In 2003, WiB again relocated its demonstration to a small, round, traffic island—named by the demonstrators "The Roundabout" ("*Ha'kikar*")—situated in a T-intersection. The horizontal road runs parallel to the mountain foot, where the mountain rises sharply, leading the eye up the steep slope to the Baha'i Shrine's golden dome and its beautiful gardens, Haifa's most famous icon. The vertical wide avenue runs perpendicular to it and gradually descends to the port. The avenue, one of the oldest European-planned streets, cuts through the heart of the "German Colony" (Ben Artzi 1996), marking the old limit of the Arab-Christian neighborhoods, and is considered one of the city's most beautiful streets—a lively leisure and shopping attraction, and a meeting place and residential area for Israeli-Arabs and Jews. The Roundabout is close to several Israeli-Arab (Catholic) schools. Students and the parents of the younger ones, and some tourists, comprise much of audience; hostility by passersby is sporadic and deliberate harassment is less frequent (Benski 2005). Five years after their relocation, veteran activists testify that they experience much greater ease, security, and support here than in the other places: "At last, we found the right place for us." None of the respondents, however, commented about the withdrawal to a less central, although busy, neighborhood-like center and the declining visibility of their demonstration.

This locational history shows that as they moved—from a regional-level visibility of the central bus station, to the metropolitan-level visibility of Hadar, to the neighborhood-like level of The Roundabout—WiB descended the hierarchy of urban centrality, hindering their ability to manipulate the local structure of opportunities. A last noteworthy point is that in Haifa (and in other large cities), WiB also dramatizes its femininity geographically, producing a geographical dissonance. They are, and are perceived to be, from a mostly privileged socioeconomic background; they are confined to the place they stand; their voices are silenced; they do not retaliate to violence; and their overall image exemplifies homogeneity, restraint, and passivity. To many Israelis they represent the feminine facet of the local hegemony, comfortably protected by the suburban order; they seem privileged women who unreasonably, and perhaps even deliberately, chose to display their "annoying" message in the heart of city against its typical hubbub. This confusing impression stems from their being out of the suburban, "domestic" place that they embody, in the city's public space where they clearly do not belong. Thus, WiB's demonstrations (con)fuse not only the gender, military, and political orders but also the spatial one, into its unitary framework of bothersome symbolism.

A Women-Only Demonstration: Geography Unfulfilled?

We conclude by analyzing how Haifa's activists met the challenge of geography, starting with two geographical oversights and moving on to examine four recognized and exercised advantages. Perhaps the activists' greatest oversight is their disregard of place symbolism: They never contemplated demonstrating where power resides and disregarded the communicative value of many known, centrally located, representational places in Haifa. Their recent relocation to a mixed, Jewish-Arab neighborhood—a different kind of symbolic place—is another example. In verification interviews, nearly five years after relocation, only two recognized a possible link between the decreased hostility and the neighborhood's demography:

> We thought of the new place as an opportunity to cut down hostile contacts and violence, we were so, so, so, tired, never thought of the Arabs there. It took me three or four years to realize that we are in their vicinity. But actually we [as a group] never discussed this consequence, certainly not then.

This quotation, revisited also later, is indicative of the activists' blindness to the multiplicity of places, political and otherwise. Although this decision dramatically changed the composition of its audience, and undermined the primacy of the publics it seeks to influence (Israeli-Jews, including right-wingers), none of the respondents referred to this negative communicative consequence. WiB have been also inattentive to the onlookers' views—how they, as demonstrators, are perceived by many ordinary viewers who do not respond violently.

Additional, more complicated, geographical oversight is the possible manipulation of a place meaning so that it also represents the demonstrators' critical message. Haifa's activists possessed two such dramatic means, their performative bodily practices that challenge the state of affairs and their opponents' aggressive response; each of these means is impressive enough to manipulate place meaning. Moreover, complementing predictability, which enables demonstrators to appropriate a place simply by representing themselves repeatedly at the same place and time, these means yield a valuable synergy. In Haifa, WiB had remained faithful to the temporal, but not to the spatial, predictability of its demonstrations. As it had reexercised relocation, it abandoned the option to spatialize its challenging practices, as a means to increase the drama it generated and to carve a place of its own. Instead, WiB absorbed itself into the place most likely to agree with its message (see Routledge 1996; Taylor 1999). A rare example of this last oversight is found in the minutes of the most recent relocation, to The Roundabout. Only one woman understood intuitively that despite media inattention, the weekly drama associating WiB with this place was a valuable asset. Her advice not to give up this place and cling to it, and suggestions for how to ease the burden of hostility and even increase WiB's visibility, were dismissed after a quick, impatient comment by another woman: "What's the point in staying there? Do we have a territory there to guard?" WiB's unanimous decision was to relocate.

Altogether, these oversights indicate that the respondents see the city as a neutral symbolic space. In other words, the respondents seemingly rejected the communicative opportunity, practiced successfully elsewhere, to interweave their criticism with place symbolism (e.g., Routledge 1996; Missingham 2002). For geographers, this spatio-political blindness by educated and experienced political activists is disappointing; yet, our lengthy conversations over a long period of time had not brought to light more than a rudimentary understanding of the functional advantages of geography.

Indeed, the respondents recognized four clear-cut geographical advantages. As elsewhere, Haifa's activists clearly identified the material public space as an important component of the public sphere and chose to place their criticism in it. A veteran activist explained the last locational choice:

> As before [previous relocations], we looked for a public place. Any place is good as long as it is public. Well, not *any* place but public also in the sense of many people, passersby. I see no difference between the [three] places, they are all public and we [WiB] must stay public or else we disappear.

Falling back on the binary of the private–public spheres (but not space), the interviewees placed great emphasis on outing their protest to keep it public. This quotation also reveals centrality as the second geographical advantage: The city space is recognized as a means of direct access to the public; however, the locational history traces a path of descending centrality and declining efficiency that terminated in the least central place and the respondents seemed oblivious to this facet of their history. Third, the power of geography is most clearly recognized through the past experience of a series of weekly demonstrations in different places. The sharp decline in the number of participants compelled the activists to cling to one place to facilitate the local network. Again, the shrinking, or at least varied, centrality typical of that period was ignored.

Relocation is the fourth, most appreciated, geographical advantage. Persistently harassed, the Haifa women have come to understand geography as a structure of opportunities to maintain their safety, evading their violent opponents. Indeed, research on women's fear of crime emphasizes that the majority of women learn that their observed femininity may put them in danger and that many, even close and familiar, public spaces are not for them. Aware of their vulnerability, women reflexively interpret signs of disorder as precursors of violence and normally adhere to practices that curtail their independence in public spaces, most commonly avoiding dangerous-looking places (Valentine 1989). Although this is not a classic case of fear of crime, it certainly exhibits some inevitable similarities with respect to being targeted by violence: Busy streets were swiftly abandoned by rushing people, aggressive opponents using verbal abuse threatened the women with physical and sexual brutality, and the police left the women unprotected, all suggesting a growing risk level. More important, Haifa's activists perceived their situation as seriously risky:

> I felt unsafe, restless, and was eager to leave before we even started [the vigil]. Violence was in the air and nobody seemed to take this threat seriously, especially not the police. From one Friday to the next I got more and more alarmed, distressed.

The activist quoted earlier recalled the resultant exhaustion and its effect on the choice of where to relocate as "an opportunity to cut down hostile contacts and violence." The respondents stressed their environmental experience of an easily noticed, very small group of women, who were threatened by many verbally aggressive opponents, and surrounded by "unimpressed bystanders and unresponsive police" (see also Benski 2005). The women activists applied a familiar practice that relieved their anxiety and responded to escalating violence with avoidance, relocating their demonstration elsewhere. It is, however, noteworthy that repracticed relocations also actualize their principle of nonviolent demonstration. Moreover, avoiding violence through relocation corresponds with WiB's political message to end the occupation by withdrawal from warfare in the occupied territories to a more peaceful existence in Israel. This complexity better explains why the women feel most comfortable where they are the least confronted.

Altogether these four advantages demonstrate that the city space is valued as a functional apparatus with a vitality that serves the activists' need to communicate their protest. Although this suggests a limited reading of the city space, it faced the activists with a paradox. As a women-only political group expressing itself almost exclusively through geography, WiB in Haifa faced a functional impasse: Were they to read the city space as a structure for communicative or for secure opportunities? Because the women activists were unable to apply both simultaneously, they resolved this spatial paradox by trading off a place of high communicative value for a highly secure one, thereby prioritizing femininity over politics. Their relocations reconciled their two noticeable, political and gendered, marginal belongings and sustained their appearance as political women. Therefore, a fair, comprehensive analysis should not be limited to the prevailing impact of their relocations. One must appreciate WiB's very long-term steadfastness to its weekly demonstrations in places of high political value and aggression (the Central Bus Station and Hadar); it represents uncountable weekly decisions by individual women to appear in public as activists and exploit the city space as political women in the face of threats to their personal security. This study demonstrates that even in the democratic world, women who wish to express themselves critically and politically need to emerge through a distinct geography and implement additional, feminine and "nonpolitical," considerations. Their right to political manifestation is not easily spatialized, and often the simple fact that they persist demonstrating as women is an achievement.

Acknowledgments

An earlier version was presented at the symposium of the IGU Commission on Gender and Geography at the National Taiwan University, Taipei (November 2007). We thank the participants and organizers, especially Nora Chiang, Lynda Johnston, Janice Monk, and Linda Peak for their valuable insights and encouragement. Tova Benski and Stanley Waterman kindly provided us with their thoughts regarding the complexities of this project, while the WiB activists patiently responded to our questions. Finally, we thank Audrey Kobayashi and the referees for their helpful comments.

Notes

1. Outside the Western liberal world, in Tiananmen Square in Beijing, the royal palace in Kathmandu, Republic Square in Beograd, the Plaza de Armas in Santiago de Chile, and the government house in Bangkok, activists demonstrated in the nation's capital city, facing central government agencies (Scarpaci and Frazier 1993; Routledge 1994; Bosco 2001; Jansen 2001; Missingham 2002).
2. In each of the last seven years we observed at least ten weekly demonstrations in the two last locations. In the verification interviews we refrained from confronting the interviewees with our own analysis and conclusions, focusing the discussions on consistent understanding of the activists' views (Jovchelovitch and Bauer 2000). Activist Hannah Safran declined anonymity, but others did not; maintaining their confidentiality matches interview protocol. Quotations from the second and third steps and from the verification interviews are unnamed; quotations from the minutes are mentioned as such.
3. Quoted from the Israeli and International WiB: http://coalitionofwomen.org/home/english/organizations/women_in_black and http://www.womeninblack.org/index.html (last accessed 7 September 2008).
4. See note 3.
5. Bystanders understand this inclusive message, and some respond by wishing for a death in the demonstrators' families (Helman and Rapoport 1997; Benski 2005).
6. We do not claim that the early activists were fully aware of the multiple meanings enfolded in their practices or that they consciously intended to disrupt so many cultural codes. Many studied the political, not geographical, context (e.g., Helman and Rapoport 1997; Shadmi 2000; Sasson-Levy and Rapoport 2003).
7. See note 3.
8. The Arab citizens of Israel have usually been referred to as Israeli-Arabs; recently some refer to them as Israeli-Palestinians. However, many retain the term Israeli-Arabs and scholars use both (Blumen and Halevi 2005, 529–30).
9. The term *anti-Zionism* refers to persons who oppose the notion of the nation (in this case, Jewish) state.
10. Respondents reported that for several years local activists discouraged men supporters from attending the demonstrations to reduce the level of violence.

References

Bar-On, D. 1999. Israeli society between the culture of death and the culture of life. *Israel Studies* 2:88–112.

Ben Artzi, Y. 1996. *From Germany to the Holy Land: Templer settlement in Palestine.* Jerusalem, Israel: Yad Izhak Ben-Zvi.

Benski, T. 2005. Breaching events and the emotional reactions of the public: Women in Black. In *Emotions and social movements*, ed. H. Flaum and D. King, 57–78. London and New York: Routledge.

Blumen, O. 2002. Women's professional choice in geography: National sentiments and national exclusion. *Women's Studies International Forum* 25:555–71.

Blumen, O., and S. Halevi. 2005. Negotiating national boundaries: Palestinian and Jewish women's studies students in Israel. *Identities: Global Studies in Culture and Power* 12:505–38.

Bosco, F. J. 2001. Place, space, networks, and the sustainability of collective action: The Madres de Plaza de Mayo. *Global Networks* 1:307–29.

Cresswell, T. 1996. *In place out of place: Geography, ideology, and transgression.* Minneapolis: University of Minnesota Press.

Garber, D. 1993. *The mass media and American politics.* Washington, DC: Congressional Quarterly Press.

Halevi, S. 1999. The premier body: Sarah, Netanyahu, Nava Barak, and the discourse of womanhood in Israel. *NWSA Journal* 11:72–87.

Helman, S., and T. Rapoport. 1997. Women in Black: Challenging Israel's gender and socio-political orders. *British Journal of Sociology* 48:681–700.

Jansen, S. 2001. The streets of Beograd: Urban space and protest identities in Serbia. *Political Geography*, 20:35–55.

Johnston, L. 2007. Mobilizing pride/shame: Lesbians, tourism and parades. *Social & Cultural Geography* 8:29–45.

Jovchelovitch, S., and M. Bauer. 2000. Narrative interviewing. In *Qualitative researching with text, image and sound: A practical handbook*, ed. M .W. Bauer and G. Gaskell, 57–74. London: Sage.

Marston, S. 1990. Who are "the people"?: Gender, citizenship, and the making of the American nation. *Environment and Planning D: Society and Space* 8:449–58.

Missingham, B. 2002. The village of the poor confronts the state: A geography of protest in the assembly of the poor. *Urban Studies* 39:647–63.

Mitchell, D., and L. Staeheli. 2005. Permitting protest: Parsing the fine geography of dissent in America. *International Journal of Urban and Regional Research* 29:796–813.

Oliver, P. E., and D. J. Myers. 1999. How events enter the public sphere: Conflict, location, and sponsorship in local newspaper coverage of public events. *The American Journal of Sociology* 105:38–87.

Pile, S. 1997. Opposition, political identities and spaces of resistance. In *Geographies of resistance*, ed. S. Pile and M. Keith, 1–32. London and New York: Routledge.

Routledge, P. 1994. Backstreet, barricades and blackouts: Urban terrains of resistance in Nepal. *Environment and Planning D: Society and Space* 12:559–78.

———. 1996. Critical geopolitics and terrains of resistance. *Political Geography* 15:509–31.

Sasson-Levy, O., and T. Rapoport. 2003. Body, gender, and knowledge in protest movements: The Israeli case. *Gender & Society* 17:379–403.

Scarpaci, J. L., and L. J. Frazier. 1993. State terror: Ideology, protest and the gendering of landscapes. *Progress in Human Geography* 17:1–21.

Shadmi, E. 2000. Between resistance and compliance, feminism and nationalism: Women in Black in Israel. *Women's Studies International Forum* 23:23–34.

Smith, J., J. D. McCarthy, C. McPhail, and A. Boguslaw. 2001. From protest to agenda building: Description bias in media coverage of protest events in Washington, DC. *Social Forces* 79:1397–423.

Staeheli, L., and D. Mitchell. 2004. Spaces on public and private: Locating politics. In *Spaces of democracy: Geographical perspectives on citizenship, participation and representation*, ed. C. Barnett and L. Murray Low, 147–60. London: Sage.

Taylor, P. J. 1999. Places, spaces and Macy's: Place–space tensions in the political geography of modernities. *Progress in Human Geography* 23:7–26.

Valentine, G. 1989. The geography of women's fear. *Area* 21:385–90.

"Foreign Passports Only": Geographies of (Post)Conflict Work in Kabul, Afghanistan

Jennifer Fluri

Geopolitical "peace-building" relies increasingly on intersections of neoliberal economies of war, violent conflict, and corruption. This article addresses U.S.-led international (post)war aid and development through a spatial examination of Kabul, Afghanistan, examining international worker epistemologies of Afghanistan and "post" conflict aid and development to investigate the spaces of privilege and power associated with political influence, (in)security, and economic and spatial inequities (2006–2008). I draw on recent scholarship in critical feminism, geography, and development studies and the work of Giorgio Agamben regarding the sovereign body and state of exception to demonstrate the spatial disparities and resource inequalities between the "international community" defined as the (un)commonwealth and "local" Afghans. I examine the sovereign status of the (un)commonwealth who manage, assist, or financially profit from international aid and development economies through four interrelated themes: economic and spatial exclusion, (in)security, mobility, and cosmopolitan auxiliary economies. *Key Words: Afghanistan, development geography, international workers, sovereignty, state of exception.*

地缘政治的"建设和平"越来越依赖于新自由主义经济战争，暴力冲突和腐败的交集。本文通过对阿富汗喀布尔地区的空间分析，讨论了在该地区以美国为首的国际战争（战后）援助和发展的情况，调查了当地国际工人对阿富汗和冲突"后"援助与发展的看法，以此调查那些具有特别权利和权力的地区，以及伴随着那些地区的政治影响力，(非)安全，经济和空间的不平等性（2006–2008)。本文作者采用了批判女权主义，地理学和发展理论方面的最近学术研究成果，以及乔治阿冈比关于主权机构和异常状态的研究结果，以此证明在"国际社会"所定义的(非）英联邦和"本地"阿富汗人之间的空间不平等和资源不平等。本文作者也考察了（非）英联邦国家的主权地位，它管理，协助，或者从国际援助和发展经济中获得经济利益，具体是通过下述四个相互关联的主题：经济和空间的排斥，（非）安全性，流动性和国际化辅助经济。*关键词：阿富汗，发展地理，国际工人，主权，异常状态。*

La "construcción de paz" geopolítica depende cada vez más de las intersecciones entre economías neoliberales de guerra, el conflicto violento y la corrupción. Este artículo se ocupa de los programas internacionales liderados por EE.UU. para proporcionar ayuda de (post)guerra y desarrollo, mediante un estudio espacial de Kabul, Afganistán, examinando las epistemologías internacionales del trabajador de ese país y los programas de ayuda y desarrollo "post" conflicto, para investigar los espacios de privilegio y poder asociados con influencia política, (in)seguridad y desigualdades económicas y espaciales (2006–2008). Mi trabajo toma en cuenta los recientes desarrollos del feminismo crítico, estudios de geografía y desarrollo, y el trabajo de Giorgio Agamben relacionado con el cuerpo soberano y el estado de excepción, para demostrar las disparidades espaciales y desigualdad de recursos entre la "comunidad internacional," definida como la "(in)comonalidad" [*(un)commonwealth*], y los afganos "locales." Examino el estatus de soberanía de la *(un)commonwealth* que maneja, ayuda o se lucra financieramente de la ayuda internacional y las economías del desarrollo, a través de cuatro temas interrelacionados: exclusión económica y espacial, (in)seguridad, movilidad, y economías auxiliares cosmopolitas. *Palabras clave: Afganistán, geografía del desarrollo, trabajadores internacionales, soberanía, estado de excepción.*

Critical geographies of peace and war increasingly discuss the interlocking spaces of neoliberal economic structures and practices of development, humanitarian aid, geopolitics, and militarism (Hyndman 2000; Power 2003; Roberts, Secor, and Sparke 2003; Katz 2005; Kothari 2005; Laurie and Bondi 2005; Coleman 2007; Sparke 2007; Vandergeest, Idahosa, and Bose 2007). Geographers and other critical researchers also link these macroscale global economic and political imbalances with racial, orientalist, and gendered discourses that influence economic, political, and legal actions (White 2002; Olds, Sidaway, and Sparke 2005; Gregory and Pred 2006; Coleman 2007; Klein 2007). I add to these studies by examining humanitarian aid and development workers in a postwar conflict zone, arguing that peace geographies of development are interlinked with neoliberal economics and imperial geopolitics that are spatially organized and embodied by international aid and development workers. Rather than "developing Afghanistan," this situation results in an extension and reproduction of hierarchical wealth and uneven development. Kabul, the capital city of Afghanistan, is a key site for the U.S.-led international system of neoliberal economics and militarized violence. In spaces such as Afghanistan, "free markets" exist within a formal system of regulation marred by extensive corruption, inconsistency, and insecurity (Johnson and Leslie 2002; Ewans 2005; Rubin 2006; Rashid 2008).

Everyday life operates in this place of inequity and corruption, and both Afghans and internationals participate in and resist this system in various ways. This is not a flawed version of an existing system but rather an outgrowth of proxy war geopolitics and neoliberal free-market economic development, which benefits *citizens of sovereignty* and ensures the continued poverty of Afghans, or *citizens of exception*. The state of exception refers to extralegal practices (such as the suspension of rights) within the context of a legal framework that is temporally bounded and often follows a state of emergency (such as 11 September 2001), and as Agamben (2003, 2) argues, the "state of exception tends increasingly to appear as the dominant paradigm of government in contemporary politics." The sovereign body that acts outside the law claims his or her application of law under the framework of legal exception to the law. This legal exception also corresponds to the sovereign body's legal home-citizenship status while placed within a space of exception. Gregory (2004) identifies locations that exhibit a continual state of legal exception and corporeal violence (i.e., Afghanistan) as *spaces of exception*. There is also an embodied mobility and privilege associated with one's location-based sovereign-citizenship or claim to "legitimate" and legal sovereignty. I define citizens of sovereignty as individuals who may live and work in places marked by a continual state of exception, while retaining benefits associated with their legitimate and "acceptable" sovereign citizenship (confirmed by one's passport or visa). Conversely, citizens of exception exist within the bounded spaces of their citizen-affiliated state, which is defined as rogue or failed by a more powerful and legitimate sovereign. The sovereign state subsequently monitors, occupies, or partially controls the "illegitimate" state. Citizens of sovereignty living within such sites of exception experience political and economic legitimacy and sovereignty not afforded to the "local" citizens of exception.

This article analyzes the sovereign bodies of international workers who manage, assist, or financially profit from international aid and development economies. A thorough examination of the complexities surrounding aid and development in Afghanistan requires more space than permitted in this article. Therefore, I focus on international workers through four interrelated themes: economic and spatial exclusion and exclusivity, spatial marginalization and (in)security, mobility, and cosmopolitan auxiliary economies. I begin with a brief review of the literature, followed by an overview of my research methods, results, summary, and conclusions.

Literature Review

In Afghanistan, the history of continued resistance to invasion is equally marked by a common unwillingness of imperialist nations to see and work with Afghans as equals. Rather, Afghans are identified as "uncivilized or traditional" and therefore force rather than diplomacy becomes the most "suitable" option for engagement (Ewans 2005; Rashid 2008). Current aid and development approaches seek to discipline the Afghan population into their peripheral position within the global neoliberal economic structure through modernization and capital-driven privatization (Atmar 2001; Barakat and Wardell 2002; Goodhand 2002; Johnson and Leslie 2002, 2004; Nawa 2006; Rubin 2006; Suhrke 2007). Modernization efforts to "liberate" Afghan women are also marred by neoliberal approaches and hegemonic feminist frameworks (Aaftaab 2005; Abirafeh 2005; Davis 2005; Zulfacar 2006; Kandiyoti 2007).

The geopolitics of humanitarian aid and development assistance has a long history of mirroring rather than subverting or critiquing neoliberal economics and their corresponding geopolitical frameworks (Baitenmann 1990; Hyndman 2000; McKinnon 2000; Fox 2001; Roberts, Secor, and Sparke 2003; Barnett 2005; Kothari 2005; Belloni 2007; Coleman 2007; Leebaw 2007; Sparke 2007). Critiques of international humanitarian aid and development highlight common themes, such as the increased role of neoliberal economic theory on aid and development praxis that strengthen donor states and further weaken recipient states (Hancock 1989; De Waal 1998; Hyndman 2000; Reiff 2002). These studies each briefly discuss the role of the international worker as an important agent in neoliberal aid and development. The racial and orientalist configuration of these spaces is a key component in shaping the development policy and in many cases a militarized response (and continual presence of foreign troops) as a necessary component of "peace-building" and humanitarianism (Woodward 2001; White 2002; Denike 2008). I focus on the international worker in an effort to "study up" (see Hyndman 2000) and within the organizations and agencies that disseminate development policy in Afghanistan.

Research Methods

This research on international workers in Kabul, Afghanistan, is based on three field site visits (summer 2006, winter 2007, summer 2008). My field methods consisted of qualitative surveys, interviews, and focus groups with a variety of international workers[1] (150 international workers, 55 percent female, 45 percent male[2]). The data include policy reports and materials provided to me by my respondents.

Research participants (RPs) identified the diversity of international workers based on the following criteria: type of work (i.e., humanitarian aid, faith-based aid, development, private investment, or security), reasons for being in the country, level of integration with Afghans, and length of time in Afghanistan (range of one month to six years), more than one's nationality or home location (North America, Europe, or Australia). Similarly my data identified that RPs' specific patterns of thought, behavior, and perceptions of Afghanistan corresponded to these categories. Also, the overwhelming majority of respondents (98 percent) reported a profound lack of training and knowledge about Afghanistan prior to their arrival.[3]

RPs identified themselves as modern/western (and used these terms interchangeably), unquestionably preferable and progressive in direct contrast to the "traditional/conservative" society of Afghanistan within which they were working (also see Kabeer 1994; Davis 2005). English was the primary language of communication among international workers, which is largely attributed to the role of the United States as the largest donor country.

I also conducted informal discussions with and observations of workers in "international" spaces such as restaurants, shopping centers, and hotels that cater to internationals. Due to the complexities of the characteristics of both the Afghan and international workers, I provide the following categorizations of the population of individuals living in Kabul city: one-third-world international workers, two-thirds-world auxiliary economy workers, and two-thirds-world local Afghans (see Table 1). One-third/two-thirds-world terminology was chosen to "represent what Esteve and Prakash [1998] call social minorities and social majorities—categories based on the quality of life led by people in both the North and the South" (Mohanty 2004, 227). This terminology also diverges from ambiguous geographic and ideological dichotomies associated with north–south, east–west, and first–third world terminology (see Mohanty 2004).

Table 1. Overview of workers living in Kabul

	One-third world	Two-thirds world	Two-thirds world
Population	International workers (professional) Expatriate Afghans	Local Afghans	International workers (auxiliary economy)
Employment options	Aid & development organizations and nongovernmental organizations (both faith- and non-faith-based) Embassies Private sector organizations: Private security, contractors, logistics, property management, war entrepreneurs	Property owner Driver Security guard Small business owner Office worker Service sector employee De-miner Domestic laborer Day laborer	Security guard Small business owner Service sector employee Sex worker
	Type of aid/development organization based on level of financial input		
Tier 1	Multimillion- to billion-dollar budget with core and continuous funding		
Tier 2	Multithousand- to million-dollar projects per year without core funding		
Tier 3	Small-scale projects without core funding and with a significant use of international volunteers		

The (Un)Commonwealth

The one-third-world workers in Afghanistan identify themselves as "the international community"; however, the term *community* does not aptly describe the diversity of individuals and their roles in reproducing, resisting, redefining, or reaping financial benefits from existing geopolitical power structures and neoliberal economic systems. I identify this group as the (un)commonwealth to describe the commonalities provided by their sovereign and one-third-world status despite significant differences in home location, occupation, knowledge, income, and so on. Despite a common understanding of what constitutes civil society, order, economic opportunity, and modernity, I included (un) as a tool to indicate that this is not a homogenous or monolithic "community" in either geographic sovereignty or ideology. Thus, (un) is intended to act as a continual reminder of the several fissures to this commonality; however, the diversity of one-third-world internationals and the many who actively resist dominant paradigms of aid and development, the (un)commonwealth retain specific opportunities, mobility, and spatial access associated with their sovereign citizenship and one-third-world status. Not all aid and development workers receive material wealth as part of their work or presence in Afghanistan. Yet, they retain a commonwealth corporeally inscribed and identified by their "legitimate" foreign or sovereign citizenship.

There is a common epistemology (with some notable exceptions) among the (un)commonwealth that highlights modernity, modernization, and neoliberal economic and "democratic" political systems as preferable, progressive, and necessary for Afghanistan's future. Conversely, they also recognize the immediate need for social programs, state-sponsored health care and education, a strong central government, and increases in Afghan government funding and oversight. This understanding, however, remains outside current neoliberal economic, geopolitical, and militarized systems for aid and development in Afghanistan.

In conjunction with these disparate and didactic characterizations for "developing" Afghanistan, the collective economies of the (un)commonwealth establish and reproduce spaces of comfort and convenience for their population that far exceed that of the two-thirds-world Afghans. The flows of capital aid within Kabul have developed temporally limited economies that include various business opportunities, such as logistics organizations, private security companies, service sector jobs, brothels, restaurants, malls, and shops. The individual and collective incomes of this (un)commonwealth also displaced much of the local Afghan population from specific sites in the central capital city due to the internationally induced increases in housing costs. As one RP noted, "The warlords own many of these properties,[4] and they rent to the Peace Lords, who have the economic means to pay" (Sue 2008). The "peace lords" in Afghanistan also benefit financially from the continued conflict and threat of war by way of economic structures of postwar aid and development (also see Japan International Cooperation Agency 2006; Dittman 2007; Issa and Sardar 2007).

These economic benefits are also used to offset some of the restrictions of employment required by many organizations. For example, the majority of RPs have large salaries that are further inflated by danger pay, housing allowances, and leave time. Danger pay exemplifies the hierarchies of global capitalism and sovereign citizenship. The sovereign body, by virtue of its legitimacy, status, and affiliation with certain aid and development organizations, receives both physical and financial insurances to maintain his or her bodily needs and personal safety. Afghan salaries are fractional by comparison, do not include danger pay, and vary widely by organization and type of work. Also, working for an international nongovernmental organization (NGO), government, or private-sector group yields a much higher salary than that of their Afghan counterparts (Dittman 2007). For example, one RP noted, "If the priority is local reconstruction and capacity building why do they pay a local such low rates, $50–100 per month when an international comes in to do the same job and is paid $200 per hour" (Sally 2006).

The quantifiable value placed on the lives and work of the (un)commonwealth with acceptable and legitimate claims to sovereignty reinforces existing global class structures that ensure bodily mobility, services, comfort, and consumption to the citizens of sovereignty in stunning contrast to the citizens of exception. In addition to these economic methods used to entice international workers, spatial divisions, physical barriers, and mobility restrictions insulate many members of the (un)commonwealth from the lives of "everyday" Afghans.

Spatial (In)Security

Personal security for many members of the (un)commonwealth (particularly employees of Tier 1 and some Tier 2 organizations) include fortified compounds that are bound by a perimeter wall, security fences, razor wire, and armed guards. This spatially bounded security provides the protected sovereign bodies inside the compound with limited spatial proximity to the "others" living outside, who are perceived to pose a continual and often undefined threat. Many compounds also employ Afghans for service sector jobs and guards are often outsourced through private security companies, which hire two-thirds-world employees from neighboring countries (Nawa 2006). International workers with strict security requirements have limited mobility; for example, they cannot enter an Afghan home, drive, or walk without a security detail.

The compound life creates a false spatial reality, which supplies both visible security for the individuals within the compound and a narrow conception of the (in)security that lies outside. It is also important to note that security is defined and experienced much differently by internationals living and working directly with Afghan communities both in and outside of the capital city (more common in Tier 2 and 3 organizations). For these individuals and organizations, security is achieved over time by building relationships with local Afghan families and community leaders. The compound dwellers of the (un)commonwealth received the highest scrutiny from their contemporaries living and working outside these security restrictions, as exemplified by the following quotes:

> I don't want to demonize the whole of the international community; however, there is a lot of arrogance. People are separated and segregated. People ride in their own prime white SUV. What can they do for the city? In a perfect world, I wish internationals thought about what they could learn from Afghans. Often they only see this place as a one-way street. (Kathy 2006)

> Internationals are rude and obnoxious. Most don't have a clue because they are stuck behind their compound walls and herded around like sheep and oblivious to the actual needs of the people here. (Jim 2006)

> I think that there is a wide diversity of people working here in Afghanistan and therefore it is difficult to generalize. There are a number of people here to make money. This group generally does not mix with the Afghan population, follows security rules strictly, and may treat the locals with less respect than they deserve. Often cultural sensitivity amongst this group is minimal. There is another group that are more compassionate and committed to the results of their work. This group is more likely to develop relationships with the local population and have increased levels of cultural sensitivity. The fact that security is so tight has a negative impact on the relationship between locals and internationals as it generates a situation of segregation and frequently mistrust of the international community. (Nat 2006)

The spatial divisions between internationals and locals are part of the epistemological processes that manifest into imagined geographies of place and people. RPs who operate within spatially bounded security spaces also identified these structural divisions as problematic and undermining their "goals." Conversely, they simultaneously identified this form of security as a necessary aspect of war-zone development.

Inside the compounds, there is little hint that you are in Afghanistan. The spatial layout of offices, homes, personal comforts, and amenities differs significantly from Afghan homes and government offices. Most two-thirds-world Afghans have "moved" to the outskirts of the city because of inflated rents or live on the mountains that surround the city where the land is free and running water and proper sanitation are scarce.

Most two-thirds-world Afghans living in Kabul have limited access to clean drinking water, electricity, Internet, and mobile phone usage due to the costs of these amenities and services. In contrast, all of my RPs (in Tier 1 and 2 organizations and most Tier 3 organizations) had daily access to hot and cold running water, bottled water for drinking and cooking, flush (Western-style) toilets, a high level of food availability, access to reliable and regular electricity (via generators), daily Internet access (in their homes, offices, or both), at least one (but more likely two or three) mobile phone, and at least one car and driver. The (un)commonwealth also experience technological efficiency and comfortable lifestyles, which occur through "imported" materials that require minimal infrastructural change in Afghanistan, whereas city-generated electricity and water accessibility remain minimal or compromised for local Afghans.

The international bodies are at times bound by security restrictions limiting their mobility within Afghanistan. Conversely, the (un)commonwealth enjoy increased mobility internationally by virtue of their sovereign citizenship, passports, and the economic means to travel. Travel outside Afghanistan provides "spatial escapes" for the (un)commonwealth and is described as essential and of high importance by the majority of RPs. International worker mobility also includes high turnover rates and short-term projects labeled by one of my RPs as "parachute-in and not quite fix development." Short-term assignments (i.e., three months to one year) provide workers a crucial line item on their résumés as part of a longitudinal career trajectory. The willingness to work in a war zone helps to secure a future expectation for more amenable job placement in North America or Europe. For example:

> People come here for six months or one-year contracts, to build their careers in the UN and NGO world. It is a kind of résumé building . . . if you come [here] it looks like you are serious and hard working. There are many people working their way up for that cushy job in Geneva or New York. (Giles 2006)

High salaries are also used to entice international workers at all levels of "experience" and "expertise" to participate in the (un)commonwealth of development. Internationals overwhelmingly believe high salaries and comfortable "first-world-like" living conditions are required because they relinquish that lifestyle to work in Afghanistan. Simultaneously, Kabul has a thriving auxiliary economy that caters to the service and consumptive desires of the (un)commonwealth.

Cosmopolitan Kabul

Local Afghans are bound by economies of desperation attributed to the aftermath of a war-induced shattered infrastructure, high unemployment, low levels of education, an unskilled labor force (or one that is perceived as such), and food and other economic insecurities. These conditions stand in direct contrast to the high salaries of international workers in the (un)commonwealth. The excessive disposable incomes of these workers and their desires for one-third-world lifestyles, entertainment, and consumption have created a quasi-cosmopolitanism in Kabul city. The disposable incomes of the (un)commonwealth developed an auxiliary economy of service sector employees, private security personnel, logistics organizations, and economies of desire, including sex workers, international restaurants, shopping malls, and hotels. The disposable incomes of this (un)commonwealth also provide economic opportunities for two-thirds-world entrepreneurs who come from neighboring countries (such as Pakistan, Iran, Uzbekistan, and Tajikistan) and open businesses to "cash in" on the excess incomes of one-third-world international workers.

The short-term work and transient nature of the (un)commonwealth exemplify sovereign citizenship and elite mobility. Members of the (un)commonwealth's behavior, desires, and personal freedoms are chosen and often determined by the temporality of their assignments and the inability of local governance to enforce the rule of law on internationals who engage in illegal activities.

Kabul city acts as a site of exception from local, national, and international juridical processes and procedures. The low-level salaries of "officials" in all branches of Afghan governance and law enforcement personnel in conjunction with the high cost of living act as a by-product of top-heavy aid and development allocations that both create the auxiliary economies of development's disposable incomes and provide the economic conditions for corruption. The enforcement of law is more often determined by the capital needs of underpaid government employees than by actual disciplinary structures of law and force of law (also see Agamben 1995).

Economically induced corruption helps to increase the perceived need among internationals for private "gun-for-hire" security services. Additionally, aid and development organizations largely hire international contract and logistics companies (rather than local Afghans) to (as stated by several RPs) "get the job done," which subsequently leads such firms to hire private security companies to ensure that projects are completed with minimal casualties. Multinational corporations in this postwar-conflict system receive the primary flow of funds through layers of top-heavy capital aid inputs, which are continually subcontracted to privately run organizations and NGOs. The level of subcontracting leads to a "lack of funds" when the project begins and a subsequent use of inadequate materials and low-paid labor. Additionally, the overwhelming majority of workers in the (un)commonwealth arrive without formal training or knowledge about Afghanistan's culture, geography, and social or political history. Local customs, norms, and laws are also dismissed by many in the (un)commonwealth.

Sovereign Privilege

The international bodies of the (un)commonwealth partake in various forms of compliance with and rejection of Afghan customs, laws, and behavioral expectations. There is a distinct spatial division in the construction and maintenance of international spaces that excludes two-thirds-world Afghans. International restaurants that serve alcohol post signs that read either "Foreign Passports Only" or "We regret we cannot serve alcohol to Afghan Nationals." This is done in an attempt to effectively "allow" internationals to oppose Afghan law, which prohibits the consumption of alcohol. Other places that cater to internationals do not physically prevent Afghans from entering; however, they secure exclusion through the costs of goods or services in these locations, which remain beyond the economic reach of most Afghans. The respondents in this study, when asked about these spaces of exclusion, also identified gendered and racialized reasons for barring Afghans from these spaces, as exemplified by the following quotes.

> Afghans aren't allowed into the restaurants because an Afghan man doesn't know how to act when he sees a woman in a bathing suit. Although, I could say the same thing about most of the Western men after a few drinks. (Paul 2006)

> Even though many expat establishments don't allow Afghans in, you can't help but "corrupt" their culture just by their presence. Some [Afghans] welcome Western influence, but other are contemptuous. When you have a place like [XX], for example, there are women sipping cocktails in their bikinis around a pool. As an expat, I think it's great because sometimes you need that escape. And it's just for that reason that Afghans aren't allowed in. It's almost an affront to their culture. But at the same time, setting up an establishment that keeps the local populations out is as well. Either way, the things that keep people like me sane, are bound to piss off the local community. (HC 2006)

The consumption of alcohol and carnival-like atmosphere of international restaurants and parties typify one aspect of the (un)commonwealth, which is not indulged in by all. The international spaces that exclude Afghans (unless they are escorted and vetted by an international) also include many international offices, businesses, and homes. Many of these places are also identified as "spatial escapes" from Afghan culture, particularly for women who are bound by greater restrictions regarding their mobility, dress, and behavioral expectations. These places exemplify the common assumptions among international workers regarding the positive and necessary aspects of consumptive modernity associated with one-third-world lifestyles as well as their identity and "sanity" while working in Afghanistan. These places also illustrate Kabul as a site of exception.

This site of exception epitomizes an economically and politically bounded space. The (un)commonwealth's actions and leisure activities are often infused with desire or fantasy, afforded by disposable incomes, and accepted through international workers' legitimate sovereignty and mobility. This allows (un)common fluidity and choice regarding one's compliance with or rejection of national or international laws and local custom. The transience of the (un)commonwealth provides a carte blanche approach to personal behaviors. For example:

> You know what we say here? "What happens in Kabul, stays in Kabul." You can drink, smoke, do drugs and of course sleep around. Stuff that you wouldn't do at home, it's okay here because the community here is always on their way elsewhere. (Simon 2006)

Citizens of sovereignty are rarely held accountable for failures, missteps, or mistakes associated with their work in postwar dissonance zones such as Afghanistan. Ordinary Afghan citizens may also act outside the law to secure a livelihood; this behavior, continually cited by aid and development workers, ensures that corruption claims rest squarely on Afghan shoulders. The Afghan citizens of exception exemplify the sacred and profane subjects of neoliberal geopolitics—targeted for underpaid modernity and the collateral bodies of militarized discipline. The site of exception and its bounded spaces of economic opportunism, income disparities, and resource disparities secure and reproduce corruption and conflict.

Conclusions

It is imperative to examine critically the spaces of aid and development at the site of geopolitical violence and neoliberal economic engagement where economies of desperation meet economies of desire. The lack of personal service amenities, health care, economic stability, and global mobility that are characteristic of postwar spaces such as Afghanistan are not experienced or embodied by most internationals who live in manufactured quasi-cosmopolitan environments (also see Hyndman 2000). Spatial removals from the daily inputs necessary to socially reproduce oneself or commute from home to work are therefore not fully understood by many in the (un)commonwealth. The (un)commonwealth, by virtue of their legitimate sovereignty and economic status, experience a flexible mobility to "escape" Afghanistan. In the country, the exclusive spaces of the aid and development elite spatially segregate the "everyday" lives of Afghans.

Spatial proximity, and lack thereof, remains an essential component of displaced development and aid; however, breaking down these physical barriers will not solve the entire epistemological processes that continue to place Afghans into a narrow and orientalist categorization of the two-thirds-world "other." For example, several popular quasi-nonfiction memoirs (such as *The Bookseller of Kabul* and *The Kabul Beauty School*) authored by one-third-world individuals who lived closely with Afghans have also orientalized and misunderstood Afghanistan due to lack of information and training as well as interpreting their personal experiences through a narrow lens of one-third-world modernity and occidental epistemologies.

The embodied experiences and movements of unequal assistance, development, reconstruction, and security reproduce cyclical benefits to many one-third-world individuals and organizations working and living in these temporally limited, postwar conflict, and capital gain zones. These sovereign bodies actively maintain aid and development ideologies that assist the structural networks of sovereign state power, militarism, and enormously unbalanced economies. For example, all respondents were eager to critique the structural problems with international aid and development and privatization; however, most removed themselves as agents of change regarding this structure and conversely identified their work as "necessary" for Afghanistan to "move forward." In these cases, geopolitical dissonance oils the engine of capital accumulation for those profiting from the economies of postwar conflict.

Conversely, the (un)commonwealth also includes fissures in the structures of neoliberal policy-driven development. There are one-third-world internationals collaborating, living, and working with Afghans, outside of the normative development and aid industry and within Afghan culture, language, and social contexts. Several have also left large aid and development organizations to start projects that counter macroscale development and aid philosophies and praxis.

Postwar conflict aid and development are crucial areas of study for geographers, because peace-building through aid and development remains within the bellicose economies and geopolitics of sovereign bodies and sites of exception. Field-based qualitative analyses such as those discussed in this article are necessary to expand critical geographies of "peace-building" and their increasingly intersected reliance on war, violent conflict, corruption, and other informal economies such as narcotics and sex trafficking. As this article argues, the places and actors in peace "building" and postwar aid and development are intricately interconnected and woven into the fabric of war's economic enterprise and opportunities. Deconstructing these geographies of power and inequity requires additional research and attention from critical geographers.

Acknowledgments

Special thanks to Audrey Kobayashi for her editorial assistance and the anonymous reviewers for their comments and suggestions. This research was possible due to a grant provided by the Nelson A. Rockefeller Center at Dartmouth College. I would also like to thank my research participants for the honest and candid information and for generously taking the time to participate in this study.

Notes

1. To ensure the confidentiality of the responses and information provided for this study, pseudonyms are used for each participant and their affiliations are not included.
2. This sample size does not represent an accurate percentage of gender within the international community. The majority of international workers are male, particularly in the private sector such as contract workers, logistics, and military and security personnel. This sample size was based on my own positionality as a female researcher and the willingness of respondents to participate in this study.
3. I define the training as profoundly lacking as respondents answered *none* or *I received a dos/don'ts sheet* to the question "Please describe (length and content of) the training you received prior to and/or after your arrival in Afghanistan." Several respondents did outside reading, which largely consisted of fiction (i.e., *The Kite Runner*).
4. Warlord ownership of property was discussed by many RPs (also see Wily 2003; World Bank 2005).

References

Aaftaab, G. H. 2005. (Re)defining public spaces through developmental education for Afghan women. In *Geographies of Muslim women: Gender, religion and space*, ed. G. W. Falah and C. Nagel, 44–67. New York: Guilford.

Abirafeh, L. 2005. *Lessons from gender-focused international aid and post-conflict Afghanistan . . . learned?* Friedrich-Ebert-Stiftung, Division of International Cooperation Department for Development Policy. Bonn, Germany: Godesberger Allee.

Agamben, G. 1995. *Homo sacer sovereign power and bare life*. Trans. D. Heller-Roazen. Stanford, CA: Stanford University Press.

———. 2003. *State of exception*. Trans. K. Attell. Chicago: The University of Chicago Press.

Atmar, M. H. 2001. Politicisation of humanitarian aid and its consequences for Afghans. *Disasters* 25 (4): 321–30.

Baitenmann, H. 1990. NGOs and the Afghan war: The politicisation of humanitarian aid. *Third World Quarterly* 12 (1): 62–85.

Barakat, S., and G. Wardell. 2002. Exploited by whom? An alternative perspective on humanitarian assistance to Afghanistan. *Third World Quarterly* 23 (5): 909–30.

Barnett, M. 2005. Humanitarianism transformed. *Perspectives on Politics* 3 (4): 723–40.

Belloni, R. 2007. The trouble with humanitarianism. *The Review of International Studies* 33:451–74.

Coleman, L. 2007. The gendered violence of development: Imaginative geographies of exclusion in the imposition of neo-liberal capitalism. *Political Studies* 9:204–19.

Davis, D. K. 2005. A space of her own: Women, work, and desire in an Afghan nomad community. In *Geographies of Muslim women: Gender, religion and space*, ed. G. W. Falah and C. Nagel, 68–90. New York: Guilford.

De Waal, A. 1998. *Famine crimes: Politics & the disaster relief industry in Africa*. Bloomington: Indiana University Press.

Denike, M. 2008. The human rights of others: Sovereignty, legitimacy, and "just cases" for the "war on terror." *Hypatia* 23 (1): 95–121.

Dittman, A. 2007. Recent development in Kabul's Shar-e-Naw and Central Bazaar districts. *ASIEN* 104: 34–43.

Ewans, M. 2005. *Conflict in Afghanistan: Studies in asymmetric warfare*. London and New York: Routledge.

Fox, F. 2001. New humanitarianism: Does it provide a moral banner for the 21st century. *Disasters* 25 (4): 275–89.

Goodhand, J. 2002. Aiding violence or building peace? The role of international aid in Afghanistan. *Third World Quarterly* 23 (5): 931–43.

Gregory, D. 2004. *The colonial present: Afghanistan, Palestine, Iraq*. New York: Wiley-Blackwell.

Gregory, D., and A. Pred, eds. 2006. *Violent geographies: Fear, terror and political violence*. London and New York: Routledge.

Hancock, G. 1989. *Lords of poverty: The power, prestige, and corruption of the international aid business*. New York: Atlantic Monthly Press.

Hyndman, J. 2000. *Managing displacement: Refugees and the politics of humanitarianism*. Minneapolis: University of Minnesota Press.

Issa, C., and M. K. Sardar. 2007. Kabul's urban identity: An overview of the sociopolitical aspects of development. *ASIEN* 104:51–64.

Japan International Cooperation Agency. 2006. *The study on the Kabul metropolitan area urban development in the Islamic Republic of Afghanistan*. Japan International Cooperation Agency Final Report. Tokyo, Japan: RECS International Inc.

Johnson, C., and J. Leslie. 2002. Afghans have their memories: A reflection on the recent experience of assistance in Afghanistan. *Third World Quarterly* 23 (5): 861–74.

———. 2004. *Afghanistan: The mirage of peace*. London: Zed Books.

Kabeer, N. 1994. *Reversed realities: Gender hierarchies in development thought*. London: Verso.

Kandiyoti, D. 2007. Old dilemmas or new challenges? The politics of gender and reconstruction in Afghanistan. *Development and Change* 38 (2): 169–99.

Katz, C. 2005. Partners in crime? Neoliberalism and the production of new political subjectivities. *Antipode*: 37 (3): 623–31.

Klein, N. 2007. *The shock doctrine: The rise of disaster capitalism*. New York: Picador.

Kothari, U. 2005. Authority and expertise: The professionalisation of international development and the ordering of dissent. *Antipode* 37 (3): 425–46.

Laurie, N., and L. Bondi, eds. 2005. *Working the spaces of neoliberalism: Activism, professionalism and incorporation*. London: Blackwell.

Leebaw, B. 2007. The politics of impartial activism: Humanitarianism and human rights. *Perspectives on Politics* 5 (2): 223–39.

McKinnon, K. 2000. Postdevelopment, professionalism, and the politics of participation. *Annals of the Association of American Geographers* 97 (4): 772–85.

Mohanty, C. T. 2004. *Feminism without borders: Decolonizing theory, practicing solidarity*. Durham, NC: Duke University Press.

Nawa, F. 2006. *Afghanistan, Inc.: A CorpWatch investigative report*. Oakland, CA: CorpWatch.

Olds, K., J. D. Sidaway, and M. Sparke. 2005. White death. *Environment and Planning D: Society and Space* 23:475–79.

Power, M. 2003. *Rethinking development geographies*. London and New York: Routledge.

Rashid, A. 2008. *Descent into chaos: The United States and the failure of nation building in Pakistan, Afghanistan, and Central Asia*. New York: Viking.

Reiff, A. 2002. Humanitarianism in crisis. *Foreign Affairs* 81 (6): 111–21.

Roberts, S., A. Secor, and M. Sparke. 2003. Neoliberal geopolitics. *Antipode* 35 (5): 886–97.

Rubin, B. 2006. Afghanistan at dangerous "tipping point." Interview with Bernard Gwertzman, Consulting Editor Council on Foreign Relations. http://www.cfr.org/publication/11620/rubin.html?breadcrumb=%2Fbios%2F115%2Fdr barnett_r_rubin (last accessed 4 August 2008).

Sparke, M. 2007. Everywhere but always somewhere: Critical geographies of the Global South. *The Global South* 1 (1): 117–26.

Suhrke, A. 2007. Reconstruction and modernization: The "post-conflict" project in Afghanistan. *Third World Quarterly* 28 (7): 1291–308.

Vandergeest, P., P. Idahosa, and P. Bose. 2007. *Development's displacements: Ecologies, economies, and cultures at risk*. Vancouver, BC, Canada: UBC Press.

White, S. 2002. Thinking race, thinking development. *Third World Quarterly* 23 (3): 407–19.

Wily, L. A. 2003. *Land rights in crisis: Restoring tenure security in Afghanistan*. Kabul, Afghanistan: Afghanistan Research and Evaluation Unit.

Woodward, S. 2001. Humanitarian war: A new consensus? *Disasters* 25 (4): 331–34.

World Bank. 2005. *Kabul: Urban land in crisis: A policy note*. South Asian Energy and Infrastructure Unit of the World Bank. http://siteresources.worldbank.org/SOUTHASIAEXT/Resources/223546-1150905429722/PolicyNote4.pdf (last accessed 12 December 2008).

Zulfacar, M. 2006. The pendulum of gender politics in Afghanistan. *Central Asian Survey* 25 (1–2): 27–59.

Territorial Tensions: Rainforest Conservation, Postconflict Recovery, and Land Tenure in Liberia

Leif Brottem and Jon Unruh

Since the cessation of civil conflict in 2003, the Liberian government is poised to expand greatly its protected area network to conserve the country's remaining rainforest. Liberia holds within its borders nearly half of the remaining Guinean rainforest, a global biodiversity hotspot. The planning process for this effort has been a central part of rebuilding Liberia's forestry sector, which helped fuel past conflict. The process, known as the Liberia Forestry Initiative, is widely considered to have been exemplary with regard to multistakeholder dialogue and policy making. Because the issue of land tenure remains widely contested, however, the initiative risks seriously aggravating land rights problems, complicating the prospects for a durable peace. This article focuses on the process through which land was zoned for strict protection and how this process is likely to exacerbate land tenure conflict in Liberia's priority conservation areas. Although the international conservation community sees postconflict scenarios as opportunities for promoting conservation initiatives, unresolved land tenure issues make for problematic outcomes, including land disputes and legal disarray. Such problems can result in significant volatility and can make the peace process and recovery much more, not less, difficult. *Key Words: conservation, forests, land tenure, Liberia, postwar.*

自 2003 年停止内战，利比里亚政府准备大幅扩大其森林保护区，以保护该国剩余的热带雨林。利比里亚在其境内拥有近一半的全球所剩余的几内亚热带雨林，是全球生物多样性的热点地区。为此举所做的规划进程已经成为重建利比里亚的林业部门的核心部分，其后果也刺激了过去的冲突。在这个过程中，即利比里亚林业动议案，被普遍认为是多方对话和决策的典范。然而，由于土地所有权的问题仍然存在广泛争议，该动议案增加了土地权利问题恶化的风险，使得持久和平的前景变得复杂。本文侧重于分析这些土地是如何被划为严格保护的区域，以及在利比里亚的优先保护区，这个过程是如何可能加剧土地使用权的冲突。尽管国际环保社会认为冲突过后的情况是促进环保动议的机会，那些尚未解决的土地使用权问题会造成有问题的结果，包括土地纠纷和法律混乱。这些问题都可以产生相当大的波动，可以使得和平进程和恢复过程产生更多，而不是更少的困难。*关键词：保护，森林，*土地使用权，*利比里亚，战后。*

Desde la terminación del conflicto interno en 2003, el gobierno liberiano mantiene su intención de ampliar al máximo su red de áreas protegidas para conservar lo que le queda al país de selva pluvial. Dentro de las frontreras de Liberia se encuentra cerca de la mitad de la restante selva guineana, uno de los puntos álgidos relacionados con la biodiversidad global. La planificación de este esfuerzo ha sido crucial en el proceso de reconstrucción del ámbito forestal de Liberia, dentro del cual se alimentó el pasado conflicto. El proceso, que se conoce como la Iniciativa Forestal de Liberia, se considera por todos como ejemplar en lo que toca al diálogo amplio entre parte interesadas y a la formulación de políticas. En la medida en que la cuestión de la tenencia de la tierra sigue siendo ampliamente debatida, no obstante, aquella iniciativa no deja de ser riesgosa para agravar los problemas de derechos a la tierra, lo cual complica la esperanza de una paz duradera. El presente artículo se concentra en el proceso a través del cual la tierra fue zonificada para darle estricta protección, y sobre cómo este proceso puede exacerbar los conflictos de tenencia de la tierra en las áreas prioritarias de conservación de Liberia. Aunque la comunidad internacional conservacionista mira los escenarios del postconflicto como una oportunidad para promover iniciativas de conservación, las cuestiones de tenencia no resueltos pueden conducir a resultados problemáticos, incluso a disputas por tierras y líos legales. Tales problemas pueden dar lugar a una significativa volatilidad y convertir el proceso de paz y recuperación en algo más difícil de lo esperado. *Palabras clave: conservación, selvas, tenencia de la tierra, Liberia, posconflicto.*

When it comes to the environment, Liberia is a shining example of conservation's role in stability in West Africa.

—Conservation International Web site[1]

The Liberian Forest Initiative (LFI) is a major component of the reconstruction effort undertaken since the end of Liberia's civil conflict in 2003. The postconflict period has provided a critical window of opportunity for conservationists to develop a transparent and equitable management plan for Liberia's critically important forest resources. Although the country's rainforests were ravaged by years of indiscriminate commercial logging, the country retained 42 percent of West Africa's remaining intact lowland rainforest, which is globally valued for its biodiversity (Republic of Liberia Forestry Development Authority 2006). At the end of the war, much of Liberia's forested areas lacked roads and people, many of whom had been displaced by prolonged violence in rural areas, making these forests significantly attractive for conservation.

When the peace accord was signed, international donor agencies, nongovernmental organizations (NGOs), and governments (e.g., Conservation International, United Nations [UN], United States Agency for International Development, World Bank) immediately began a stabilization and reconstruction effort in Liberia. Following nearly two decades of violence and instability, ongoing reconstruction and reform has taken place in every sector of Liberian society, from physical infrastructure to governance (Sawyer 2005). The end of the war was a historic moment when Liberia would be redefined and rebuilt as rapidly as possible, and international agendas regarding peace, economic opportunity, and conservation would be brought to bear.

Land tenure emerges as a central part of the challenge. As a primary factor contributing to the war (Richards 2005; Unruh 2009), land tenure in Liberia has been the basis of violently antagonistic social relations between the state and rural inhabitants, between several ethnic groups, and between generations within chiefdoms. The current president of Liberia has "expressed fear that the issue of land reform, if not swiftly redressed by the government and its international partners, could crop up into another war in the country" (Daygbor 2007). In particular she has noted in recent speeches "that land reform is needed now to contain future troubles," and that "land disputes are a major hurdle in the wake of attaining genuine peace in the country." It was recently estimated that there are several hundred thousand rural youth in Liberia and neighboring Sierra Leone currently vulnerable to militia recruitment, and that "[y]oung people without secure tenancy rights will continue to float in the countryside without stable social commitments, and thus remain vulnerable to both chiefs and militia recruiters" (Richards 2005, 587).

The LFI is a major land reform mechanism encompassing virtually every aspect of the forestry sector through a framework known as the "three Cs": commercial timber, community forestry, and nature conservation (Republic of Liberia Forestry Development Authority [FDA] 2006). The LFI has successfully coordinated the forest sector interventions of an array of Liberian and external governments and agencies. It also included a consultation process with local communities, nongovernmental authorities, and other members of civil society. This article concerns the LFI-produced land use plan, which has delimited the entire country into zones designated to meet the conservation, community, and commercial objectives defined through the LFI process. We focus specifically on the planned creation of protected areas that will comprise the core of Liberia's conservation strategy. We argue that the conservation component of the LFI land use plan sows the seeds for future land tenure uncertainty and rural injustice, which were major contributors to the war (Richards 2005; Unruh 2009). We focus specifically on conservation and its linkage to renewed conflict for two reasons: (1) conservation organizations working in Liberia frame their work in contradictory terms: as significant contributors to peace and stability in the country; and (2) conservation, more than other land uses, demands territories that are free of people. This demand poses unique problems in Liberia. Large areas of the country that hold conservation value may be perceived as lacking human settlement, but this perception opportunistically ignores key aspects of social history and conflict in rural Liberia, notably, large-scale displacement, contested land claims, and fragmented political authority. Liberia represents a confluence of postconflict land tenure uncertainty, high-value forest resources, and the combined use of geospatial technologies and environmental discourses to articulate and legitimize a large-scale exclusionary land use plan.

The article is organized as follows: A methodological and theoretical overview is followed by three analytical sections covering (1) the role played by remote sensing-based (RS) information and land use modeling in the LFI conservation strategy; (2) how biodiversity and climate change discourses legitimize the land use zones designated by the plan; and (3) the volatility of postconflict land tenure in Liberia. This volatility will greatly complicate attempts by the central government to impose territory-based land use prescriptions. Such top-down impositions, including those with seemingly benign conservation goals, may aggravate rather than reduce the risk of renewed conflict in Liberia.

Methodology

The field research in Liberia included a series of individual and group interviews and focus group discussions totaling 210 people in the months of December 2006 and February 2007. Those interviewed included smallholder farmers, large landholders, ministry officials, university researchers, NGOs, lawyers, UN personnel, commercial agriculture associations, bilateral and multilateral donors, international legal and development organizations, and a former president of the country (Unruh 2009). The discourse analysis focused on publicly available LFI documents, notably the World Bank Forest Resource Assessment conducted in 2004 and the land suitability analysis carried out in 2006. Interviews with conservation and forest-sector NGOs provided supplemental information.

This article draws from the theoretical tradition in geography of examining the social transformation of nature (Castree 1995; Escobar 1998; Haraway 1997). According to Swyngedouw (2004, 130), "nature becomes a sociophysical process infused with political power and cultural meaning." In postconflict Liberia, this process is manifested through a territorializing project (Vandergeest and Peluso 1995) at a specific scale that is produced through geospatial technology and scientific discourse. Liberia today is a unique case because its specific historical and socio-spatial conditions (shattered political structures and displaced rural populations) that made the LFI project possible also contain the seeds of its undoing. Further, it demonstrates how new scientific discourses, namely, climate change mitigation, compound existing ones like biodiversity conservation, to strengthen territorial claims to natural resources without addressing the political legacy of the conflict.

Geospatial Technologies

Geospatial technologies (RS tools and geographic information systems [GIS]) played a critical role in the creation of Liberia's land use plan. By representing and classifying Liberia's landscape in a specific way, these technologies were instrumental in the delineation of future protected areas and other zones in which land use will be prescribed and controlled. Our argument concerning geospatial technologies is based on the following premises: (1) By representing Liberia's forest cover as a gradient of intact dense forest, RS classification implicitly perpetuates the notion of the pristine forest ecosystem, a notion of questionable historical veracity in West Africa (Fairhead and Leach 1998); (2) classification itself is a political act, the power of which is connected to its visibility and to the degree to which it can be contested as a form of knowledge (Bowker and Star 1999); and (3) RS land cover information privileges spatio-environmental analysis based on certain types of knowledge and information, namely, that which is visually measurable (Turner 2003).

The land use planning component of the LFI relied heavily on a land cover and forest resource assessment produced for the World Bank in 2004. The assessment was based on standard RS analysis of six Landsat scenes from the years 2001 to 2004. We do not question the technical rigor of the geospatial analysis, but we do question the legitimacy of basing a countrywide land use plan on a set of satellite imagery from a contemporary four-year period. The use of contemporary land cover, as classified through RS and scientific discourse, belies a multicentury history of complex population fluctuations and land use patterns in Liberia. As Fairhead and Leach (1998) describe, certain parts of the country were widely assumed to be uninhabited as recently as the eighteenth century; however, at least one scholar (Mayer 1951) has compellingly argued that as early as the seventeeth century far higher population and far less mature forest cover characterized Liberia.

As Turner (2003, 262) argues for the case of the Sahel, the use of a single set of RS information provides a "snapshot of particular times." In the Sahelian case, RS has been used during a time of rapid population expansion that began in the 1950s, leading to neo-Malthusian interpretations of land use change and resource degradation. In the Liberian case, reliance on contemporary RS data provides a land cover snapshot from a period of extreme social dislocation and much reduced population densities in many areas, thereby reinforcing notions of empty and intact forests that are under threat from agriculture and other types of land use.

The classification scheme used in the 2004 assessment reflects this notion of pristine versus human-disturbed areas in its use of three forest categories: closed dense, open dense (distinguished by signs of recent logging disturbance), and agriculture disturbed. This classification aimed to provide information for the following objectives: country-level land use planning and management, the optimization of forest inventories, and a baseline for land cover monitoring (Bayol and Chevalier 2004). These classes also reflect the objective of measuring forest cover against a previous 1979 inventory known as the Liberia Forest Reassessment to ascertain change. A compelling aspect of the 1979 inventory, which goes unmentioned in the Bayol and Chevalier (2004) report, is that it was criticized as flawed by a subsequent study conducted in 1985 by the FDA. The FDA report describes inaccuracies in the 1979 inventory that led to exaggerated estimates of forest loss (Fairhead and Leach 1998). Nonetheless, the 1979 inventory by the Food and Agricultural Organization (1981) was and continues to be deeply influential as demonstrated by its use in subsequent reports and in the LFI.

The 2004 classification and inventory reproduce a narrative of unidirectional forest loss in Liberia with significant flaws and inaccuracies that become invisible, particularly to those without the proper training, knowledge, or social connections to access and comprehend its contents, as an uncontested baseline of what is recognized as forest cover, exploitable forest resources, and of who may exploit them as dictated through the LFI process. This type of RS classification also privileges visual information that can be detected through satellite imagery (Turner 2003; Walker and Peters 2007). Bayol and Chevalier (2004) acknowledge the difficulty in distinguishing closed and open dense forest at the spatial scale of their analysis, so they relied on the identification of roads as a visual proxy of logging disturbance used to designate areas as open dense forest. This classification decision, based largely on a single visual feature, carries enormous implications for rural Liberians: More than 2.4 million hectares, roughly 25 percent of Liberia's total area, were classified as closed dense forest, which translates into strict protection and land uses from which smallholders will largely be excluded.

The open–closed forest distinction is also critical to the designation of nearly 1 million hectares for future protected areas. In a 2006 land suitability analysis conducted by a U.S. Forest Service consultant, distance from roads is used as a proxy for "no human settlement" (Nebel 2006, 5). Land located more than three kilometers from roads is assumed to be uninhabited by people and more than eight kilometers to be free of agriculture. According to the consulting report, Liberian forestry officials confirmed this critical assumption, which is problematic given the oppressive history of state forest services in Liberia and the rest of West Africa (Fairhead and Leach 1998). Although a GIS overlay of human settlements was used in the model, UN refugee data show high levels of displacement around certain areas classified as closed dense forest (Internal Displacement Monitoring Center 2007) and officials from two international agencies working in Liberia acknowledged that many forest areas appearing to be uninhabited are likely to be that way as a result of the war. Further, customary land tenure in Liberia, like much of West Africa, extends far beyond the limits of settled areas. Thus, although smallholder agriculture, small-scale timber extraction, and use of nontimber forest products occur in such areas, more important and pervasive is that virtually all land, settled or not, is claimed by indigenous chiefdoms.

Nonetheless, distance from roads and distance to edge of closed dense forest are two of the key variables used to define suitability for conservation because they positively correlate with biodiversity (Nebel 2006, 9). The suitability analysis ultimately produced a map containing 919,930 hectares of zones "optimally" suitable for conservation as twenty-three new protected areas. Strikingly, the conservation suitability analysis mimicked the land cover classifications produced by Bayol and Chevalier (2004) "to a large degree" (Nebel 2006, 9). Although the consultant emphasizes that the land use model is a tool that does not provide definitive answers, the land use zones produced by the suitability analysis are preserved without change in subsequent forest management documents.

The use of RS land cover classes to zone vast areas for protection (and human exclusion) is the fundamental element in the link between conservation and renewed conflict in Liberia. In the final section, we argue that the application of a comprehensive land use plan in the context of Liberia's volatile postwar land tenure situation is a very risky strategy. First, however, we examine how RS information and knowledge gain power through discourses of biodiversity and climate change mitigation.

Conservation Discourse and Policy

Globally dominant conservation strategies focus on maximizing the area of land that is (1) designated as intact habitat (i.e., closed dense forest) and (2) protected through the exclusion of people (see Brandon, Redford, and Sanderson 1998; Brooks et al. 2002). By producing national-scale maps of land cover that include classifications such as closed dense forest, RS technologies provide an effective tool that, in conjunction with indicators such as species richness capacity (Defries et al. 2005), is used toward these strategic goals. The policy argument is simple: more intact habitat, more biodiversity.

Climate change mitigation plays an increasingly important role in this logic as national and international policymakers seek large tracts of intact forest to sequester atmospheric carbon. Carbon sequestration through "avoided deforestation"[2] has been internationally recognized and Liberia is poised to become one of the first countries to benefit financially through the World Bank-led Forest Carbon Partnership Facility. Avoided deforestation dovetails with conservation efforts that seek to maximize *permanent* intact forest cover. International biodiversity and climate change discourse and policy merge as habitat becomes carbon sink. As the political importance of climate change mitigation increases, the growing attraction of financial capital to forested developing countries such as Liberia only compounds the future risk of land tenure conflict in the country. Advocates for smallholder forest inhabitants in Africa already cite the potential for increased land alienation as a primary risk associated with the UN-led Reducing Emissions from Deforestation and Forest Degradation in Developing Countries Programme (Nengo 2008).

The influence of this international conservation policy in Liberia is profound. As part of its goal to maintain the integrity of its forest cover over a twenty-five-year time horizon, Liberia's National Forest Management Strategy of 2007 (Republic of Liberia Forestry Development Authority 2007) merged the open dense and closed dense forest classes produced by Bayol and Chevalier (2004) into one category labeled "permanent forest" (Republic of Liberia Forestry Development Authority 2007, 11). The new category totals 4.39 million hectares, an equivalent of 47 percent of the nation's territory, including the 919,930 hectares deemed suitable for strict protection, which represents nearly 10 percent of Liberia's territory as mandated by the UN Convention on Biological Diversity (UNCBD). The UNCBD, which the Liberian government signed in November 2000, provides the legal anchor for national conservation strategy and the central role played by territorial land use zoning in it.

Liberia's conservation strategy, guided by the UNCBD article on strict territorial protection, reflects the reality that protected areas are one of the "cornerstones" of nature conservation (DeFries et al. 2005, 19). As the previous section demonstrated, the locations and extent of Liberia's future protected areas are heavily influenced by a land cover map that is at best decontextualized and at worst part of a flawed narrative of regional deforestation. Despite long-standing efforts to better incorporate people into protected areas (Brechin et al. 2002), the "protectionist paradigm" of conservation is resurging globally and is strongly reflected in Liberia's strategy. According to Hutton, Adams, and Murombedzi (2005, 347) this resurgence reflects: "urgency of action, calls to better incorporate biological science in conservation, and the completion of global-scale analysis of biodiversity hotspots." Brandon, Redford, and Sanderson (1998) argue that "protected areas are the last safe havens for large tracts of tropical ecosystems" (Hutton, Adams, and Murombedzi 2005, 348). As previously noted, justifications for protecting large tracts of land that are classified as intact relate to species richness capacity, which is an accepted measure of a given area's potential to conserve biodiversity (DeFries et al. 2005).

Although grounded in conservation science, the application of species richness capacity in the policy process depends on the highly political and very problematic act of erasing the social history and human–environment interactions from landscapes targeted for conservation (West, Igoe, and Brockington 2006, 148). This is a necessary step in a twofold process that transforms landscapes into conservation territories. In Liberia's case, the second step is the creation of a stable land cover category of permanent forest. Although stakeholders can dispute the terms of forest use, its definition and its spatial extent are nonnegotiable. Landscapes that were sites of historic struggle, conflict, and dislocation become naturalized spaces to be defined as intact habitat, free of people and strictly protected.

Postwar Land Tenure

Because customary land tenure cannot be legislated out of existence, the classification approach sets up extremely contentious relationships about land between the state and rural inhabitants. Because land tenure in postwar Liberia is highly volatile, such relationships detract significantly from efforts within the peace process. Currently the root land tenure problem in the country is the massive confusion that exists on a range of legal, administrative, boundary, claim, and ownership issues. For example, there is confusion about which laws apply where and to whom; there is contradiction within and between laws; there were multiple changes in subnational boundaries and definitions and roles of administrative areas; and large land areas are simultaneously claimed by chiefdoms, the state, private interests, and informal groups. The link between such confusion and wide-ranging land tenure insecurity and conflict is explicit (Bruce and Migot-Adholla 1994; Unruh 2008), and this link plays a large role in postwar Liberia (Unruh 2009). With little clarity regarding rights, authority, and dispute resolution, a large-scale attempt to redefine land tenure and rezone land use, particularly through territorial exclusion, is perilous.

Approximately half of the country was under the control of armed factions during the 1990s, and the government is just now coming to understand what has transpired with regard to private, political, and administrative change in land units. The fieldwork revealed that the activities of ex-combatants are a particular problem, some still reporting to militia commanders who engage in a variety of extractive activities (Unruh 2009). The administrative and judicial systems required to handle land matters in the postwar context are currently extremely underdeveloped, nonfunctional, or overstretched. Implementing the wide-reaching LFI land use plan in this sociopolitical and institutional environment sets up a potentially volatile dilemma.

Postconflict land tenure and the LFI are embedded in a long history of conquest, claims, and violence that dates to the arrival of the first Americo-Liberians. Armed conflict over land occurred as early as 1822, the year Liberia was established, due to profound conceptual differences in land tenure between Americo-Liberians and indigenous inhabitants (Bruce 2007). With severe suppression of the interior in the early half of the twentieth century, the resulting aggrieved sociopolitical situation grew over time as extensive land appropriation grew into acute grievance, uncertainty, and conflict. By the eruption of the war in 1990, the means of acquiring land had developed into a volatile crisis (Governance Reform Commission [GRC] 2007; Unruh 2009). Riddell and Dickerman (1986, 102), quoting Liberian expert Gus Liebnow in a personal communication, note that during the Doe regime (1980–1990), "[l]and tenure is at the heart of rural discontent." Subsequent to the war, the Liberian GRC (2007) concluded that access to land was a root cause of the armed conflict. Rural people continue to be skeptical of central authority and it is unsurprising that they are now contesting new territorial claims. Land rights in Liberia after the war are therefore an acute concern (P. Banks, personal communication, 6 December 2006, conversation; Daygbor 2007; K. Johnson, personal communication 12 February 2007, conversation) and the rapid reassertion of centralized control is risky. An additional concern is the potential reinstatement of certain abuses by customary leadership, particularly with regard to land access for migrant youth. Extremely exploitive arrangements existed prior to the war, which fed youth into the militias (IRIN 2007). Indeed "reform of rural rights seems as urgent an issue as tracking the gun-runners or diamond- and timber-smugglers" (Richards 2005, 588).

The general lack of clarity in land rights is aggravated by the existence of the dual tenure system in the country (statutory and customary). There are constant and persistent clashes in Liberia involving customary versus statutory rights over the management, authority, and control of land resources (GRC 2007). The way that this duality functioned was a significant contributor to the war (Unruh 2009).

The fieldwork revealed that effective boundary demarcation is a large and confusing problem. In a number of cases how much land exists in the various counties and concessions is unknown. In others, mistaken numbers are used to calculate such areas (Unruh 2009). In addition, the fieldwork observed that under Liberia's deed system, only the number of acres and approximate boundaries are recorded, and there exists no registry. Complicating this situation is the significant purposeful destruction and alteration of deeds during the war. These issues are all fundamentally important for conservation planning and land inventory. Interviews with members of the GRC indicate that there is currently overlap and jurisdictional ambiguity between the state-supported customary units of Clan and Paramount Chieftaincies, and the townships and cities subject to the statutory system. The overall situation is that many boundaries exist in a state of extreme confusion and contestation.

For example, members of the GRC and practicing lawyers estimate that currently up to 90 percent of all cases in all statutory courts are land and property related. Moreover, land disputes are the most frequent cases in local courts (Richards 2005). This is a serious problem after a war when access to justice and the promotion of the rule of law are priorities in the country's peace process (Unruh 2009). At the same time there is a deep distrust of customary courts and their capacity to decide land issues in a fair manner. Interviews with government and customary communities indicate that there is no legal or institutional mechanism whereby disputes and other issues can be resolved between the statutory and customary tenure systems, and that the interface between commercial holdings and smallholder farmers is the focus of increasing conflict (Unruh 2009). A particular example involves whether the Mandingo ethnic group in the north of the country are to be considered citizens of Liberia and thereby legitimately able to claim and own land. Conflicts are emerging between adverse possession claims via statutory law by Mandingos and traditional claims by other ethnic groups.

In short, customary land rights that exist within the national-scale land use zones are still highly contested. Significantly problematic with regard to the LFI is that key forest management decisions were finalized in 2007. Although the public consultation phase of the LFI had concluded, important elements of Liberian civil society did not fully support the management plan, as indicated in an open letter signed by fourteen Liberian and seven international NGOs on 25 September 2006. The letter states clearly that the Forestry Reform Law, adopted on 19 September 2006, failed to address community land tenure, access and user rights, or meaningful public participation in forest management and sector reform. Nearly two years later, the NGO coalition of Liberia issued a statement in July 2008 identifying three timber contracts issued by the FDA that violate the property rights of local communities. Allegedly the government was fully aware of existing community land claims but chose to ignore its own policy. The statement further describes a larger problem: The reform process and rule of law in the forestry sector are at risk due to opaque procedures.

Conclusion

Liberia's land use planning process was made possible by the presumed "clean slate" that was the country's rural landscape following the end of hostilities. Although rural Liberia was dangerous and unstable, it was a clean slate in the sense that no customary authorities could counter the territorializing effects of postwar land use planning. Actors seeking to control Liberia's vast and valuable forest resources, including its biodiversity and potential to sequester carbon, have used geospatial technologies and scientific discourse as tools to establish a system of territory-based authority covering the entire country. We have argued, however, that this project ignores issues relating to a problematic political and scientific discourse, technology application, and land tenure at considerable peril.

Notes

1. http://www.conservation.org/explore/africa_madagascar/liberia/pages/liberia.aspx (last accessed 30 April 2009).
2. *Avoided deforestation* is the conventional term referring to the ensemble of multilateral initiatives that aim to prevent the conversion of tropical forests into other land covers to reduce greenhouse gas emissions.

References

Bayol, N., and J. Chevalier. 2004. *Current state of the forest cover in Liberia*. Washington, DC: World Bank.

Bowker, G., and S. L. Star. 1999. *Sorting things out: Classification and its consequences*. Cambridge, MA: MIT Press.

Brandon, K., K. Redford, and S. Sanderson. 1998. *Parks in peril: People, politics, and protected areas*. Washington, DC: Island Press.

Brechin, S., P. Wilshusen, C. Fortwangler, and P. West. 2002. Beyond the square wheel: Toward a more comprehensive understanding of biodiversity conservation as social and political process. *Society & Natural Resources* 15 (1): 41–64.

Brooks, T., R. Mittermeier, C. Mittermeier, B. da Fonseca, A. Rylands, P. Konstant, and J. Flick. 2002. Habitat loss and extinction in the hotspots of biodiversity. *Conservation Biology* 16:910–23.

Bruce, J. 2007. *Insecurity of land tenure, land law and land registration in Liberia: A preliminary assessment*. Washington, DC: The World Bank.

Bruce, J., and S. E. Migot-Adholla. 1994. *Searching for land tenure security in Africa*: Dubuque, IA: The World Bank and Kendall/Hunt.

Castree, N. 1995. The nature of produced nature: Materiality and knowledge construction in Marxism. *Antipode* 27: 12–48.

Daygbor, J. 2007. Land dispute—Country's next war trigger. *The Analyst* (Monrovia) All Africa Global Media 21 March. http://allafrica.com/stories/200703210858.html (last accessed 1 May 2009).

DeFries, R., A. Hansen, A. Newton, and M. Hansen. 2005. Increasing isolation of protected areas in tropical forests over the past twenty years. *Ecological Applications* 15 (1): 19–26.

Escobar, A. 1998. Whose knowledge, whose nature? Biodiversity, conservation, and the political ecology of social movements. *Journal of Political Ecology* 5:53–82.

Fairhead, J., and M. Leach. 1998. *Reframing deforestation: Global analysis and local realities: Studies in West Africa*. London and New York: Routledge.

Food and Agricultural Organization. 1981. *Forest resources of tropical Africa: Country briefs*. Rome: Food and Agricultural Organization.

Governance Reform Commission (GRC). 2007. *The way forward: Land & property right issues in the Republic of Liberia*. Monrovia: Government of Liberia.

Haraway, D. 1997. *Modest-witness@second-millennium. FemaleMan-Meets OncoMouse*. London and New York: Routledge.

Hutton, J., W. Adams, and J. Murombedzi. 2006. Back to the barriers? Changing narratives in biodiversity conservation. *Forum for Development Studies* 32 (2): 341–70.

Internal Displacement Monitoring Center. 2007. *Facilitated return of Liberian IDPs by county*. Geneva: UNHCR. http://www.internal-displacement.org/8025708F004-CE90B/httpCountry_Maps?ReadForm&country=Liberia&count10000 (last accessed 31 August 2009).

IRIN. 2007. *Liberia: Donor fatigue threatening DDR process*. New York: United Nations Office for the Coordination of Humanitarian Affairs.

Mayer, K. 1951. *Forest resources of Liberia*. Washington, DC: U.S. Department of Agriculture.

Nebel, M. 2006. *GIS for land use planning in Liberia: Delineation of land use areas*. Monrovia: Liberia Forest Initiative. http://www.fao.org/forestry/29662/en/ (last accessed 30 April 2009).

Nengo, E. 2008. *Dialogue between the World Bank and indigenous peoples in Central and East Africa on the forest carbon partnership facility*. Bujumbura, Burundi: UNIPRPOBA-IPACC.

Republic of Liberia Forestry Development Authority. 2006. *National forestry policy and implementation strategy: Forestry for communities, commerce, and conservation*. Monrovia: Forest Development Authority of Liberia.

———. 2007. *National forest management strategy* (draft). Monrovia: Forest Development Authority of Liberia.

Richards, P. 2005. To fight or to farm? Agrarian dimensions of the Mano River conflicts (Liberia and Sierra Leone). *African Affairs* 105:571–90.

Riddell, J., and C. Dickerman. 1986. Land tenure profile: Liberia. In *Country profiles of land tenure: Africa 1986*, ed. J. Riddell and C. Dickerman, 98–103. Madison: Land Tenure Center, University of Wisconsin-Madison.

Sawyer, A. 2005. *Beyond plunder: Toward democratic governance in Liberia*. London: Lynne Rienner.

Swyngedouw, E. 2004. Scaled geographies: Nature, place, and the politics of scale. In *Scale & geographic inquiry: Nature, society, and method*, ed. E. Sheppard and R. McMaster, 129–53. Oxford, UK: Blackwell.

Turner, M. 2003. Reflections on the use of remote sensing and geographic information science in human ecological research. *Human Ecology* 31 (2): 255–79.

Unruh, J. 2008. Toward sustainable livelihoods after war: Reconstituting rural land tenure systems. *Natural Resources Forum* 32:103–15.

———. 2009. Land rights in postwar Liberia: The volatile part of the peace process. *Land Use Policy* 26:425–33.

Vandergeest, P., and N. Peluso. 1995. Territorialization and state power in Thailand. *Theory and Society* 24 (3): 385–426.

Walker, P., and P. Peters. 2007. Making sense in time: Remote sensing and the challenges of temporal heterogeneity in social analysis of environmental change—Cases from Malawi. *Human Ecology* 35 (1–2): 69–80.

West, P., J. Igoe, and J. Brockington. 2006. Parks and peoples: The social impact of protected areas. *American Review of Anthropology* 35:251–316.

Halfway to Nowhere: Liberian Former Child Soldiers in a Ghanaian Refugee Camp

Lucinda Woodward and Peter Galvin

This study utilizes Kunz's kinetic model of refugee displacement to interpret the placelessness experienced by Liberian former child soldiers in the Buduburam refugee camp in Ghana. From August to December 2007, a clinical psychologist and a geographer interviewed ten Liberian former child soldiers to determine spatial and social barriers to successful resettlement and the prospects for overcoming these obstacles. Based on the interviews, five areas of intervention were suggested: (1) geographic desegregation and relocation, (2) education and employment, (3) psychological counseling, (4) societal acceptance and reintegration, and (5) security and protection. *Key Words: child soldiers, forced migration, Liberians, peace studies, refugees.*

本研究使用了昆斯的难民流动动力学模型，对位于加纳的布杜布拉姆难民营中利比里亚的前儿童兵所经历的流离失所情况进行了分析解释。从2007年的8月至12月，由一名临床心理学家和一名地理学家对十名利比里亚前儿童兵进行了采访，以确定成功安置所面临的空间和社会障碍，以及克服这些障碍的前景。基于采访的结果，对下述五个领域提出了调解建议：（1）地理上废除种族隔离和种族搬迁，（2）教育和就业，（3）心理咨询，（4）社会接受和重返社会，（5）安全和保护。*关键词：儿童兵，被迫迁移，利比里亚，和平研究，难民。*

En este estudio se utiliza el modelo kinético de Kunz sobre desplazamiento de refugiados para interpretar el sentido de desubicación que experimentaban los niños soldados liberianos en el campo de refugiados de Buduburam, en Ghana. Entre agosto y diciembre de 2007 fueron entrevistados por un psicólogo clínico y un geógrafo diez de aquellos niños soldados, para determinar tanto los obstáculos espaciales y sociales a un exitoso proceso de reasentamiento, como las posibilidades de remover tales barreras. Con base en los resultados de las entrevistas se sugirieron cinco áreas de intervención, a saber: (1) desegregación geográfica y relocalización, (2) educación y empleo, (3) consejería psicológica, (4) aceptación social y reintegración, y (5) seguridad y protección. *Palabras clave: niños soldados, migración forzada, liberianos, estudios sobre la paz, refugiados.*

The trauma experienced by war refugees is complex and profound. An abundance of geographical research has addressed the impact of forced migration on individuals and social groups (Black 2001), and a growing body of literature has embraced the concepts of place and space in interpreting the refugee experience (Black and Robinson 1993; Brun 2001). Particular hardships confront internationally displaced persons having served as child soldiers in their home countries. Such individuals are prone to posttraumatic stress and face unique obstacles to repatriation, resettlement, and social integration (Thompson 1999). Geographic isolation, social segregation, and psychological stress severely limit their ability to reintegrate. Spatial and social barriers to resettlement have relegated these young men and women to a state of geographical and psychological limbo. Despondent and desperate, some return to soldiering, the only occupation they have ever known. Rerecruited as mercenaries in new or continuing regional conflicts, they help perpetuate the cycle of war. The placelessness experienced by these individuals is both geographical and psychological, and combined spatial and psychological interventions are necessary if they are to be successfully resettled. The goal of this study, conducted by a geographer and a psychologist, was to identify barriers to resettling former child soldiers through personal interviews with individuals and observation of their physical placement and social interaction within the camp. The spatial patterns and processes indentified by E. F. Kunz (1973) in his kinetic model of refugee displacement were applied to this end and to formulate recommendations for effective psychosocial reintegration and resettlement of youth ex-combatants.

Background: The Liberian Civil Wars

The civil conflict in Liberia spanned fourteen years and killed an estimated 250,000 Liberians (Refugees

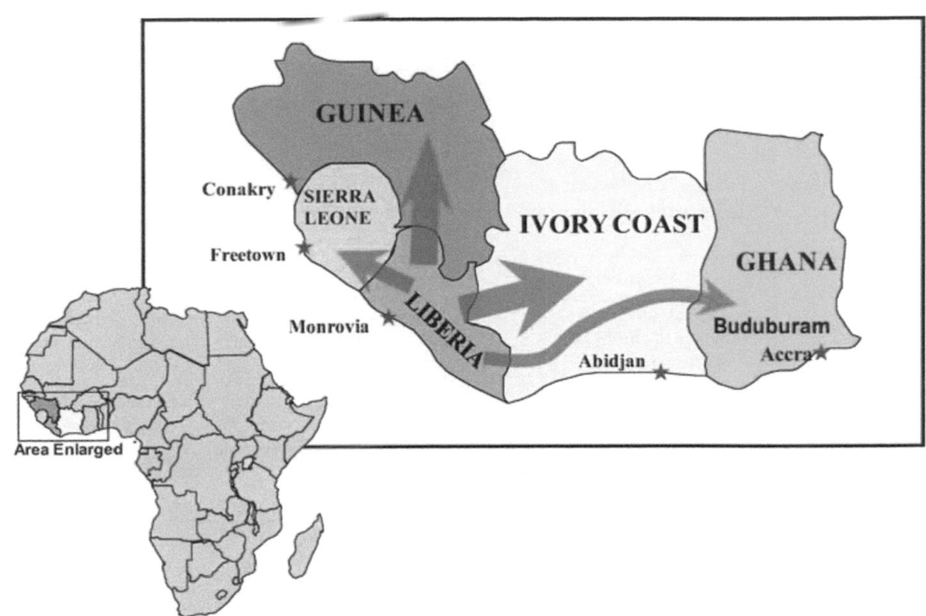

Figure 1. Flow of Liberian Civil War refugees after 1989.

International 2005). Hundreds of thousands became internally displaced and, by 2004, more than 770,000 refugees had fled to nearby countries in West Africa (Figure 1), especially Côte d'Ivoire, Guinea, Sierra Leone, and Ghana (United Nations High Commissioner for Refugess [UNHCR] 2006, 379). According to Wood (1994), Liberia had the second highest rate of forced migration in the world, second only to Bosnia. The direct causes of this flood of refugees included ethnic and tribal conflict as well as insurgency and governmental persecution.

By 2004, Liberian refugees numbered 70,000 in Côte d'Ivoire, 41,000 in Ghana, 127,000 in Guinea, and 65,000 in Sierra Leone (UNHCR 2006, 379). Among these refugees are thousands of former child soldiers. The United Nations defines a child soldier as "any person under 18 years of age who is part of any kind of regular or irregular armed force or armed group in any capacity, including but not limited to cooks, porters, messengers and anyone accompanying such groups, other than family members. The definition includes girls recruited for sexual purposes and for forced marriage" (United Nations Children's Fund [UNICEF] 1997).

Child soldiers figured prominently throughout Liberia's civil wars. Children were conscripted for many reasons. Easily bullied, they were naïve and likely to follow adult orders without question. They were plentiful. Many child recruits were orphaned, abandoned, homeless, and hungry—desperate for the food, companionship, and protection that marauding armies could supply. Often they formed the vanguard of attacks, living shields for older, more valued, seasoned fighters (Thompson 1999).

Two civil wars in Liberia, from 1989 to 1996 and 1999 to 2003, brought over 40,000 refugees from that country to Buduburam, near Accra, Ghana. By 2007 nearly 30,000 remained, many of them former child soldiers having languished for years without gainful employment, food and shelter allowances, medical care, or social services. Unable to return to Liberia, resettle abroad, or enter vocational training, and unable to find sustaining work in Ghana, these men and women remain in what some observers have described as a legal and political "limbo" (Mountz et al. 2002). Kunz (1973) described this state as "midway-to-nowhere," when refugees can neither repatriate to their country of origin nor move on to a new country of asylum. Many former child soldiers report feeling emotionally stunted—neither children nor adults—segregated, with no sense of spatial or social belonging. Theirs has become, as well, a psychosocial and geographical limbo. Through qualitative deconstructionist methodology, this study will contribute to a better understanding of the spatial and psychosocial dimensions of acute, intermediate forced displacement on former child soldiers. This study expands on Kunz's kinetic model of displacement by examining the personal backgrounds and attitudes of

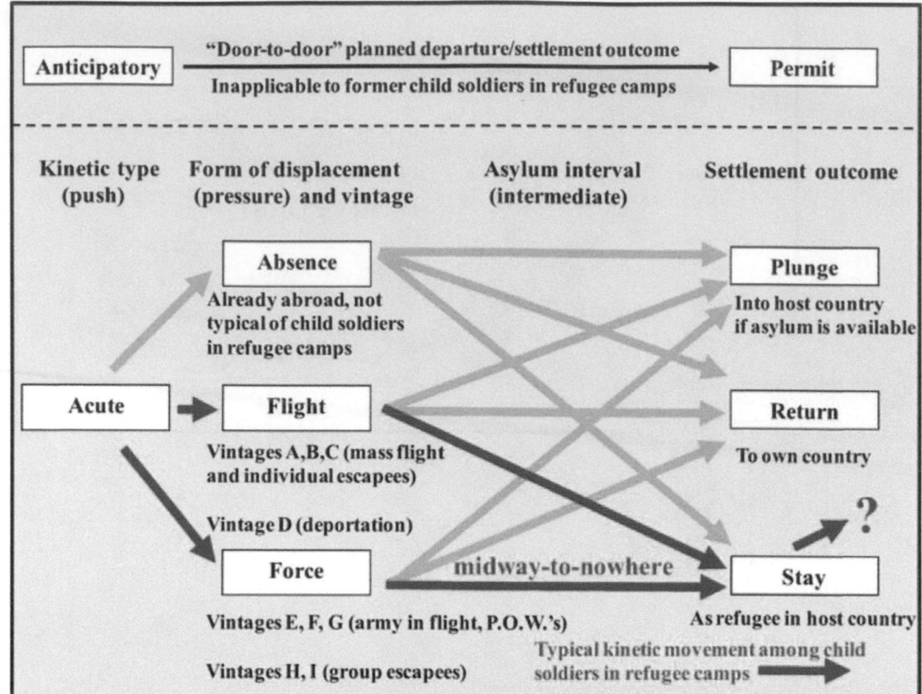

Figure 2. Flow diagram of refugee settlement patterns (based on Kunz 1973).

former Liberian child soldiers as they affect settlement outcomes.

The Kinetic Model of Refugee Movement

Kunz's model (1973; see Figure 2) explains how various push factors and processes (*kinetic types*) lead to distinctive rationales and timing of refugee migration (*form of displacement* and *vintage*). As forced migrants, refugees are pushed, not pulled. Safety and survival are the hope of all refugees, but particular combinations of kinetic type, vintage, and form of displacement result in different refugee settlement outcomes. Kunz identifies two kinetic types: *anticipatory* motivation (émigrés responding to early declining conditions in their home country) and *acute* motivation (émigrés pushed by military action or political persecution to seek refuge in another country). The three main forms of acute displacement are *flight* (characterized by voluntary flight of groups or individuals), *force* (usually associated with military or political action), and *absence* (refusal to return to one's country of origin after a peaceful departure). The timing of refugee movement, or *vintage*, can be assigned to one of nine categories, designated A through *I* (earliest to latest). The four possible settlement outcomes (*permit*, *plunge*, *return*, or *stay*) are governed by kinetic type, form of displacement, and vintage. It was hypothesized that the former child soldiers interviewed in this study would best fit into vintages E, F, and G—for example, prisoners of war (POWs) and soldiers—and that the resulting kinetic movement would be acute—force and flight—stay with an intermediate, midway-to-nowhere asylum interval.

Methods

From August 2007 to December 2007, ten former Liberian soldiers living in Ghana's Buduburam refugee camp were interviewed privately and individually using a structured set of open- and closed-ended questions. Participants were recruited through a connection with Veteran Child Soldiers of Liberia (VECSOL), a nongovernmental organization (NGO) promoting the needs of Liberian former child soldiers. Interviews (each about two hours long) were conducted off camp to ensure confidentiality. Participant ages ranged from twenty-three to thirty-seven years. Interviewees received the equivalent of US$3 (approximately one day's wages). Following deconstructionist methodology, after data collection, the initial interpretation of the interviews was submitted to the original participants for comments and corrections prior to final write-up.

Interviewees were selected as a representative sample of sex, age, service positions, and military factions that participated in the Liberian war. Our sample thus included nine men and one woman. The mean reported age of recruitment was 12.4 years, with a range from ten to seventeen years of age. According to participant report, the mean number of years of military service was five years, ranging from 2.5 to seven years of combat spanning the two civil wars.

Results

Refugee Characteristics

According to Kunz (1973), demographic characteristics consistent with vintage E/F/G refugee movement as seen in POWs or conscripted military agents would include high masculinity, active age groups, and a diverse section of socioeconomic and educational backgrounds. The data from our sample of child soldiers were consistent with the predictions in his model.

All ten participants described their lives as "normal" before the war. Six reportedly lived within metropolitan Monrovia. Four reported coming from more rural areas of Liberia. All reported attending school before the civil conflict. On average, the highest grade completed was seventh (although some reportedly never completed sixth grade and one reported returning to school to earn a high school diploma after fleeing the war). Participants indicated their parents and families represented a wide range of socioeconomic and educational backgrounds. Several indicated that occupational ties to the existing government put their parents in danger once the rebels attacked.

One male ex-combatant explained:

> I was very small before the war, going to school. Everything was fine with us before the war. When the war came everything turned around. I was sixteen. I was in school and I liked to study math and science and history. I lived with my mother and father. I had brothers and sisters. My parents wanted me to finish school and go to college and learn to become a doctor.

Another former child soldier lamented:

> We had a good standard of life. There was no hunger and we had a high school that trained us to go on to college or get vocational training to work for the town company. . . . I wanted to become a civil engineer.

Refugee Movement Factors

Kinetic type and form of displacement. Seven of the interviewees reported having been orphaned or separated from their families by the fighting. This was frequently cited as partial motivation for joining the military service because there was reportedly little or no organized care provided to such children. One participant described the day he was separated from his family:

> I was very scared at that time because we knew soldiers were coming. Sometimes I would ask, "Why? Why are they coming?" but my mother could not explain. The war first reached our neighborhood in 1990. At that time I had passed ten years old. There were children all over, there was shooting. People came to our house. They said there was a curfew. No one should wear jeans or white t-shirts because they would think we were rebels. Some brothers were passing by and saw us and said, "Everybody come out of the house. If I come in the house and I find anyone I will shoot them." All the other boys came out but I hid. I walked on my knees to a window and broke out and ran. But there was still shooting—dead bodies everywhere. Soldiers pulled me aside and said, "You boys get off the streets or I will shoot you." So I went to empty buildings and slept. For four days I didn't eat and the only thing I could do was loot. But if you were not careful if they saw you looting they would kill you. There was no understanding, no one trusted each other . . . I never found out anything about my mother—it has been eleven years and I haven't heard from her.

Of those who were separated from loved ones, most had reportedly heard no news of their families since their recruitment. Three individuals stated that they knew their family's whereabouts or had had contact with their parents or siblings. According to one young man, his role as a soldier was critical to his family's survival, as he provided not only protection but sustenance:

> I was always in contact with my family. Whenever I had money, food, I went to see my family. My family all survived the war, but I don't know where my father is, since they took him away. I don't know where he is. I haven't heard from my mother or sister since I came to Ghana.

Recruitment and Training. Six interviewees indicated they were recruited through coercion or force. It was reportedly common for rebel soldiers to kill entire families, leaving just the second oldest son to choose between dying with his family or living as a recruit. By account, one young man watched his older brother and father executed by rebel soldiers; his mother was then forced to butcher and cook his brother's body for the soldiers to eat. He was reportedly forced to join the soldiers after that to protect his mother and remaining family members.

Another's report described forced relocation prior to conscription, a common practice used to separate families:

> I can remember there was heavy firing and the [soldiers] were wearing wigs. It was evening time and we couldn't see the bullets. We couldn't take anything. We locked the door and we went where the soldiers told us. My five-year-old sister was crying. We went to a place outside Monrovia in a house where we slept on the floor. I lived there six months before joining the soldiers. [One day] I went to a well to get water and the soldiers took us. We were all children, boys and girls. The chief blindfolded us and told us we would be in the service. I was underage (11) and I didn't know what to think. Some tried to escape and they were killed. Some of my friends tried and they got killed.

Most participants indicated feelings of anger and resentment about their treatment during conscripted military service. According to the female ex-combatant, rape and abuse were common experiences for the girl child soldiers. Four individuals indicated that recruitment often took the form of gradual indoctrination through friendship or peer pressure. Two such stories are described here in the participants' words:

> My sister befriended one of the rebels. He used to come to our house and used to bring food. She deemed it [prostitution] necessary because of the family. He wanted to give me an opportunity so I visited him. He wanted me to hold his arms [guns]. [He] was not a child soldier—twenty or twenty-five. At first, I did not go on missions, just helped out around town. I thought if I'm with this guy, nothing can happen. He was in [my town] three to four months and was reassigned . . . and took me along with him. He kept me in his yard to protect his family. He left me with an AK-47 but in the home I had an Uzi. He taught me how to use it.

> Johnson's forces came and I lost my family. At that time I met a friend who said he would help me. He gave me some baggage and I worked for him [as a porter]. And he told me I should join the forces. They gave me a gun and taught me to advance and shoot. Then they taught me to kill animals. Soon I had to kill people. . . . My friend taught me to take cover and things. I had an MK3. My friend got shot in one of the battles. We were in the area when the enemy came. He got shot in the leg and it turned bad and it got cut off.

Qualitative Analysis of Common Themes

This study set out to identify geographical and social barriers to resettlement among former child soldiers and to formulate recommendations for effective spatial and psychosocial reintegration. Their responses revealed a placelessness consistent with Kunz's (1973) description of similar refugees who have "arrived at the spiritual, spacial, temporal and emotional equidistant no man's land of midway-to-nowhere" (133). When asked to describe the problems they face currently as a consequence of their young military service, respondents cited difficulties that encompassed five overriding issues.

Geographical Isolation. The former child soldiers interviewed for this study were residents of the Buduburam refugee camp in Ghana, about forty kilometers west of Accra. The camp was opened by the UNHCR in 1990. Housing more than 40,000 Liberians at its height, the camp grew from a scattering of tents into a sprawling maze of mud brick dwellings with shops, churches, schools, restaurants, and a market. Open sewers and garbage heaps abound (Figure 3). Water is available but expensive. Conditions are difficult for any refugee but particularly bad for former child soldiers. As exemplars of vintages E/F/G—late-arriving former combatants—they have settled and subsist on the spatial and social fringe of Buduburam, primarily in Zone 9 (Figure 4). Within that peripheral zone, most have congregated in one section, the Area E GARP (AEG) or "ghetto" described by one of the study's participants:

> The GARP is a terrible place to be . . . but very helpful to Former Child Soldiers (FCS) . . . They gamble in . . . pool . . . cards . . . and dice . . . [and] sometimes fight . . . usually [over] money or stolen items. The AEG . . . is crowded because of the smoking of marijuana. You don't have to get money before you can smoke, friends can easily share with you. Selling marijuana is a basic source of income for FCS. . . . There are rooms in which drugs like cocaine are smoked . . . also gambling with the dice and the practice of prostitution. Most of the FCS survive by stealing, selling marijuana, and gambling. The GARP is also . . . their home. They sleep outside whether it rains or it is cool. The GARP is both heaven and hell for FCS.

Consistent with Kunz's model, the child soldiers comprised a wave of refugee migration. Most fled in the years just prior to Charles Taylor's election as president of Liberia, thus marking them as a cohort of political outcasts and late arrivals. Their ghetto lies on the outskirts of camp, far from the social and economic heart

Figure 3. Open sewer in Buduburam.

of the settlement. Its geographic and social isolation became clear from a map of Buduburam sketched by an older vintage, noncombatant refugee. His mental map of the camp and community ended well before Zone 9 (see Figure 4). It is also likely that apprehensiveness of earlier vintage groups, whose political credos might differ from the supporters of Taylor's regime, has contributed to the inability of child soldiers to integrate into expatriate Liberian society, just as the greater population of refugees' internment on camp has prevented them from integrating into Ghanaian society.

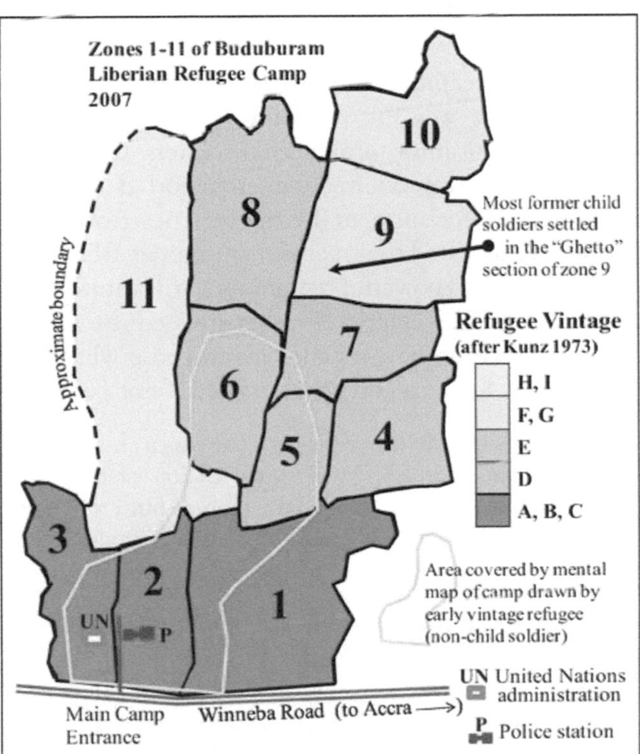

Figure 4. Spatial-temporal settlement patterns at Buduburam.

According to Hyndman (2000, 100), "spatial organization and *segregation* of [refugee] camps shape the social routines of its residents." She labels such camps "noncommunities" and the refugees "subcitizens" (xxv, 113). In that context the spatial and social isolation of former child soldiers in Buduburam renders them *sub*-subcitizens. If camp society does not constitute a true community, it presents, nevertheless, the first and perhaps most crucial opportunity for social reintegration of former child soldiers.

An integral link between the psychology of human interaction and geographic location suggests that variables such as perceived liking (Rhodes, Halberstadt, and Brajkovich 2001) and ingroup–outgroup bias (Yehuda 1969) can be impacted by planned proximity and familiarity. To rectify psychological isolation induced by spatial variables, it is recommended by Carmalt (2007) that geographic approaches to human rights work be increasingly implemented. At Buduburam, specifically, service organizations, schools, and NGOs dedicated to serving child soldiers should be constructed in zones of higher population activity such as Zones 2, 4, or 6, where greater contact with the general population would contribute to desegregation of the veteran population.

Education and Employment. Each of the ten participants described a powerful desire to continue his or her education and to secure employment either in a host country or on return to Liberia. The words *school* and *education* were repeated an average of eight times per interview, making it overwhelmingly the most common issue that emerged. All the participants indicated frustration with the limited employment opportunities for ex-combatants on the refugee camp. Most stressed that they were unofficial refugees with no papers enabling them to receive United Nations support or services. Likewise, as unofficial refugees, they are unable to gain employment in the Ghanaian workforce. Hence, typical employment mentioned during the interviews included doing laundry, making mud bricks, digging gardens, and working unskilled construction with an average reported pay of three cedis per day (approximately US$3). Perhaps a guarantee of amnesty, at least while in camp, would encourage former child soldiers to seek official refugee status and its accompanying benefits.

As stated by one young man:

> My life on camp, I'm not too pleased with that anymore. I don't have a personal way of staying. Sometimes I work. Someone will call you to plaster a house or lay blocks of sixty bricks. Your wages are four cedis. To eat is one cedi, just for one plate of food. Even to take a shower is two cedis. Those are problems.

Another participant described local prejudice against any entrepreneurial attempts made by refugees. "Now I go deep into the woods and make bricks. If you are unlucky, people will come along and beat you up, take your shovel, your water." All of the interviewees reported that better international support and financial aid for educational programs would be crucial for successful retraining and employment of former child soldiers.

Psychological Counseling. Next, an expressed need for psychological services and counseling emerged from the interviews. Eight of ten respondents indicated anger at politicians (often Charles Taylor, by name) for their exploitation as child soldiers and their current refugee status. Five participants specified psychological problems such as recurrent nightmares, hypervigilance, emotional numbness, lack of interpersonal trust, and even psychotic symptoms as a consequence of their war experience. One young man described auditory hallucinations that haunt him nightly in a stream-of-consciousness manner that suggested disorganized thought processes:

> When I am laying down, I think about it [the war] all the time. Sometimes I hear a group of individuals [who aren't there]. I don't know them. I can't see them. Maybe they are people who have died, who I've killed. I dream about them.

Drug abuse and addiction are commonly reported psychological sequelae of child soldiering practices. As reported by four of our participants, drugs such as marijuana, cocaine, alcohol, and even gun powder were frequently used to suppress fear and make them "braver." Two young men indicated that addictions counseling services will need to play a role in the treatment and retraining of former child soldiers.

> I'd take drugs like marijuana, alcohol, tobacco. Also, GP—more people do that—they take drugs to stop fear.

> They gave us pieces of paper and put gun powder in it and forced us to eat it. I didn't want to eat it but they beat us if you didn't. It makes you crazy. It made you want to go fight.

Psychological interventions geared to healing trauma experiences, promoting interpersonal communications, and treating substance abuse would be critical to any permanent resettlement programs for former Liberian child soldiers.

Societal Acceptance and Reintegration. Another common issue was the need for societal acceptance and reintegration for former child soldiers. Of the ten ex-combatants interviewed, seven reported attempts of military re-recruitment by mercenaries from Côte d'Ivoire. One stated that recruitment money from mercenaries can be a powerful temptation to destitute child soldiers in refugee camps. Another stressed the importance of teaching new social skills to those who have learned that toting a gun brings power, if not peace.

> Most of my friends have gone back to being a child soldier. A man came to camp to recruit with lots of money to fight in Ivory Coast . . . I need the money, but I said no to fighting and I won't go back. I put a lot of thought into this.

All ten interviewees had heard of mercenary recruitment of child soldiers for combat in foreign wars. Most indicated that active attempts at social reintegration (teaching new ways of being and behaving) and subsequent societal acceptance are requisite to preventing such predatory practices.

Security and Protection. To overcome the push–pressure–stay dynamic of the intermediate immigrant, as describe by Kunz—a situation in which the refugee wishes to move but is unrepatriatable or unacceptable to another country—international efforts must be made to assure the security and protection of the former child soldier. For those who attempted escape, running was often a life-changing event that precluded ever returning to Liberia for fear of retribution either from their former commanders or from families of those they had hurt.

All of the participants expressed fear of retribution as war criminals either from fellow refugees in Ghana or from the Liberian government should they return. As one former child soldier explained:

> Seven months ago in exile, someone came to me and said, "I know who you are and you killed my father and I will seek revenge." I know they are right. I find people chasing me everyday, saying, "You killed my father." But I need resettlement because there can be no peace. I don't believe that home is safe for me.

The need for official amnesty for young veterans became clear from one man's perspective:

> When I was in Monrovia, a guy saw me and said, "I know you fought, you are not so big, but you fought." Before I could talk, he hit me over the head with the butt of a gun . . . See here [tilting head and showing a long jagged scar].

Retribution reportedly often takes the form of eye-for-an-eye vengeance: "I cannot go back to Monrovia because they would kill my mother, if I killed a mother." For these men and women, there is no safe haven. "I have no intention of going back to Liberia because of what would occur," said another participant. "The people who are there, these people would get even with me." All ten participants in the sample expressed a desire for repatriation in another country or in their host country to assure their security and protection from both official and unofficial retribution.

Discussion

This study utilized Kunz's kinetic model of displacement and deconstructionist data collection techniques to examine common themes among ten former Liberian child soldiers residing in a refugee camp in Ghana, West Africa. Overall, analysis of the demographic and experiential variables described by the Liberian child soldiers in this study suggests that former child soldiers can be classified as vintage E/F/G migrants whose acute, forced displacement has resulted in an intermediate push–pressure–stay settlement outcome. Our study empirically confirmed aspects of Kunz's model with a population of former Liberian child soldiers and added to the existing theory by examining psychosocial factors and formulating resettlement recommendations derived from the interview data.

Five principal issues emerged from these interviews:

1. Attention to geographic location: Spatial isolation and social segregation at the Buduburam camp have impaired the reintegration of former child soldiers into Liberian and Ghanaian society.
2. The need for education and employment: Universally, participants expressed the desire to continue their interrupted education and become contributing members of society through either trade school or college training.
3. The desire for psychological counseling: Many of the former child soldiers in this study reported experiencing symptoms consistent with post-traumatic stress disorder following their combat experiences. To date, none had received formal treatment for this or any other psychological condition (including substance abuse or dependence) from a trained professional. A majority of interview participants expressed an interest and desire to participate in therapy for their mental health issues.
4. The longing for societal acceptance and reintegration: Host communities must be educated and counseled about the process and importance of accepting former child soldiers. It is proposed that spatial integration would promote social acceptance through day-to-day contact. Re-recruitment of former child soldiers reportedly remains a continuing problem in refugee camps where employment opportunities are scarce and human services are nonexistent for unregistered refugees. Psychological disarmament is the primary suggestion for preventing the re-recruitment of ex-combatants.
5. Issues of security and protection: Fear of retribution and punishment as former war criminals haunted all of the interview participants in this study. Most had experienced some type of threat from a victim and all expressed reticence to return to Liberia. Solutions that emerged included formal amnesty from the Liberian government as

well as resettlement or permanent legal residency in the host country.

As a final caveat, all the former child soldiers interviewed felt that their participation in the war shaped their personalities in significant ways. "When you are a child, you have the love of a mother and father. When you are a soldier you have none of that. You are a child, but you are forced to do things you shouldn't even see as a child—killing, molesting, raping." Yet, the capacity of these young men and women to adapt remains astonishing. As one young man stated, "What has happened, has already happened. Man changeth. You can't stay the same." The stories presented here are a testament both to the human consequences of war and the power of human resiliency.

Acknowledgments

We thank the Office of International Programs, Indiana University, for an International Enhancement Grant, and Indiana University Southeast for an Improvement of Teaching grant, both of which partially funded this research. Additional thanks go to Matthew Decker, Western Michigan University, for helping to organize the project. We thank Morris Matadi, Brocks Pokai, and Francis Mulbah from the Initiative for the Development of Former Child Soldiers (IDEFOCS), Monrovia, Liberia, for their invaluable advice and support. Above all, we express heartfelt thanks to all the former child soldiers we interviewed, who, by necessity, must remain unnamed.

References

Black, R. 2001. Fifty years of refugee studies: From theory to policy. *International Migration Review* 35:57–78.

Black, R., and V. Robinson, eds. 1993. *Geography and refugees: Patterns and processes of change*. New York: Belhaven.

Brun, C. 2001. Reterritorializing the relationship between people and place in refugee studies. *Geografiska Annaler* 38:15–25.

Carmalt, J. C. 2007. Rights and place: Using geography in human rights work. *Human Rights Quarterly* 29:68–85.

Hyndman, J. 2000. *Managing displacement: Refugees and the politics of humanitarianism*. Borderlines, Vol. 16. Minneapolis: University of Minnesota Press.

Kunz, E. F. 1973. The refugee in flight: Kinetic models and forms of displacement. *International Migration Review* 7:125–46.

Mountz, A., R. Wright, I. Miyares, and A. J. Bailey. 2002. Lives in limbo: Temporary protected status and immigrant identities. *Global Networks* 2:335–56.

Refugees International. 2005. Liberia. http://www.refugeesinternational.org/content/article/detail/2921/ (last accessed 4 January 2009).

Rhodes, G., J. Halberstadt, and G. Brajkovich. 2001. Generalization of mere exposure effects to averaged composite faces. *Social Cognition* 19:57–70.

Thompson, C. 1999. Beyond civil society. *Review of African Political Economy* 26:191–206.

United Nations Children's Fund. 1997. Children and armed conflict. http://www.unicef.org/emerg/index_childsoldiers.html (last accessed 4 January 2009).

United Nations High Commissioner for Refugees. 2006. *2004 UNHCR statistical yearbook*. http://www.unhcr.org/statistics/STATISTICS/ (last accessed 4 January 2009).

Wood, W. B. 1994. Forced migration: Local conflicts and international dilemmas. *Annals of the Association of American Geographers* 84:607–34.

Yehuda, A. 1969. Contact hypothesis in ethnic relations. *Psychological Bulletin* 71:319–42.

Reconciliation in Conflict-Affected Societies: Multilevel Modeling of Individual and Contextual Factors in the North Caucasus of Russia

Kristin M. Bakke, John O'Loughlin, and Michael D. Ward

Over the past two decades, there has been a growing interest in reconciliation in societies emerging from conflict. The North Caucasus region of Russia has experienced multiple and diverse conflicts since the collapse of the Soviet Union, and violence continues, although at a lower level than a decade ago. We examine willingness to forgive members of other ethnic groups for violence that they have perpetuated as an indicator of the potential for reconciliation in the region. Using data from a large representative survey conducted in five ethnic republics of the North Caucasus in December 2005, we analyze responses to the forgiveness question in relation to social–psychological models of reconciliation, and we add a key geographic measure, distance to violent events, to the usual theories. Using the survey data ($N = 2{,}000$) and aggregate data for the eighty-two sampling points, we use a multilevel modeling approach to separate out the effects of individual and contextual factors. We find little support for the social identity theory expectations as ethnic hostility is not an important factor, except for in the case of the Ossetians, a mostly Orthodox minority disproportionately affected by multiple conflicts and the Beslan school killings. Instead, personal experiences of violence and terrorism, the impacts of military actions against communities, differences in general trust of others, and the extent to which the respondent's life has been changed by violence negatively influence the willingness to forgive. Conversely, respondents in ethnic Russian communities and those relatively close to violence are more willing to engage in postconflict reconciliation. *Key Words: contextual effects, multilevel modeling, North Caucasus, postwar reconciliation, Russia.*

在过去的二十年里，人们对摆脱冲突的社会和解的兴趣与日俱增。自苏联解体以来，俄罗斯的北高加索地区经历了多次不同类型的冲突，尽管比十年前的水平要低，暴力仍在继续。我们研究了某个族群愿意原谅其他族群成员暴力的程度，以此作为该地区和解的潜在指标。利用 2005 年 12 月在五个北高加索民族共和国进行的大规模代表性调查的统计数据，我们分析了调查对象对宽恕问题的答复，这些问题是基于有关社会和解的社会心理模型而设计的，相对于通常的理论，我们在研究中添加了一个重要的地理因素，即相距暴力事件的距离因素。利用调查数据（N=2,000）和 82 个采样点的汇总数据，我们使用了多层次模型方法分离出个人和环境因素的影响。我们发现社会身份理论在此缺乏支持，民族仇恨并不是一个重要的因素，除了奥塞梯人的个例，他们主要是信仰东正教的少数民族，在多次的冲突和伊斯兰学校屠杀事件上受到了不相称的影响。相反，关于暴力和恐怖主义的个人经验，针对社区的军事行动的影响，一般性的对他人的信任差异，作为受访者个人生活受到暴力的影响的负面程度，影响了他们宽恕的意愿。与此相对的是，俄罗斯族社区的受访者和那些与暴力相对较近的人们更愿意从事冲突后的和解。*关键词：语境效果，多层次模型，北高加索，战后和解，俄罗斯。*

Durante las dos últimas décadas ha surgido creciente interés por la reconciliación en las sociedades que emergen de conflictos. La región del Cáucaso Norte en Rusia ha experimentado múltiples y diversos conflictos desde el colapso de la Unión Soviética y la violencia continúa allí, aunque con niveles de intensidad menores de los de hace una década. Examinamos la condición de buena voluntad de perdonar a miembros de otros grupos étnicos por la violencia que ellos han perpetuado, a título de indicador del potencial de reconciliación de la región. Con datos de una encuesta representativa amplia realizada en cinco repúblicas étnicas de Cáucaso Norte, en diciembre de 2005, analizamos las respuestas a preguntas sobre perdón en relación con modelos socio-psicológicos de reconciliación, agregando a las teorías usuales una medida geográfica clave, la distancia a sucesos violentos. Utilizando los datos de la encuesta (N=2.000) y datos agregados para los ochenta y dos puntos de muestra, aplicamos un enfoque de modelaje de nivel múltiple, para separar los efectos de factores individuales y contextuales. Hallamos poco soporte a las expectattivas sobre la teoría de identidad social por cuanto la hostilidad étnica no es un factor importante, con la excepción del caso de los osetios, una minoría con predominio ortodoxo afectada de modo desproporcionado por múltiples conflictos, y la masacre de la escuela de Beslan. Por el contrario, las experiencias personales de violencia y terrorismo, los impactos de acciones militares contra comunidades, las diferencias en el nivel de confianza general en otros, y el grado con que la vida del entrevistado ha sido cambiada por la violencia, influyen negativamente en la voluntad de perdonar. A la inversa, los entrevistados en comunidades étnicas rusas y quienes viven relativamente cerca de la violencia, están más inclinados a comprometerse en reconciliación posconflicto. *Palabras clave: efectos contextuales, modelaje de nivel múltiple, Cáucaso Norte, reconciliación de posguerra, Rusia.*

Restoring peace in postconflict and conflict-affected societies is a complex process. The warring parties need to lay down their weapons, agree on an institutional division of power and resources, and begin the material reconstruction of properties and infrastructure damaged during the fighting—with or without the help of the international community (e.g., Roeder and Rothchild 2005; Collier 2006). Our study focuses on an equally important element of postwar reconstruction, conflict reconciliation that involves the emotional and cognitive processes that help former adversaries to live together in peace. We investigate why some individuals in conflict-affected societies are more inclined to forgive the perpetrators of violence than are others. We do so by examining individual and district-level indicators likely to affect reconciliation in the North Caucasus region of Russia, employing survey, census, and violence data. Since the end of the Cold War, the North Caucasus region of Russia has been the scene of different types of violent conflict: interethnic, religiously motivated, and separatist struggles. Based on a large public opinion survey carried out in December 2005 and original data that pinpoint the locations of violent incidents in the region between 1999 and 2005, this is the first study that systematically examines intergroup forgiveness in the North Caucasus. Unlike much previous works of this genre, we specifically examine whether there is a "geography" to reconciliation, beyond that explained by variations in the characteristics of the people in conflict zones. Does it matter in which community a person lives in understanding the ability to forgive? Does the community's relative level of violence produce a climate of forgiveness or of blame and accusation?

Conflict in the North Caucasus

Our study region in the North Caucasus is an ethnically diverse area of the Russian Federation, consisting of six republics (Chechnya, Ingushetia, Dagestan, Kabardino-Balkaria, Karachaevo-Cherkessia, North Ossetia) and the large Russian-dominated territory of Stavropol' (see Figure 1). The most destructive conflict has taken place in Chechnya, where civil war broke out in 1994 when Moscow responded to Chechen separatist demands with military force. In 1992, North Ossetia was the scene of a violent interethnic conflict, when informal militias representing the Ingush population concentrated in the region's Prigorodnyy *rayon* clashed with North Ossetian militias, both sides laying claim to the territory. The violent phase of the conflict, although short-lived, resulted in a large outflow of Ingush settlers from North Ossetia. Although unresolved and still a very sensitive matter (O'Loughlin, Ó Tuathail, and Kolossov 2008), this conflict has not resulted in large-scale violence since November 1992.

By 1999 the Chechen conflict began to spill over into the neighboring regions, in particular Dagestan, Ingushetia, and Kabardino-Balkaria, each of which also faces its own internal domestic conflict(s). Although fighting in Chechnya has diminished since 2002 as the rebellion has been quashed by Chechnya's new pro-Moscow president Ramzan Kadyrov, violence is increasing in other parts of the North Caucasus. By one estimate, at least seventeen insurgent organizations of varying sizes (50–2,000 members) were active in the Northern Caucasus in 2005 (Lyall 2006). Readily available weapons, unemployment, radical Islamist forces, and religious discrimination are contributing factors to the violence (Matsuzato and Ibragimov 2005). Overall, the North Caucasus has been characterized by violence directed at Russian military targets, local police, and government officials rather than at civilians (Lyall 2006; O'Loughlin and Witmer forthcoming), although there has been a considerable number of civilian kidnappings, both at the hands of the Russian security forces and the militias under the control of local leaders. Perhaps the most well-known of these attacks was the tragic Beslan (North Ossetia) school hostage crisis of September 2004. Estimates of the total killed in the

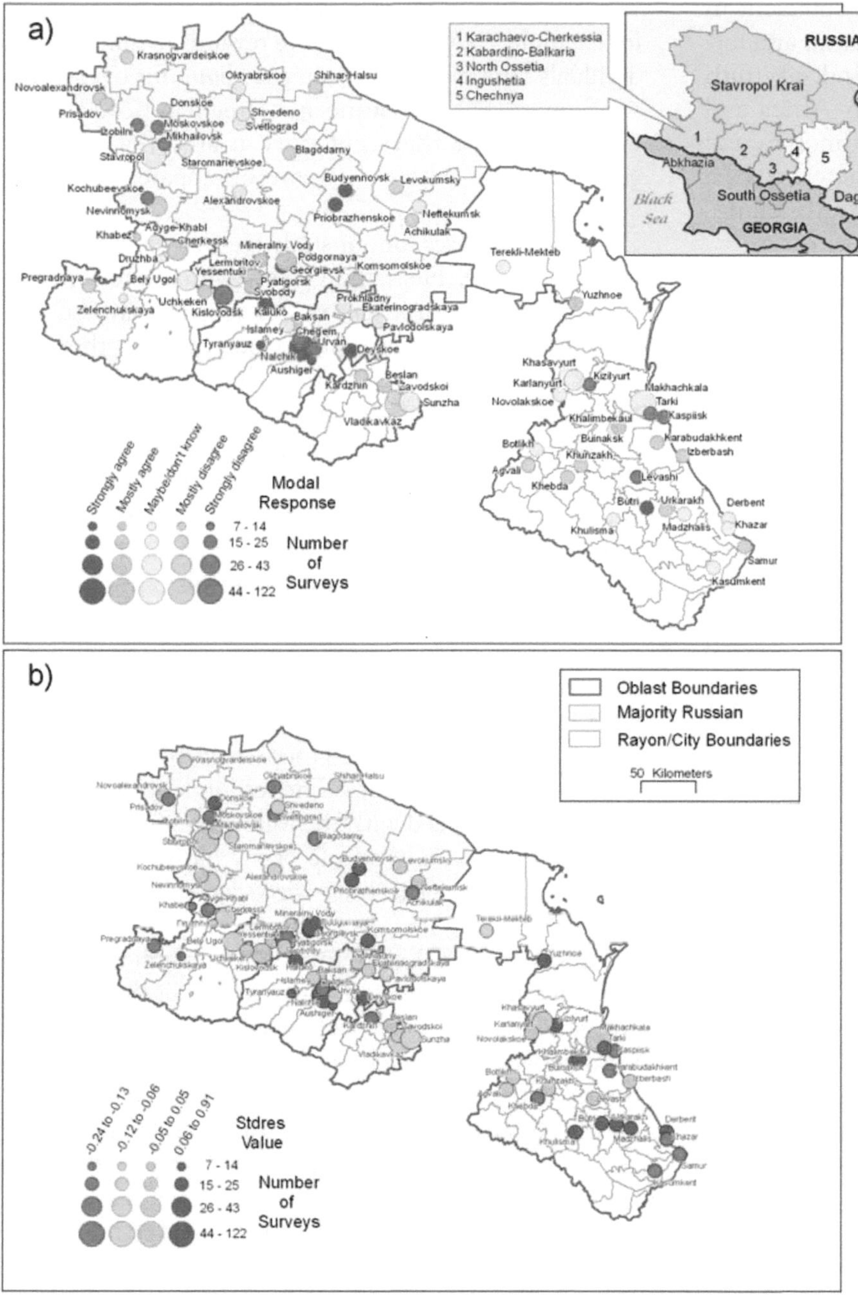

Figure 1. Modal responses to (a) the forgiveness question and (b) standardized residuals from the Generalized Linear Latent and Mixed Models (GLLAMM) model with locational inset of the North Caucasus region.

various intermeshed North Caucasian conflicts over the past fifteen years range from 75,000 to 100,000.

Reconciliation After Conflict

Conflict reconciliation is distinct from conflict settlement and resolution. Central to reconciliation is the removal of the negation of "the other" in people's identities (Kelman 2008). As such, reconciliation goes beyond conflict settlement, which concerns the interests at stake in a conflict, and conflict resolution, which concerns pragmatic changes in the relationship between former adversaries. Reconciliation is about internalizing and integrating the changed relationships into one's identity. More generally, social psychologists define intergroup reconciliation as "a process that leads to a stable end to conflict and is predicated on changes in the nature of adversarial relations between the adversaries and each of the parties' conflict-related needs, emotions, and cognitions" (Nadler, Malloy, and Fisher 2008, 4). Although reaching and implementing a settlement are critical for lasting peace in conflict-affected societies, such formal steps might not be sufficient in the absence of empathy, trust, understanding, and forgiveness among the former adversaries. Indeed, truth and reconciliation commissions in South Africa and elsewhere begin by hearing personal testimonies and applications for amnesty in societies characterized by a violent past through reconciliation and "truth telling" (e.g., Gibson and Gouws 1999; Ross 2004).

A critical step toward reconciliation is intergroup forgiveness, which is not about forgetting the past but about trying to come to terms with the past and creating a shared vision of the future by learning new aspects about oneself and one's own group and exploring the world from other groups' points of view (Noor, Brown, and Prentice 2008). Forgiveness can help prevent collective memories of violent events feeding into a recurring cycle of violence. Forgiveness is often thought about in terms of interpersonal relationships, but in societies where members of different ethnic groups have fought one another, a growing body of research in social psychology suggests that forgiveness is conceptualized as a group concern (Hewstone et al. 2008). We assess intergroup forgiveness based on a question asked of respondents in the North Caucasus in December 2005 that (indirectly) probed whether they could forgive people of other nationalities for the violence they have committed in the post-Soviet years.

Social Psychology and Intergroup Forgiveness

Several of the empirical findings on intergroup forgiveness are based on studies of Catholic and Protestant communities in Northern Ireland. There, researchers have found that identity with one's own group (ingroup identity), trust in members of other ethnic groups (outgroup trust), and contact with members from other ethnic communities (the contact hypothesis) are key determinants for intergroup forgiveness (Hewstone et al. 2006; Noor, Brown, and Prentice 2008). These studies draw on social identity theory, which views identity as central to both conflict emergence and reconciliation. Social identity theory assumes that people have both personal and social identities, and social identity comes from group membership. Because people seek a positive social identity, they compare their own group (the ingroup) with relevant other groups (outgroups). In experiments, people tend behaviorally to favor their own group, the ingroup, even if they are randomly assigned to a group with no substantive bonds holding the group members together. Indeed, even when people are assigned to groups based on some arbitrary and minimal criteria, they will favor members of their own group if, for example, they are to allocate rewards to different individuals (the minimal-group paradigm; for overviews, see Brown 2000; Hewstone and Greenland 2000). The implications are that conflicts can arise out of intergroup relations where there are no apparent material conflicts of interest.

Two hypotheses for our study are suggested by social identity theory. First, individuals who express strong pride in their ethnic group (ingroup) might be less likely to forgive perpetrators of violence from other ethnic groups (outgroups) than are those who express less pride in their ingroup. In our analysis, we include a survey question that asks the respondents the following: "To what extent to you feel proud to be a member of your national group?" The answers are given on a scale ranging from 1 (*very proud*) to 5 (*not proud at all*). Second, individuals who express low levels of interethnic trust might be less likely to forgive perpetrators of violence from other ethnic groups than those who are more trusting of outsiders. To consider this hypothesis, we include a survey question that measures individuals' trust in their own ethnic group—and by implication, their trust in other groups. The respondents were asked whether they agreed with the following statement: "It's possible to trust only people of my nationality," using a scale ranging from 1 (*strongly agree*) to 5 (*strongly disagree*).

Horowitz (1985) explained ethnic conflicts as noninstrumental competition. He pointed out that because ethnicity cannot easily be changed, intergroup comparisons between ethnic groups become even more salient than when groups are randomly assembled. Focusing on the difference between dominant and subordinate groups, Horowitz found that economically poorer groups are most often the initiators in ethnic conflicts because they perceive that they are less developed or inferior to more advanced groups. As such, it is plausible that individuals who are poorer or who have experienced ethnic discrimination are less likely to forgive perpetrators of violence from other ethnic groups than are individuals who have not experienced ethnic discrimination. To account for such a possibility, we include a survey question that asks the respondents the following: "Have you been discriminated against because of your ethnicity or religion?" The answers were given on a scale ranging from 1 (*yes, often*) to 4 (*never*).

More generally, inclusion, empathy, and respect for others are factors that both constitute and further foster conflict reconciliation (Kelman 2008). For our analysis, we include a question that asks the respondents, "Generally speaking, can most people be trusted, or do you need to be careful?" Our expectation is that individuals who are generally trusting of others are more likely than untrusting individuals to forgive perpetrators of violence.

Experiences of Violence

We hypothesize that an important determinant of forgiveness is the respondents' experiences of violence because research on postconflict societies often assumes that violent conflict damages interpersonal trust (e.g., Posner 2004; Widner 2004). Yet not all individuals living in conflict-affected societies personally experience violence. We would expect, following the research of Hewstone et al. (2006) for Northern Ireland, that someone who lives in an area that has been the target of frequent attacks is less likely to forgive perpetrators of violence than is a person without such experiences. To assess this hypothesis, we include two individual-level indicators and one district-level indicator for experiences of violence. First, we include a question (with a binary response) that asks whether the respondents' lives have significantly changed due to violence and danger in the North Caucasus. Our assumption is that changes caused by violence are likely to make individuals less forgiving. We also include a question that asks whether the respondent's community was targeted by military operations or police actions in the conflicts in the North Caucasus (binary response), expecting that those living in communities targeted in such actions would be less forgiving.

In addition to examining the effect of individual—and, thus, subjective—experiences of violence, we also assessed whether survey respondents residing in violence-ridden areas were less likely to forgive perpetrators of violence than respondents living in more peaceful areas. Based on original data (O'Loughlin and Witmer forthcoming), we aggregated violent events for eighty-two sampling points across the North Caucasus region and thus created an indicator that counts the number of violent incidents between 1999 (the start of the second Chechen war) and 2005 (the time of the survey) within a 50-km radius of each survey respondent. Our expectation is that respondents residing in areas characterized by violent incidents are less likely to forgive perpetrators of violence. We used a threshold of 50 km because a plot of violence by 25-km distance bands shows a substantial decrease in the occurrence at this distance.

Socioeconomic Status and Ethnic Composition

As control variables in the model, we include each individual's self-reported material status, based on a question that assesses the degree to which he or she can purchase the things he or she needs and wants. The expectation is that economic hardship might cause more negative assessments of others, which can affect individuals' abilities or willingness to forgive perpetrators of violence. We also include an indicator for the share of Russians in the sample districts. There has been a significant redistribution of Russians in the North Caucasus since the conflicts began in the early 1990s with big drops in the Russian proportions in the ethnic republics (especially Chechnya, Dagestan, and Ingushetia) and a growth in Stavropol' due to flight from conflicts elsewhere (Belozerov 2005). We therefore expect less forgiveness in communities with a high Russian ratio. Finally, we include a predictor of whether the respondent was Ossetian. A small, mostly Orthodox population in a predominantly Muslim region, Ossetians have traditionally been allied to Moscow and have fought three recent wars with their Ingush and Georgian neighbors over territory. The horrific violence at the Beslan school occurred just over a year before our survey and caused an enormous shock to locals. We expect that the nature of this terror, which killed about 330, including more than 180 children, will foster low levels of forgiveness among the Ossetians.

Contextual Effects in Postwar Reconciliation

It is now widely acknowledged by geographers that "places matter"; that is, that individual-level predictors do not fully account for variation in political and social behavior between communities. Long a debating point between political geographers and political scientists (e.g., Agnew 1996; King 1996; O'Loughlin 2000), recent methodological advances have allowed geographers to clarify the relative importance of contextual effects and to point to the limitations of place-free approaches in the behavioral sciences. Work in electoral geography especially has indicated the significant influence of community interactions (titled the "friends and neighbors" effect) on voters' choices (Johnston et al. 2004). A growing recognition of the value of multilevel modeling approaches for separating out the individual (first-level) and community (second-level) effects has generated dozens of studies that indicate modest (5 to 15 percent) but important contributions of contextual effects (Jones and Duncan 1996). Given the variability in the local geographies of civil wars (Kalyvas 2006), our expectation is that the reconciliation process is also geographically variable and related to violence experiences. A multilevel approach allows us to measure the social-psychological effects on the individual, as reviewed earlier, and the geographic effects of community exposure to violence.

Data, Methods, and Results

We assess intergroup forgiveness based on the responses of 2,000 individuals in four North Caucasus republics (North Ossetia, Dagestan, Kabardino-Balkaria, and Karachaevo-Cherkessia) and one territory (Stavropol') to the question, "There are people who are convinced that they could never forgive people of other nationalities for the violence they have committed in the last fifteen years. Are you among those people?" The sensitive nature of this question required an indirect wording, a tactic that was confirmed in the pilot testing of the survey instrument. Due to missing data, seventy-eight responses were dropped from the analysis. Because of the dangers involved in doorstep interviewing in the most violent republics, Ingushetia and Chechnya had to be excluded from the analysis. The survey was distributed in proportion to population and ethnicity in the sample regions, and it is the most comprehensive survey carried out in the region to date. A geographically stratified sampling strategy captured the locational variation within the republics: urban–rural, mixed and homogenous communities, mountains–piedmont–plains location, and material wellbeing. There are eighty-two sample points, ranging from large cities (e.g., Stavropol', Mineralnyy Vody, Makhachkala, Vladikavkaz) to isolated rural settlements (see Figure 1). Details on the survey are available in Bakke et al. (2009).

The dependent variable, forgiveness, is an ordered, categorical variable, and the 1,922 responses are ordered from "definitely no" ($n = 312$), "mostly disagree" ($n = 527$), "maybe" ($n = 536$), "mostly agree" ($n = 328$), to "definitely yes" ($n = 219$). The map of the distribution of modal responses in Figure 1a shows both dramatic variation across the region and sizable differences within each republic. "Strongly agree" and "mostly agree" (with the proposition) indicate a reluctance to forgive others for the violence they committed. These views characterize samples in or near places that experienced significant hostage-taking and consequent loss of life (Budyennovsk-Priobrazenskoye in 1995 and Beslan in 2004) or large-scale attacks and massive loss of life (Sunzha in 1992, Yessentuki-Mineralnyy Vody in 2003, and Nal'chik-Ausiger in 2005). Communities furthest from the zones of greatest violence, in northern Stavropol', Karachaevo-Cherkessia, and most of Dagestan, have the highest levels of propensity to forgive.

The model's independent predictors are personal attributes and self-reported attitudes of respondents as well as characteristics of each of the eighty-two *rayoni* (counties) in which the sampling points are located. Key contextual variables are the Russian proportion of the population in each *rayon* (from the 2002 Russian census) and the number of violent events that have occurred within 50 km of the sampling point. This combination of individual and district-level data results in a statistical model that is somewhat unusual. We use a multilevel, ordered probit specification to capture the individual and aggregate variable elements of our model. Multilevel models are now common in the social sciences, and so are ordered probit specifications; however, only a few examples of multilevel, ordered probits have been developed, mostly in the context of fitting conditional models across diverse societies (King and Wand 2007).

To represent the observed categorical variable, we define a latent response variable ($Y_{i,d}$) for the ith individual in the dth district such that the range of the latent variable is divided into $k + 1$ categories that correspond to the observed data by estimated parameters. The latent model for forgiveness is

$$y_{i,d} = \begin{cases} 1 \text{ (definitely no)} & \Leftrightarrow -\infty < y^*_{i,d} < \tau_1 \\ 2 \text{ (mostly disagree)} & \Leftrightarrow \tau_1 < y^*_{i,d} < \tau_2 \\ 3 \text{ (maybe)} & \Leftrightarrow \tau_2 < y^*_{i,d} < \tau_3 \\ 4 \text{ (mostly agree)} & \Leftrightarrow \tau_3 < y^*_{i,d} < \tau_4 \\ 5 \text{ (definitely yes)} & \Leftrightarrow \tau_4 < y^*_{i,d} < \infty \end{cases}$$

This latent model has no analytical solution and is tackled by a normal quadrature approach (Rabe-Hesketh and Skrondal 2008), available via the Generalized Linear Latent and Mixed Models package in STATA, used to produce the estimates.

We specify our ordered probit model of forgiveness as significantly related to Ossetian ethnic membership, level of targeting of the community by the Russian military, level of ethnic pride, the scale of changes in the respondent's life due to violence, the respondent's level of general trust, and two aggregate measures, share of

Table 1. Multilevel ordered probit model estimates of forgiveness—GLLAMM estimates for the North Caucasus sample

Predictor	Coefficient	Standard error	Z-score (Probability)
Ossetian	0.3235	0.121	2.69 (0.007)
Military target	0.3651	0.093	3.92 (0.000)
Pride in ethnic group	0.0708	0.061	1.18 (0.239)
Changed life	0.2848	0.054	5.31 (0.000)
General trust	0.1676	0.067	2.51 (0.012)
Russian share of population	−0.0004	0.001	−2.86 (0.004)
Violence within 50 kilometers	−0.0002	0.0001	−3.56 (0.000)
$\tau 1$	−1.147	0.108	−10.64 (0.000)
$\tau 2$	−.225	0.107	−2.11 (0.035)
$\tau 3$	0.623	0.106	5.87 (0.000)
$\tau 4$	1.425	0.109	13.08 (0.000)

Notes: Number of respondents (Level 1 units) = 1,922. Number of Level 2 units (survey points) = 82. Log likelihood = −2785.83. GLLAMM = Generalized Linear Latent and Mixed Models.

Russians in local population and incidents of violence within 50 km of the respondent's home. We checked other possible sociodemographic controls (gender, occupation, age, etc.) but do not include them because they are insignificant. More important, five factors that we expected to be related to forgiveness from the social psychology of reconciliation literature cited earlier—perception of discrimination, material wealth status, pride in the respondent's own ethnic group, nature of intergroup contacts, and trust in other ethnicities—are not statistically significant when considered in the same ordered probit regression as the significant relationships reported in Table 1; therefore, they are dropped from the model. We retain the pride measure (a binary variable separating those who express strong pride or pride in their ethnic group) because of its central importance in the social-psychological theories of reconciliation.

In the null multilevel model (with no predictors), the Level 1 (respondents) variance accounts for 71.3 percent and the Level 2 (sample points) for 28.7 percent of the total variance; however, after the ordered probit model is fitted, the map of the standardized residuals for the eighty-two survey points (Figure 1b) shows no evident spatial or contextual patterning, nor any correlation with the original dependent variable values. Only in Dagestan is there any evident clustering of residuals but the overall range is small. The sample level variances for the survey points in the model, however, are significant at 0.470 with a standard error of 0.076.

In examining the relationships, as expected, Ossetians are less likely to forgive than are other ethnic groups, an outcome of the involvement of this community in an unresolved territorial conflict with the Ingush and the attack on the Beslan school a year before the survey. People perceiving that they live in communities that have experienced disproportionate attacks from the Russian military and allied paramilitary forces are also less willing to forgive perpetrators of violence. Although the ethnic pride predictor shows a coefficient in the expected (positive) direction, the relationship is not significant. Respondents who feel that their lives have been significantly changed by violence in the region are also less likely to forgive than those who have not experienced such violence-induced life changes. Persons who are more cautious (picking the prompt "you cannot be too careful") are less likely to forgive than are respondents with a higher level of general trust ("most people can be trusted"). Both of our aggregate indicators are negatively related to the dependent variable, indicating a higher level of forgiveness. The value for the Russian share of the population is undoubtedly related to the geographic distribution of this ethnicity to the north and west of the North Caucasus and thus farther from the locales of highest violence. The only unexpected result was for the measure of exposure to violence; people living in communities with more violent incidences within 50 km are more likely to forgive than those in more peaceful locales.

To illustrate the results graphically in Figure 2, we present the predicted value of the latent representation of forgiveness (minus any random effects) in two different scenarios. The blue line shows the density of predictions for all 1,922 respondents for the model. The brown line portrays a similar calculation, except that we removed a key contextual variable, the share of the Russian population in the model. The comparison of these two distributions illustrates clearly that inclusion of this variable changes the model predictions, increasing substantially the number of predictions of higher levels of forgiveness. Figure 2 shows that ignoring the multilevel context yields results that are far too optimistic about forgiveness while underestimating the number of respondents who self-report lower levels of forgiveness.

Our results do not strongly support the theoretical propositions of the social-psychological theories of reconciliation after extended violence, but they are in line with other studies of interethnic attitudes in former

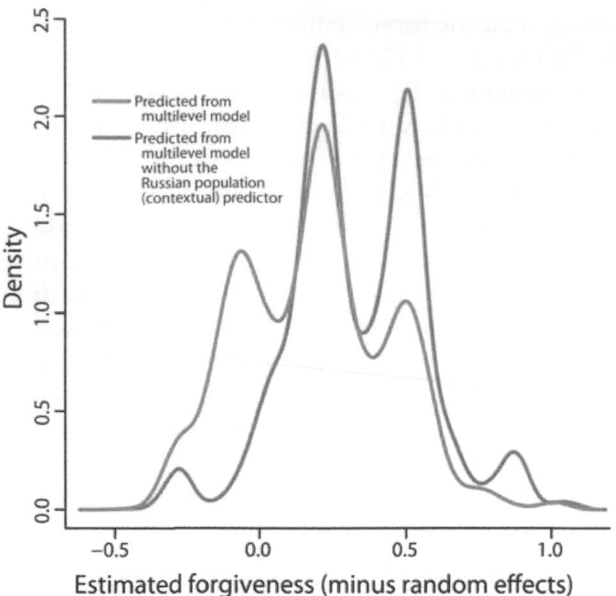

Figure 2. Effect of the removal of contextual variable (Russian percentage) from the model of forgiveness.

Communist states. Whitt and Wilson (2007), in an experimental game in Bosnia with different ethnicities, find that although there is a preference for the ingroup, the bias against the outgroups (other ethnicities) is less than expected; they conclude that this higher norm of fairness bodes well for reconciliation. For the former Soviet Union, Hale (2008) showed that the cognitive process of using ethnicity as an "uncertainty-reducing" process is associated with peaceful and cooperative ethnic relations in some regions but activated in movements for secession and conflict in others. With the exception of the Ossetians, whose experiences of violence are unique in the region, we did not find that ethnic group membership, or other sociodemographic categories, helped us understand people's willingness to forgive others for violence. Instead, like the study of long-term trauma among Bosnians exposed to violence by Ringdal, Ringdal, and Simkus (2008), we find that those more directly affected by violence in their daily lives—those who consider themselves targeted and who feel forced to adapt to changed circumstances due to violence—experienced the most significant long-term effects. In our case, those with personal experiences of violence have a lower propensity to forgive those who have perpetrated it. In the most extreme example of war trauma's effects, a study of thirty-four suicide terrorists in Chechnya (based on interviews with the terrorists' family members, friends, and hostages), Akhmedova and Speckhard (2006) found that the terrorists' own experiences of years of violence led to personal trauma and a wish to avenge the deaths of relatives or friends.

Conclusions

The study of intergroup reconciliation and forgiveness is a relatively new research agenda. Our aim in this study is not to develop a theory of intergroup forgiveness. Rather, building on the existing literature, our study empirically examines intergroup forgiveness in a society that has been the scene of several intergroup conflicts. The current situation in the North Caucasus is *not yet* a postwar one where truth and reconciliation through national or international commissions can be promoted. Instead, multiple conflicts drag on in many localities, with occasional outbursts of dramatic violence, and their geographies shift from year to year. Although the fighting in Chechnya, between federal forces and their local allies on one side and separatist rebels on the other, has weakened in the past few years, low-level violence continues to diffuse to more communities outside Chechnya. The impact of this spread on communities new to the conflicts will result in more forced adjustments by people previously (barely) unaffected. This diffusion is also expected to postpone the possibility of reconciliation in the region.

Our study emphasized the importance of the level of personal exposure to violence and its effects on one's daily activities. Moreover, the usual sociodemographic categories are not very helpful in understanding the propensity toward forgiveness, which undermines any easy categorization of certain groups (ethnic, age, gender, socioeconomic status) as more willing and able to engage in reconciliation. Because violence in civil wars is both localized and temporally shifting, our modeling highlights the geography of willingness to forgive across communities, after accounting for its variation due to the sociodemographic composition of residents.

Unlike postwar developments in countries like Northern Ireland, Liberia, South Africa, Rwanda, and Guatemala, reconciliation in the North Caucasus of Russia is a work in progress. The multiple and shifting lines of conflict there, sometimes involving local communities but usually involving some combination of state authorities and religious, ethnic, or regional opponents, makes any assumptions about the permanence of a downturn in violence subject to rapid reversal.

Acknowledgments

We are grateful to Alexei Grazhdankin of the Levada Center in Moscow for organizing and successfully carrying out the contract for the large and complex surveys in the North Caucasus; to Vladimir Kolossov of the Russian Academy of Sciences for his key insights and fieldwork leadership, his contribution to the design and wording of the questionnaire, and his continuing friendship and collegiality; to Gerard Toal of Virginia Tech University for his contribution to the survey instrument design and fieldwork comradeship in multiple trips to the Caucasus; to Frank Witmer and Nancy Thorwardson for (carto)graphical assistance; and to the *Annals* reviewers and editor, Audrey Kobayashi, for comments that improved the article. This research was supported by the National Science Foundation (Grant 0433927), via its Cross Directorate Initiative in Human and Social Dynamics.

References

Agnew, J. 1996. Mapping politics: How context counts in electoral geography. *Political Geography* 15 (2): 129–46.

Akhmedova, K., and A. Speckhard. 2006. A multi-causal analysis of the genesis of suicide terrorism: The Chechen case. In *Tangled roots: Social and psychological factors in the genesis of terrorism*, ed. J. I. Victoroff, 324–55. Amsterdam: IOS Press.

Bakke, K. M., X. Cao, J. O'Loughlin, and M. D. Ward. 2009. Social distance in Bosnia-Herzegovina and the North Caucasus region of Russia: Inter- and intra-ethnic attitudes and identities. *Nations and Nationalism* 15 (1): 229–55.

Belozerov, V. S. 2005. *Etnicheskaya karta Serverogo Kavkaza* [Ethnic map of the North Caucasus]. Moscow: OGI.

Brown, R. 2000. Social identity theory: Past achievements, current problems and future challenges. *European Journal of Social Psychology* 30 (6): 745–78.

Collier, P. 2006. Post-conflict economic recovery. Department of Economics, Oxford University. http://users.ox.ac.uk/~econpco/research/pdfs/IPA-PostConflictEconomic Recovery.pdf (last accessed 20 September 2009).

Gibson, J. L., and A. Gouws. 1999. Truth and reconciliation in South Africa: Attributions of blame and the struggle over apartheid. *American Political Science Review* 93 (3): 501–17.

Hale, H. E. 2008. *The foundations of ethnic politics: Separatism of states and nations in Eurasia and the world*. New York: Cambridge University Press.

Hewstone, M., E. Cairns, A. Voci, J. Hamberger, and U. Niens. 2006. Intergroup contact, forgiveness, and experience of the "Troubles" in Northern Ireland. *Journal of Social Issues* 62 (1): 99–120.

Hewstone, M., and K. Greenland. 2000. Inter-group conflict. *International Journal of Psychology* 35 (2): 136–44.

Hewstone, M., J. B. Kenworthy, E. Cairns, N. Tausch, J. Hughes, T. Tam, A. Voci, U. von Hecker, and C. Pinder. 2008. Stepping stones in Northern Ireland: Intergroup contact, forgiveness, and trust. In *The social psychology of intergroup reconciliation: From violent conflict to peaceful coexistence*, ed. A. Nadler, T. E. Malloy, and J. D. Fisher, 199–226. New York: Oxford University Press.

Horowitz, D. L. 1985. *Ethnic groups in conflict*. Berkeley: University of California Press.

Johnston, R., K. Jones, R. Sarker, C. Propper, S. Burgess, and A. Bolster. 2004. Party support and the neighbourhood effect: Spatial polarisation of the British electorate, 1991–2001. *Political Geography* 23 (4): 367–402.

Jones, K., and C. Duncan. 1996. People and places: The multilevel model as a general framework of the quantitative analysis of geographical data. In *Spatial analysis: Modelling in a GIS environment*, ed. P. Longley and M Batty, 79–104. New York: Wiley.

Kalyvas, S. N. 2006. *The logic of civil war violence*. New York: Cambridge University Press.

Kelman, H. C. 2008. Reconciliation from a social-psychological perspective. In *The social psychology of intergroup reconciliation: From violent conflict to peaceful coexistence*, ed. A. Nadler, T. E. Malloy, and J. D. Fisher, 15–32. New York: Oxford University Press.

King, G. 1996. Why context should not count. *Political Geography* 15 (2): 159–64.

King, G., and J. Wand. 2007. Comparing in-comparable survey responses: Evaluating and selecting anchoring vignettes. *Political Analysis* 15 (1): 46–66.

Lyall, J. M. K. 2006. Landscapes of violence: A comparative study of insurgency in the Northern Caucasus. Paper presented at the annual meeting of the Midwest Political Science Association, Chicago.

Matsuzato, K., and M.-R. Ibragimov. 2005. Islamic politics at the subregional level in Dagestan: Tariqa brotherhoods, ethnicities, localism, and the Spiritual Board. *Europe-Asia Studies* 57 (5): 753–79.

Nadler, A., T. E. Malloy, and J. D. Fisher. 2008. Intergroup reconciliation: Dimensions and themes. In *The social psychology of intergroup reconciliation: From violent conflict to peaceful coexistence*, ed. A. Nadler, T. E. Malloy, and J. D. Fisher, 3–14. New York: Oxford University Press.

Noor, M., R. Brown, and G. Prentice. 2008. Prospects for intergroup reconciliation: Social-psychological predictors of intergroup forgiveness and reparation in Northern Ireland and Chile. In *The social psychology of intergroup reconciliation: From violent conflict to peaceful coexistence*, ed. A. Nadler, T. E. Malloy, and J. D. Fisher, 97–114. New York: Oxford University Press.

O'Loughlin J. 2000. Geography as space and geography as place: The divide between political science and political geography continues. *Geopolitics* 5 (1): 126–37.

O'Loughlin, J., G. Ó Tuathail, and V. Kolossov. 2008. The localized geopolitics of displacement and return in the Eastern Prigorodnyy Rayon of North Ossetia. *Eurasian Geography and Economics* 48 (6): 635–69.

O'Loughlin, J., and F. Witmer. Forthcoming. The geography of violence in the North Caucasus, 1999–2007. *Annals of the Association of American Geographers*.

Posner, D. N. 2004. Civil society and the reconstruction of failed states. In *When states fail: Causes and consequences*, ed. R. I. Rotberg, 237–55. Princeton, NJ: Princeton University Press.

Rabe-Hesketh, S., and A. Skrondal. 2008. *Multilevel and longitudinal modeling using Stata*. 2nd ed. College Station, TX: Stata Press.

Ringdal, G. I., K. Ringdal, and A. Simkus. 2008. War experiences and war-related distress in Bosnia and Herzegovina eight years after war. *Croatian Medical Journal* 49 (1): 75–86.

Roeder, P. G., and D. Rothchild. 2005. *Sustainable peace: Power and democracy after civil wars*. Ithaca, NY: Cornell University Press.

Ross, A. 2004. Truth and consequences in Guatemala. *Geojournal* 60 (1): 73–79.

Whitt, S., and R. K. Wilson. 2007. The dictator game: Fairness and ethnicity in postwar Bosnia. *American Journal of Political Science* 51 (3): 655–68.

Widner, J. 2004. Building effective trust in the aftermath of severe conflict. In *When states fail: Causes and consequences*, ed. R. I. Rotberg, 222–36. Princeton, NJ: Princeton University Press.

“Post”-Conflict Displacement: Isolation and Integration in Georgia

Beth Mitchneck, Olga V. Mayorova, and Joanna Regulska

The Abkhaz civil wars and continuing territorial conflicts in Georgia have resulted in the long-term displacement of more than 200,000 people since the early 1990s. Although the international and local discourse is about integrating internally displaced persons (IDPs), little research has documented the meaning of isolation or integration for the daily lives of the IDPs or the local population. We engage the discourse about integration and isolation by analyzing the composition, size, and density of social networks in the “post”-conflict environment and the socio-spatial characteristics of social interactions and social networks. We combine a formal social network analysis with a daily path analysis to explore how socio-spatial patterns are formative of social networks and explore how various demographic factors, including gender, dwelling status, and employment status, may be related to the nature of social interactions and social networks. Our results are initially puzzling and suggest the need to rethink the meaning of isolation and integration within postconflict situations. We had expected to find greater diversity of social interactions in both populations, especially IDPs in private accommodations, because they are generally thought to have more diverse social interactions. The social network and daily path analyses, however, suggest evidence of social isolation within social networks among the entire population, not only among IDPs. We find a high degree of social isolation in two ways: (1) the persistent dominance of family and kin in all social networks and (2) highly dense (or closed) social networks in the entire population across gender, dwelling, and migrant status. The only demographic factor that appears to distinguish patterns is whether an individual engages in income-generating activity. Finally, using narrative interviews, we also explore the meaning of integration and isolation during displacement in the Georgian context. *Key Words: Abkhazia, displacement, Georgia, integration, internally displaced persons, social networks.*

阿布哈兹内战和格鲁吉亚持续的领土冲突自 90 年代初起已经造成 20 多万人长期流离失所。尽管国际和地方的言论是要是使国内流离失所者（IDPs）一体化，但是很少有对国内流离失所者或当地居民日常生活的隔离或一体化的意义的研究记载。我们通过分析在“后”冲突环境中的社会网络的组成、大小、和密度以及社会交往和社会网络里的社会空间特点，从而涉及一体化和孤立的言论。我们结合日常路径分析和正规社会网络分析，探讨社会经济空间格局是如何影响社会网络的形成的，并探讨各种人口因素，包括性别，居住状况和就业状况是如何与社会交往和社会网络相联的。我们的结果最初令人费解，也提示我们在“后”冲突形式中对反思孤立和一体化意义的必要。我们本来期望在这两个群体中发现更大的社会交往的多样性，特别是住在私人住所中的国内流离失所者，因为一般认为他们有更多样的社交活动。但是，我们的社会网络和日常路径分析表明社会孤立的证据存在于整个人口，而不仅仅是国内流离失所者群体中。我们发现两种方式的高度的社会孤立：（1）家庭和亲人在所有的社会网络中持久的主导（2）跨性别、住所、和移民身份的全部人口的高密度的（或闭塞的）社会网络。唯一的能区分不同模式的人口因素是一个人是否有从事创收活动。最后，使用说明采访，我们还探讨了在格鲁吉亚人口流离失所过程中一体化和孤立的意义。*关键词：阿布哈兹，流离失所，格鲁吉亚，一体化，国内流离失所者(IDP)，社会网络。*

Las guerras civiles de Abjasia y los continuados conflictos territoriales de Georgia han dado por resultado el prolongado desplazamiento de más de 200.000 personas, desde principios de los años 1990. Si bien en el discurso internacional y el local se clama por integrar a las personas desplazadas internamente (PDI), muy poca es la investigación hecha para documentar el significado de aislamiento o integración en la vida cotidiana de las PDI, o en la de la población local. Nosotros abordamos el discurso sobre integración y aislamiento analizando la composición, tamaño y densidad de las redes sociales en el entorno “pos”-conflicto, y las características socio-espaciales de las interacciones sociales y redes sociales. Combinamos un análisis formal de redes sociales con un análisis de trayectorias diarias para explorar de qué manera contribuyen los patrones socio-espaciales a la formación de redes sociales, y para indagar cómo pueden relacionarse

factores demográficos que incluyan género, estatus residencial y estatus de empleo, con la naturaleza de las interacciones sociales y de las redes sociales. A primera vista, los resultados obtenidos son intrigantes y sugieren la necesidad de repensar el significado de aislamiento e integración en situaciones pos-conflicto. Habíamos esperado encontrar una mayor diversidad de interacciones sociales en ambas poblaciones, especialmente en las PDIs en albergue privado, porque se suele creer que éstas tienen interacciones sociales de mayor diversidad. Sin embargo, los análisis de redes sociales y de trayectorias diarias parecen detectar el aislamiento social en las redes sociales en toda la población, y no solamente entre las PDIs. Encontramos un alto grado de aislamiento social expresado de dos maneras: (1) el dominio pertinaz de familia y parentela en todas las redes sociales, (2) redes sociales de alta densidad (o cerradas) en toda la población, por encima de consideraciones de género, residencia y estatus del migrante. El único factor demográfico que parece distinguir patrones tiene que ver con que el individuo se involucre en actividades que generen ingreso. Por último, en el contexto de Georgia, también exploramos el significado del aislamiento y la integración durante el desplazamiento, por medio de entrevistas narrativas.
Palabras clave: Abjasia, desplazamiento, Georgia, integración, personas desplazadas internamente, redes sociales.

As the August 2008 war between Georgia and Russia illustrates, ethno-territorial conflicts occur in the presence of a peacekeeping force and even when a political discourse suggests that the conflict is "frozen." Territorial conflicts are not unique; at the end of 2007, the United Nations High Commissioner for Refugees (UNHCR) estimated that there were 26 million internally displaced persons (IDPs) worldwide (UNHCR 2008). A key reason for the large number of IDPs is the nature and extent of protracted conflicts. In Georgia, as in many other countries around the world, displacement is a constant state due to violence surrounding territorial disputes and desires for territorial autonomy or independence.

International and local discourses concern integrating IDPs into the local populations, but little research has documented what isolation or integration mean in the context of the daily lives of IDPs, or of the local population, in a "post"-conflict environment. As we discover the limitations of applying insights from the governance, social network, and international refugee literatures to internal displacement, we are engaging in a theory-building project. Using a conceptual framework that centers both the migrant and the social network in the governance environment of conflict situations,[1] we explore how social networks and interactions can be formative or indicative of social isolation or integration. We look specifically at the composition, size, and density of social networks in the postconflict environment for both the IDPs and the local population and explore how the socio-spatial characteristics of social interactions relate to isolation and integration. Ager and Strang (2008) conceptualize the various domains in which integration can occur and suggest that employment and housing types are markers and means of integration, and that integration is often used but understood in a variety of ways. We argue that meanings of integration vary across participants in the governance environment and that illustrating these meanings shows how a dominant integration discourse contributes to masking isolation and to developing policy directions that have limited chances at success. In the following sections, we briefly discuss the international discourse about integrating IDPs, the Georgian background, our research design, and then patterns of social interaction and networks.

The Governance Environment and the Integration Discourse

Forced migrants are part of a *governance environment* that surrounds the postconflict situation and frames the social context for management and interaction. Those actors and agencies involved in managing the IDPs are viewed here as constituting the governance environment. Governance suggests that forced migration is neither the sole domain of institutional actors nor of individuals but rather a political-economic process in which many individuals and institutions participate, including national and international nongovernment organizations (NGOs) and supranational state and nonstate actors as well as the migrants (e.g., Leitner 1997; Silvey and Lawson 1999; Nagel 2002). Participants in the governance environment are engaged in developing the discourse around integration of forced migrants. Others have researched the governance environment for refugees (see Hyndman 2000; Loescher 2001; Helton 2002), but little research has focused on the nexus linking daily practice with social networks and the governance environment for IDPs.

The Integration Discourse

The UNHCR has long been a key player in managing forced migrants.[2] The original statute charging the

UNHCR sets a goal of assimilating refugees (UNHCR 2007). The terminology has evolved and currently the UNHCR cites local integration of refugees as a key goal (UNHCR 2006). UNHCR documents about Georgia also cite the importance of promoting integration into the current place of residence, at the same time as allowing for return to Abkhazia (UNHCR 2002). Although the UNHCR is one player in the larger governance environment, it helps frame the discourse and the focus on integration found in materials of many other organizations. The Norwegian Refugee Council, a longtime player worldwide and in Georgia, frames the issues as promoting integration into local communities and targeting reintegration to origin (Internal Displacement Monitoring Centre 2006). Walter Kalin (2006), the UN Secretary-General's Representative on the Human Rights of IDPs, notes that Georgian IDPs in collective centers are highly isolated and that the Georgian government had planned to facilitate "economic and social integration of the displaced into local communities, including through the privatization of collective centres to the benefit of IDPs" (67).

The Georgian government has consistently maintained that return is the only viable option. The ministry tasked with managing IDPs is called the Ministry for Refugees and Accommodations (MRA). The Government of Georgia has created a system of separate educational and medical facilities for the IDPs as well as a separate government called the Legitimate Government of Abkhazia (for those IDPs from Abkhazia). The separate facilities were thought to meet the needs of the IDPs better than integrating them into the national educational and medical structures because such facilities simultaneously provide employment for the IDPs trained in these sectors and could provide services better suited to meet their specific needs. The separate structures have also been interpreted as reinforcing the social isolation of the IDP population and shaping the desire to return to Abkhazia.

The Georgian government recently dramatically changed its approach to management of IDPs and the subsequent discourse. As a result of an international coalition of governmental and nongovernmental agencies, the MRA managed the large-scale and cooperative development of the State Strategy for Internally Displaced Persons–Persecuted (Government of Georgia 2007).[3] A large number of local NGOs also participated in the development of the State Strategy as a means of including IDPs in the process. Prior to this policy document, the Georgian government itself had not clearly articulated the goal of integration in addition to safe return.

The strong international discourse on the isolation of IDPs and the need for social integration shape our analysis of social networks. The discourse, conventional wisdom, and limited research on refugees and social networks suggest that IDPs in collective centers are highly isolated relative to and from the local population. Because collective centers are often on the outskirts of towns or in isolated locations, we expected that social interactions and thus social networks would be more limited in size and would encourage the development of highly dense (closed) networks, more so than IDPs in private housing or than the local population who live spatially more integrated. In addition, we expected that IDPs in collective centers would have social networks comprised of both relatives and close friends living in the collective centers who would take on the characteristics of fictive kin because of the social isolation within the collective center (see Ebaugh and Mary 2000 for discussion of fictive kin among immigrant communities).

There is also a very strong discourse among the Georgian government and humanitarian aid organizations to move IDPs from collective centers, in part because they believe those living in private accommodations do not experience the physical or social isolation of collective centers. We expected to find that IDPs in collective centers have smaller and denser networks than those of other populations. We also expected to find that women's social networks were larger and less dense among the displaced because of the discourse surrounding the greater entrepreneurial activity and social engagement of women in displacement, both territorial and social. A previous study of Indonesian IDPs found that women's networks were larger and more fully established than men's after displacement (Okten and Osili 2004). Another study of Bosnian refugees in New York and Vienna also found men's networks were small and more focused on the refugee community than women's networks and women's networks were more integrated into the local population (Franz 2003). We expected to find similar results in Georgia, in part because a woman's daily path is more varied due to housework and child care activities.

The Georgian Case

Since gaining its independence in 1991, Georgia has experienced a number of dramatic transformations, including the fall 2003 final dissolution of ties with the past communist regime (i.e., the Rose Revolution) when a former Soviet leader was replaced with current President Saakashvili. There have been historically rooted civil wars (Suny 1994), persistent economic crises (Demetriou 2002), and continual political crises with the Russian government. Several territorial conflicts have arisen, including those in Abkhazia, South Ossetia, and Adjaria. Armed conflict in Abkhazia began in 1992 and continued until a ceasefire in 1994, yet some were displaced as recently as the late 1990s.[4]

The majority of the IDPs live in the capital city of Tbilisi, Kutaisi, or Zugdidi, a city adjacent to the conflict zone. More than fifteen years after the initial conflict, about 45 percent lived in what are called collective centers and the remaining 55 percent lived in private accommodations either on their own or with a host family. These figures have decreased in the last two years due to an effort on the part of the Georgian government to resettle IDPs into private accommodations that includes a privatization plan for collective centers that are considered viable. The drive to move IDPs from collective centers is related to the horrible living conditions in the centers and a discourse in Georgia that the best way to improve IDP living standards is to move them from collective centers into private accommodations (Government of Georgia 2007). The collective centers are located in former hotels, sanatoria, hospitals, kindergartens, and office buildings.

IDPs remain temporarily settled and estimates suggest that unemployment is between 35 and 45 percent for IDPs, substantially higher than the estimated 15 percent for the local population (Holtzman and Nezam 2004). These high unemployment rates underscore the importance of livelihood strategies and access to information about employment. A recent report by the Georgian Institute for Polling and Marketing[5] (2005) finds that IDPs use informal livelihood strategies to a greater degree than do the general population. Despite the prevalence and necessity of informal strategies for livelihood, part of the discourse around IDPs is the "syndrome of dependence on assistance and lack of initiative" (Government of Georgia 2007).

Because the majority of the literature on IDPs in the region is written as reports to support or influence policymaking and humanitarian aid institutions (e.g., Dershem and Sakandelidze 2002; Sumbadze and Tarkhan-Mouravi 2003), there is little on which to base a formal analysis of social networks. Most of these reports note that networks can play an important role in the development and realization of livelihood strategies, but no study before ours has looked specifically at the construction and use of the social networks in relation to dwelling type. Dwelling type is important because, as noted earlier, much of the discourse on integration centers on the geographical imperative of moving IDPs from collective centers to private accommodations as a means toward achieving integration. Underlying the discourse is an assumption that living among the local population will lead to social and economic integration.

Research Design

The research design is interdisciplinary, combining insights and methodologies from gender studies, geography, sociology, and anthropology. The single survey instrument includes structured questions, narrative questions, a formal social network analysis, and a daily path analysis. The narrative questions collect departure histories as well as personal perspectives on their lives. The interviews were conducted face-to-face with trained Georgian interviewers from June through August 2007.

Because we want to compare the social interactions and networks among and between the local and displaced populations, it is critical to include a sample from the local population in the three case study sites. Without the comparison, we would not know how IDPs' social lives compare to those of the local population. We chose three cities where the majority of IDPs from Abkhazia live and interviewed a total of 180 people, roughly half men and half women, who included 120 IDPs and 60 from the local population. We interviewed the IDPs in proportion to those living in private accommodation and collective centers at the time that the interviews took place. The sample was purposive in that we included only people of working age distributed across dwelling type and gender.

The daily path analysis assesses the spatiality of living in displacement and provides a mechanism by which we link social interaction and social network formation to spatial behavior. Questions asked for the daily path analysis include the following: Would you please describe the path you followed last Thursday as you went about your daily routines? Please describe with whom you interacted, where, and in what context you interact with people, what daily tasks you try to accomplish, and how frequently you visit these places and see these people. Respondents could give us up to two different days because many respondents have multiple means of generating income that might lead to different daily paths.

The social network analysis asked respondents to identify up to fifteen individuals and then to provide detailed information about each individual's role in the

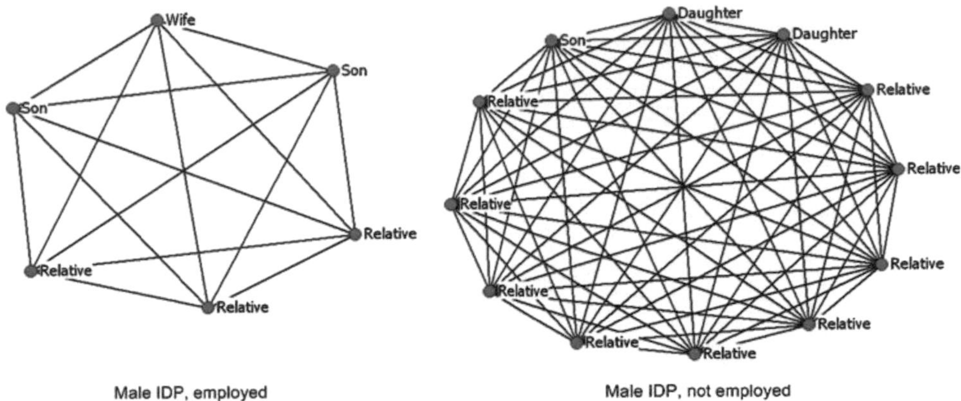

Figure 1. Social isolation. IDP = internally displaced person.

social network, whether the individual is engaged in income-generating activity,[6] where and in what context the respondent met the individual, how they communicate, whether the individual is an IDP, and, if so, if the respondent knew her or him in Abkhazia. We also asked the respondents to identify which members of the social network not only know one another but also interact independently (see McGrady et al. 1995; Breiger 2003). From this latter question, we calculate measures of density (or how closed the network is). In addition, we use standard social network mapping to visualize the social networks.

What Does Isolation Look Like?

In this section we look at what isolation looks like and investigate the role that dwelling type might play in shaping isolation as well as integration. Figure 1 provides examples of social network density maps of two individuals from our sample that display social isolation. This is a typical pattern of fully connected and highly dense social networks for the mean and large-sized networks, irrespective of gender, IDP status (meaning whether they are IDPs), or dwelling type. This means that everyone who is identified as being part of an individual's social network knows everyone else in the social network. We find the pattern for individuals with small, medium, or large networks irrespective of city, type of dwelling, or gender. A tightly connected social network signifies social isolation because the lack of social connections outside the trusted circle reduces an individual's access to information and reduces the potential for building social capital (Marsden 1987). The reduction of access to information has the potential to impact individuals negatively, especially in postconflict environments where information is critical for access to income-generating activities as well as information that can lead to access to housing, education, and medical services.

Isolation is also found in the composition of the social networks. We expected to find friends and neighbors in the social networks (e.g., fictive kin) in nearly all the networks, irrespective again of status, but network members are all household or family members (Figures 1 and 2), suggesting a reliance on family and extended family to a much greater degree than neighbors that likely results in a similar lack of information or a lack of a variety of sources of information. Although there is a possible debate about which came first, the conventional wisdom among humanitarian aid workers and Georgian government officials is that many IDPs lack entrepreneurial behavior and many note that it is likely because of a lack of integration. Yet, we find this social network pattern in the local and IDP populations. This finding calls into question whether or not the perceived social isolation of IDPs has any causality related to reduced initiative if the networks are similar for the local population (Renzulli, Aldrich, and Moody 2000).

Social isolation within social networks is evident by examining additional maps that display the size and composition of the networks in terms of gender, social relationship (family or friend), and IDP status. Figure 2 maps the average or the median-sized network for two women, one from the local population and one IDP from a collective center. Both are not engaged in income-generating activity. The social isolation is evident in the closed nature or high density of the networks as well as the relative isolation of the IDP and local populations from one another. Even in the case of a woman from a collective center who has one unconnected nonfamily member in her social network, the friend is still an IDP.

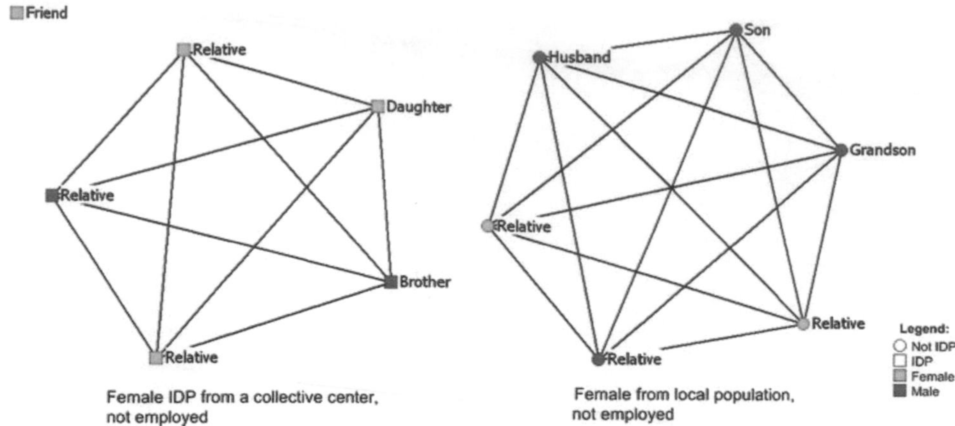

Figure 2. Women's networks. IDP = internally displaced person.

Data on network size and density demonstrate the similarity in network composition across the population while suggesting some potentially important differences between IDPs in collective centers and private accommodations. Table 1 shows that there is neither a statistically significant difference between the network size of IDPs and the local population nor a statistically significant difference between IDPs in different dwelling forms.[7] Roughly about 50 percent of the total sample had 100 percent density measures, whereas more IDPs in our sample had the lowest density, evidence of greater range of social integration. The IDPs in collective centers were more similar to the local population in terms of the density of their social networks than IDPs in private accommodations, who had the least dense networks of all subgroups. We suggest that the colocation in physical and social space of IDPs in private accommodations is formative of the slightly less dense and larger social networks.

The IDP and local populations might have similar degrees of social isolation; the median size and density of social networks are the same for the local population and IDPs in collective centers. It appears that, in general, social interactions in Georgia occur within small to medium-sized networks made up of mixed gender groupings that are highly connected and generally among relatives.[8] The social network literature suggests that this nature of social interaction is what might influence the perception of reduced entrepreneurialism and dependency, in large part because of the limited access to information and social engagement. So, from a social network perspective, the location of the collective

Table 1. Network size and density for the local population and internally displaced persons

	Displacement status		IDPs by dwelling type	
Network characteristics	IDPs (%)	Local population (%)	IDPs in CCs (%)	IDPs in PH (%)
Network size				
Small (0–4)	32.20	27.42	34.92	29.09
Medium (5–9)	40.68	53.23	42.86	38.18
Large (10–15)	27.12	19.35	22.22	32.73
Total	100 (n = 118)	100 (n = 62)	100 (n = 63)	100 (n = 55)
Pearson chi-square	2.729		1.663	
Network density				
Low (0–50%)	24.14	14.75	18.03	30.91
Medium (51–80%)	18.10	24.59	19.67	16.36
High (81–99%)	9.48	14.75	13.11	5.45
Fully connected (100%)	48.28	45.90	49.18	47.27
Total	100 (n = 116)	100 (n = 61)	100 (n = 61)	100 (n = 55)
Pearson chi-square	3.542		3.973	

Notes: IDPs = internally displaced persons; CCs = collective centers; PH = private housing.

Table 2. With whom do you interact on a daily basis?

	Displacement status		IDPs by dwelling type	
Network member	IDPs (%)	Local population (%)	IDPs in CCs (%)	IDPs in PH (%)
Relationship to respondent				
Immediate family or relative	71.32	75.80	76.61	66.07
Not a relative	28.68	24.20	23.39	33.93
Total	100 (n = 781)	100 (n = 405)	100 (n = 389)	100 (n = 392)
Pearson chi-square	2.711		10.594*	
IDP status				
IDP	77.49	6.17	87.09	67.70
Not an IDP	22.51	93.83	12.91	32.30
Total	100 (n = 782)	100 (n = 405)	100 (n = 395)	100 (n = 387)
Pearson chi-square	545.055*		42.132*	

Notes: IDPs = internally displaced persons; CCs = collective centers; PH = private housing.
$^{*}p < .001$.

center or the conditionality of displacement are less determinative of social isolation than the nature of the social network and social interactions.

The daily path analysis and questions to respondents about with whom and where they interact also suggest a high level of social isolation within networks and a low level of spatial mobility and interaction (Tables 2 and 3) and suggest similar forms of social interactions across IDPs and the local population. IDPs in private housing tend to interact more with nonrelatives than the local population and IDPs in collective centers (note that interaction is different from inclusion in the social network).

Socio-Spatial Signs of Isolation

Social isolation of IDPs is further seen through the constancy of the social network over time. IDPs noted that of the IDPs with whom they interact, over 90 percent were known to them in Abkhazia prior to displacement (Table 3). This suggests that there has been relatively little change in their social networks despite territorial displacement. Given that the majority of members of social networks are relatives, this is not surprising. It does, however, raise the question of the territorial limits to social networks in Georgia and the potential for distance decay or spatial limitations of social networks despite the prevalence of cell phones and other forms of technological means of overcoming physical distances. McPherson, Smith-Lovin, and Cook (2001) find that social contacts and thus the formation of social networks are more likely to occur with individuals closer located in geographical space. In addition to the long-term and space-bound nature of membership in social networks for IDPs in collective centers, IDPs in private accommodations report that despite not knowing some members of their current social networks in Abkhazia, they still are IDPs from Abkhazia. Following McPherson et al., this suggests a convergence of the importance of physical location and social space (living in territorial displacement) for the formation of new social networks.

When asked where people spend their days, the local population as well as IDPs note remarkably similar spatial patterns, except for those engaged in income-generating activity (Table 4). There are few statistically significant differences between groups with respect to

Table 3. Where did you meet people with whom you interact daily?

	Displacement status		IDPs by dwelling type	
Network member	IDPs (%)	Local population (%)	IDPs in CCs (%)	IDPs in PH (%)
Know this person from Abkhazia	93.25	30.56	96.03	89.59
Know this person from other place	6.75	69.44	3.97	10.41
Total	100 (n = 622)	100 (n = 72)	100 (n = 353)	100 (n = 269)
Pearson chi-square	220.558**		10.065*	

Notes: IDPs = internally displaced persons; CCs = collective centers; PH = private housing.
$^{*}p < .01.$ $^{**}p < .001.$

Table 4. Where do you spend your day and where do you go on a daily basis?

	Displacement status		IDPs by dwelling type	
Places	IDPs (%)	Local population (%)	IDPs in CCs (%)	IDPs in PH (%)
Yard, garden, birja, street corner				
Does not go to	80.51	88.71	80.95	80.00
Goes to	19.49	11.29	19.05	20.00
Total	100 (n = 118)	100 (n = 62)	100 (n = 63)	100 (n = 55)
Pearson chi-square	1.968		0.017	
Kins', friends', neighbors' homes				
Does not go to	78.81	83.87	76.19	81.82
Goes to	21.19	16.13	23.81	18.18
Total	100 (n = 118)	100 (n = 62)	100 (n = 63)	100 (n = 55)
Pearson chi-square	0.664		0.557	
Work				
Does not go to	63.56	46.77	65.08	61.82
Goes to	36.44	53.23	34.92	38.18
Total	100 (n = 118)	100 (n = 62)	100 (n = 63)	100 (n = 55)
Pearson chi-square	4.694*		0.135	
Public places besides work				
Does not go to	62.71	61.29	65.08	60.00
Goes to	37.29	38.71	34.92	40.00
Total	100 (n = 118)	100 (n = 62)	100 (n = 63)	100 (n = 55)
Pearson chi-square	0.035		0.324	

Notes: IDPs = internally displaced persons; CCs = collective centers; PH = private housing.
*$p < .05$.

how and where they spend their days, again, unless they report income-generating activity.

Integration and the Social Network

The discourse suggests that integration of the IDPs into the local population will resolve problems associated with the physical and social isolation of IDPs, particularly those in collective centers. When members of the governance environment talk about integration, they refer to blending IDPs into the local population for housing and access to and use of social services including health, education, and social assistance. Prior to our analysis, we expected to find strong evidence of integration into the local population of IDPs in private accommodation. Again, this is a major part of the discourse; indeed, the Urban Institute managed a housing voucher program in 2006 to move IDPs out of collective centers and into private accommodations "to effectively integrate into their host communities" (Golda 2007, 4). We expected to find evidence of integration in the social networks of IDPs in private accommodations, including more varied, less dense, and larger social networks.

The configuration of social interactions and networks for IDPs in private accommodations provides some evidence for greater social integration. In terms of size and density, there appear to be no statistically significant differences for IDPs in collective centers or private accommodations, despite the discourse. Other aspects of social interactions do vary and are statistically significant across social groups. IDPs in private accommodations note interacting more on a daily basis with nonrelatives than either IDPs in collective centers or the local population, and they interact more with non-IDPs than do those in collective centers. Additional information about the IDPs with whom they interact suggests that although they interact more with people they have met since displacement, they are still interacting with IDPs from Abkhazia (Table 3). Place and displacement still figure prominently into choice of social interactions.

Additional findings suggest that physical proximity along with engaging in income-generating activity might increase the likelihood of more diverse and integrated social interactions and networks (Table 5). Figure 3 displays the social network maps for a male IDP in private accommodation engaged in income-generating activity and a female IDP in private accommodation not engaged in income-generating activity. Both maps indicate greater diversity in terms of whether the members of the social network are IDPs,

Table 5. Network size and density by engagement in income-generating activities for internally displaced persons in private housing

Network characteristics	Respondent engages in income-generating activities	
	Yes (%)	No (%)
Network size		
Small (0–4)	12.00	44.83
Medium (5–9)	36.00	37.93
Large (10–15)	52.00	17.24
Total	100 (n = 25)	100 (n = 29)
Pearson chi-square	9.763**	
Network density		
Low (0–50%)	52.00	13.79
Medium (51–80%)	16.00	17.24
High (81–99%)	4.00	6.90
Fully connected (100%)	28.00	62.07
Total	100 (n = 25)	100 (n = 29)
Pearson chi-square	9.807*	

*$p < .05$. **$p < .01$.

but the male IDP engaged in income-generating activity has a lower density network with a more diverse set of members including friends, neighbors, and relatives.

Implications

The analysis of social networks and social interactions suggests a highly compartmentalized society based in part on the formation of social networks around family members. A major part of the discourse suggests that IDPs live in different spaces and once they live among the local population, integration will occur. Our findings suggest otherwise. We argue that the disconnect might arise from the various competing conceptualizations of integration among participants in the governance environment. We view these competing views as stemming from variable institutional agendas and interpretations of the root causes of isolation and integration. Several agendas and programs promote movement out of collective centers as a key solution to social integration (e.g., Urban Institute voucher program and National Strategy); yet, our findings suggest that such integration does not result from simply changing geographic location. Evidence from the narrative interviews suggests that IDPs themselves see integration in a more complex way that is more about socio-spatial relations than about geographical proximity to the local population.

In discussing whether separate IDP schools should exist, a female IDP living in private accommodation in Zugdidi noted:

> At one of the schools . . . a child of my neighbor was not accepted this year only because the child was an IDP. . . . And they told the parents of the child that the local kids were a priority and only if there were places left, they would accept the IDP. This is a kind of disintegration.

A male IDP also living in private accommodations in Tbilisi noted that collective centers have slowed down the process of integration but suggested that the separate educational facilities contribute to the creation of a separate population category:

> It should not exist as such like it is today . . . and when they have relationships with local children they would have the same problems, the same level of education . . . but on top of that I mean friendly relations with one another. Their

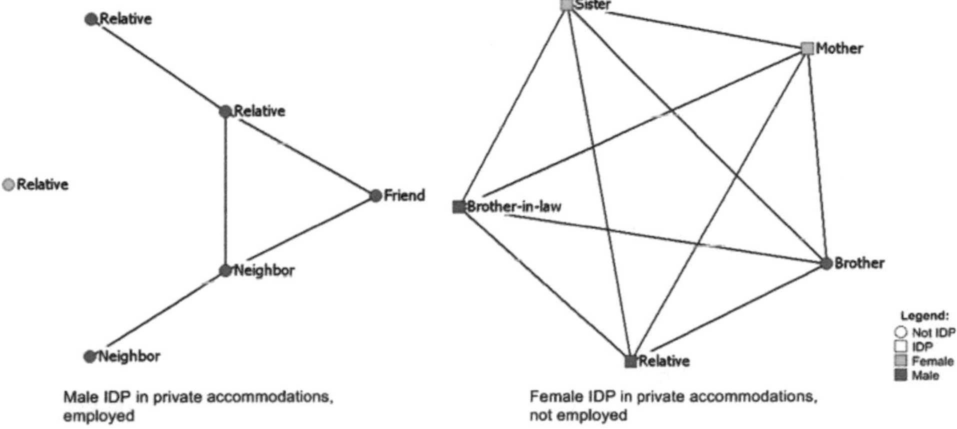

Figure 3. Social networks of internally displaced persons in private accommodations. IDP = internally displaced person.

> integration will take place with the local society and that would create a better foundation for the future.

Both these IDPs live in private accommodations and they both point to the daily practice of separate facilities as discouraging integration from the perspective of social interaction—not from the perspective of where they live. They seem to implicitly recognize the importance of sustained social interaction outside the dwelling place to facilitate social integration.

Our findings about the slight difference in composition of the social networks of those IDPs living in private accommodation seem to support the view of the IDPs that there must be places or spaces where relationships can form—perhaps social interaction that becomes formative of social networks. Our findings also suggest, however, that living in private accommodations is not the determining factor for less dense or larger social networks. If dwelling type or spatial colocation does not largely shape social networks, then the issue becomes the specific mechanisms or spaces that can change patterns of social interaction. The finding that engaging in income-generating activity appears related to larger and less dense networks suggests the importance of combining sites and forms of social interaction to change what appear to be stable patterns of social isolation into family networks. As noted earlier, Georgian culture is focused on family networks, so changing those patterns is neither simple nor necessarily recommended as a policy direction.

Is an apparent or perceived lack of initiative on the part of IDPs in collective centers related to the structure or composition of their social networks? We conclude that low levels of initiative are not due solely to the combination of IDP status and location in a collective center. The discourse about dependency does not flow to the local population, which has remarkably similar patterns of social interaction to IDPs in collective centers. So what do integration and isolation mean in the post-conflict situation? The short answer is different things to different people. In terms of spending time in one's dwelling, we find that IDPs are less spatially isolated from or just as isolated as the local population in terms of daily practice and that irrespective of dwelling type and social group, individuals spend about the same amount of time in various places. It is clear that integration cannot be achieved by simply changing dwelling type or location. Also clear is that policy prescriptions based only on changing geographical locations will not result in changing patterns of social interaction, given the stability of the social networks in displacement and the prevalence of similar social groupings. Greater social integration is associated with participation more broadly in income-generating activity. Although the policy framework and the current discourse are about integrating access to social services, increasing access to income-generating activity might result in greater social interaction and integration than any other single change.

Acknowledgments

The authors wish to acknowledge gratefully the support of the National Science Foundation Human and Social Dynamics Program and insights from an anonymous reviewer. In addition, we thank our Georgian collaborators, Nana Sumbadze and George Tarkhan-Mouravi, the interviewers at the Institute for Policy Studies, and numerous members of the international humanitarian aid community who spent countless hours with us.

Notes

1. In using the concept of the governance environment, we use Jessop's (1998) view of heterarchic governance and the self-organization of individuals or groups not required to work together and Rhodes's (1996) view of interactions across network participants. We prefer governance rather than management because it allows for the inclusion of IDPs in the process. Governance environment also relates more directly to social network approaches.
2. Forced migrants include both refugees and IDPs. The legal terminology is that refugees cross an international boundary and IDPs do not.
3. The National Strategy was adopted as a new policy direction in 2007 and the associated action plan was adopted in summer 2008.
4. The August 2008 war resulted in an estimated additional 200,000 IDPs and refugees—the refugees are in Russia and the IDPs in Georgia.
5. This polling organization is widely used by Western governmental agencies and NGOs as local implementing partners.
6. We use the term *income-generating activity* in Georgian and English because employment as a category is hard to define. Engagement in the informal economy rather than the formal one might be more important to many people as a form of income or revenue generation. Also, many individuals have multiple "jobs" and the one in the formal sector might not actually provide income. Receiving monetized wages from positions in the formal economy has been a casualty in many former Soviet countries.
7. The ranges used for both density and size measures are not determinative of the statistical significance. In other words, the density ranges illustrate low density versus fully connected and the size ranges were based on prior work on social networks in Georgia by Sumbadze (1999).

8. We do have some probability data, however, that suggest that IDPs in collective centers have the highest probability of all groups of having neighbors in their social networks.

References

Ager, A., and A. Strang. 2008. Understanding integration: A conceptual framework. *Journal of Refugee Studies* 21:66–91.

Breiger, R. I. 2003. Emergent themes in social network analysis: Results, challenges, opportunities. In *Dynamic social network modeling and analysis: Workshop summary and papers*, ed. R. Breiger, K. Carley, and P. Pattison, 19–35. Washington, DC: National Academies Press.

Demetriou, S. 2002. Rising from the ashes? The difficult (re)birth of the Georgian state. *Development and Change* 33 (5): 859–83.

Dershem, L., and I. Sakandelidze. 2002. *The status of households in Georgia 2002*. Tbilisi, Georgia: Save the Children.

Ebaugh, H. R., and C. Mary. 2000. Fictive kin as social capital in new immigrant communities. *Sociological Perspectives* 43 (2): 189–210.

Franz, B. 2003. Transplanted or uprooted? Integration efforts of Bosnian refugees based upon gender, class and ethnic differences in New York City and Vienna. *European Journal of Women's Studies* 10 (2): 135–58.

Georgian Institute for Polling and Marketing. 2005. *Technical report for IDP opinion survey in Georgia*. Tbilisi, Georgia: Georgian Institute for Polling and Marketing.

Golda, A. 2007. *Georgia IDP housing voucher program pilot phase II program report, February 2007–October 2007*. Washington, DC: The Urban Institute.

Government of Georgia. 2007. On approving of the state strategy for internally displaced persons–persecuted. Unofficial translation. Tbilisi: Government of Georgia. http://www.internal-displacement.org/8025708F004CE90B/(httpDocuments)/0860F04B3162B38CC12572950056DBED/$file/State+Strategy+for+IDP+−+ENG.pdf (last accessed 19 September 2008).

Helton, A. C. 2002. *The price of indifference*. New York: Oxford University Press.

Holtzman, S. B., and T. Nezam. 2004. *Living in limbo: Conflict-induced displacement in Europe and Central Asia*. Washington, DC: World Bank.

Hyndman, J. 2000. *Managing displacement: Refugees and the politics of humanitarianism*. Minneapolis: University of Minnesota Press.

Internal Displacement Monitoring Centre (IDMC). 2006. *Georgia: IDPs' living conditions remain miserable, as national strategy is being developed: A profile of the internal displacement situation*. Geneva, Switzerland: Norwegian Refugee Council.

Jessop, B. (1998). The rise of governance and the risks of failure: The case of economic development. *International Social Science Journal* 50 (155): 29–45.

Kalin, W. 2006. Georgia must act on promises to end displacement crisis. *Forced Migration Review* 25:67.

Leitner, H. 1997. Reconfiguring the spatiality of power: The construction of a supranational migration framework for the European Union. *Political Geography* 16: 123–44.

Loescher, G. 2001. *The UNHCR and world politics*. New York: Oxford University Press.

Marsden, P. V. 1987. Core discussion networks of Americans. *American Sociological Review* 52:122–31.

McGrady, G. A., C. Marrow, G. Myers, M. Daniels, M. Vera, C. Mueller, E. Liebow, A. Klovdahl, and R. Lovely. 1995. A note on implementation of a random-walk design to study adolescent social networks. *Social Networks* 17:251–55.

McPherson, M., L. Smith-Lovin, and J. M. Cook. 2001. Birds of a feather: Homophily in social networks. *Annual Review of Sociology* 27:415–44.

Nagel, C. R. 2002. Geopolitics by another name: Immigration and the politics of assimilation. *Political Geography* 21 (8): 971–88.

Okten, C., and U. O. Osili. 2004. Social networks and credit access in Indonesia. *World Development* 32 (7): 1225–46.

Renzulli, L., H. Aldrich, and J. Moody. 2000. Family matters: Gender, networks, and entrepreneurial outcomes. *Social Forces* 79 (2): 523–46.

Rhodes, R. 1996. The new governance: Governing without government. *Political Studies* 44:652–67.

Silvey, R., and V. Lawson. 1999. Placing the migrant. *Annals of the Association of American Geographers* 89 (1): 121–32.

Sumbadze, N. 1999. *Social web: Friendships of adult men and women*. Leiden, The Netherlands: DSWO Press.

Sumbadze, N., and G. Tarkhan-Mouravi. 2003. Working paper on IDP vulnerability and economic self-reliance. United National Development Program, Tblisi, Georgia.

Suny, R. G. 1994. *The making of the Georgian nation*. 2nd ed. Bloomington: Indiana University Press.

United Nations High Commissioner for Refugees (UNHCR). 2002. Georgia. In *2002 global report*, ed. A. Encontre. Geneva, Switzerland: UNHCR. http://www.unhcr.org/4a0c24cf6.html (last accessed 19 September 2009).

———. 2006. *Global appeal 2007: Strategies and programmes*. Geneva, Switzerland: UNHCR. http://www.unhcr.org/static/publ/ga2007/ga2007toc.htm (last accessed 19 September 2008).

———. 2007. *Statute of the Office of the United Nations High Commissioner for Refugees*. Geneva, Switzerland: UNHCR. http://www.unhcr.org/protect/PROTECTION/3b66c39e1.pdf (last accessed 19 September 2008).

———. 2008. *2007 global trends: Refugees, asylum-seekers, returnees, internally displaced and stateless persons*. Geneva, Switzerland: Field Information and Coordination Support Section, Division of Operational Services, UNHCR. http://www.unhcr.org/statistics/STATISTICS/4852366f2.pdf (last accessed 19 September 2008).

Satellite Data Methods and Application in the Evaluation of War Outcomes: Abandoned Agricultural Land in Bosnia-Herzegovina After the 1992–1995 Conflict

Frank D. W. Witmer and John O'Loughlin

The devastation of wars is most often measured in terms of the number of dead and missing people, but other conflict effects are long-lasting and far-reaching. The 1992–1995 war in Bosnia-Herzegovina resulted in almost 100,000 killed and almost half of the population displaced. This article analyzes the war's effects by evaluating impacts on the postwar agriculture environment from which most Bosnians derive their livelihoods. The war's impacts showed significant geographic variability, with localities near the frontlines and in eastern Bosnia-Herzegovina particularly affected. Thirty-meter Landsat imagery from before, during, and after the war was used to identify abandoned agricultural land in two study areas (northeast and south) within Bosnia-Herzegovina, characterized by different climates, soil, and vegetation. In the image analysis methodology, multiple change detection techniques were tested, and ultimately a supervised classification was chosen. Ground reference data collected during the spring seasons of 2006 and 2007 show the remote sensing methodology is effective in identifying abandoned agricultural land for the northeast study region but not for the southern one. The differential success rates were due primarily to variations in climate and soil conditions between the two regions, but also point to contrasts due to the different nature of the war in the two study regions. The study has important implications for the use of remote sensing data in tracking the course of conflicts and evaluating their long-term impacts. *Key Words: Bosnia-Herzegovina war, environmental outcomes, GPS, ground-referencing methods, Landsat imagery.*

死亡和失踪人数是最常见的战争破坏程度衡量指标，但其它冲突的影响是长期和深远的。1992—1995 年，波斯尼亚和黑塞哥维那的战争造成了近 10 万人死亡和几乎一半的人口流离失所。农业是大多数波斯尼亚人获得生计的主要途径，本文通过评估战后农业环境来分析战争的影响。战争的影响有着显著的地域差异，特别是在前线附近，以及受到影响的波斯尼亚一黑塞哥维那东部地区。利用战争之前，期间和战后的 30 米分辨率的地球资源卫星图像，在波斯尼亚和黑塞哥维那的两个研究区域内（东北部和南部），以确定那些被放弃的农业用地。这两个研究区域各自具有不同的气候、土壤和植被特征。在图像分析方法上，本文测试了多种变化监测方法，最后采用了监督分类方法。地面参考数据采集于 2006 年和 2007 年的春季，研究表明，确认被放弃的农业用地，遥感方法在东北研究区是行之有效的方法，但不适用于南部研究区。成功率的差异主要是由于两个区域之间气候和土壤条件的差异，同时也指出了因发生在这两个研究地区的战争的不同性质而造成的对比。这项研究对使用遥感数据跟踪冲突进程和评估其长期影响具有重要意义。*关键词：波斯尼亚和黑塞哥维那的战争，环境后果，全球定位系统，地面参照方法，地球资源卫星图像。*

Muy a menudo la devastación de las guerras se mide en términos del número de muertos y desaparecidos, pero otros efectos de los conflictos son duraderos y de largo alcance. La guerra de 1992–1995 en Bosnia-Herzegovina resultó en cerca de 100.000 muertes y en el desplazamiento de casi la mitad de la población. Este artículo analiza los efectos de la guerra, evaluando los impactos sobre el entorno de la agricultura en la posguerra, de la cual depende la vida de la mayoría de los bosnios. Los impactos de la guerra exhiben una variabilidad geográfica significativa, notándose particularmente afectadas las localidades cercanas a los frentes de guerra y en la parte oriental de Bosnia-Herzegovina. Se utilizó un conjunto de imágenes Landsat de antes, durante y después de la guerra para identificar la tierra agrícola abandonada en dos áreas de estudio dentro de Bosnia-Herzegovina (nordeste y sur), caracterizadas por tener diferentes climas, suelos y vegetación. En la metodología del análisis de las imágenes se probaron técnicas de detección de cambio múltiple, y en últimas se escogió una clasificación supervisada. Los datos de referencia del terreno, recogidos durante las primaveras del 2006 y 2007, muestran que la metodología de percepción remota es efectiva en la identificación de tierra agrícola abandonada en la región de estudio del nordeste, pero no en la del sur. Las tasas diferenciales de éxito se debieron primariamente a variaciones de las condiciones de clima y suelo entre las dos regiones, pero apuntan también a contrastes debido a la diferente naturaleza de la guerra en las dos regiones estudiadas. El estudio tiene importantes implicaciones sobre el uso de datos de percepción remota para hacerle seguimiento al curso de los conflictos y evaluar sus impactos a largo plazo. *Palabras clave: guerra de Bosnia-Herzegovina, efectos ambientales, GPS, métodos de referenciación del terreno, imágenes Landsat.*

There are approximately 110 million mines and other unexploded ordnance (UXO) scattered in sixty-four countries on all continents, remnants of wars from the early twentieth century to the present. Africa alone has 37 million landmines in at least nineteen countries. Angola is by far the most affected zone with 15 million landmines and an amputee population of 70,000, the highest rate in the world. Despite the belated awareness of the toll of landmines due to the 1997 Nobel Peace Prize for the International Campaign to Ban Landmines (ICBL), the removal of mines proceeds at a glacial pace due to the danger, cost ($300 to $1,000 per mine removed), and lack of international agreement on targeting priorities. Efforts to prohibit the future use of mines by militaries have foundered on their low price ($3–$10), easy accessibility, and effectiveness in military campaigns (Mather 2002). In mine-ridden countries such as Mozambique and Angola, which suffered protracted civil wars after independence in 1975, notions of sustainable development are illusory as agriculture, transportation, resource access, and general international investment are hindered (Unruh, Heynen, and Hossler 2003). In Bosnia-Herzegovina (BiH), only about 4 percent of the mines laid over a decade ago have been removed and at present rates of removal, it will take over a century to complete the process. Up to 4,000 km^2 still harbor antipersonnel mines, antitank mines, and UXO (ICBL 2002). In the first nine postwar years, 1,522 people were killed or severely injured by landmine accidents across BiH (United Nations High Commissioner for Refugees 2004).

The War in Bosnia-Herzegovina and the Two Study Areas

Bosnia-Herzegovina is the most mine-afflicted country in Europe, with an estimated 1.3 million people, roughly one third of the population, living in 1,366 mine-affected communities. In 2007, there were more than 12,000 locations requiring clearance (Fitzgerald 2007). About 1 million mines, mostly antipersonnel, still remain in Bosnia-Herzegovina and only about 60 percent of mined areas have been identified (Bolton 2003; Mitchell 2004). As seen in Figure 1, the minefields are found in all regions of the country with a concentration near the war's frontlines, the line of demarcation (Inter-Entity Boundary Line [IEBL]) between the Republika Srpska and the Federation of Bosnia-Herzegovina as agreed in the Dayton Peace Accords in November 1995. Because the region was a strategic narrow strip connecting the two main bodies of Serb-controlled territory during the war, mines are most concentrated in the northern municipality of Brčko.

Although the presence of minefields in all parts of the country is a daily reminder of the war's legacy, the localized distribution of the dead and missing people shows a more uneven distribution. The eastern boundary of the country along the Drina, including the massacre site of Srebrenica and the site of massive ethnic cleansing (Zvornik), shows a disproportionate loss of life (Ó Tuathail and Dahlman 2006). Sarajevo was besieged for almost three years and suffered the single biggest loss of life for any community in the war. Other major regional centers with ethnic divides (Prijedor, Doboj, Mostar, and Foča) also experienced high losses, although every municipality registered dead or missing (Figure 2). The moves of refugees and internally displaced persons (IDPs), mostly due to ethnic cleansing during the war, meant that the mixed communities evident on the map of census data from 1991 in Figure 2 were mostly changed to nearly homogenous communities. The provisions of Annex VII of the Dayton Accords guarantee the safe returns of refugees and IDPs to their former homes, but despite a wave of returns between 1999 and 2004, the process is far from complete and fear of retribution keeps many potential returnees away (Ó Tuathail and O'Loughlin this issue).

The outcomes of wars are multifaceted and include a range of social, economic, political, and environmental consequences from fighting. Consociational political

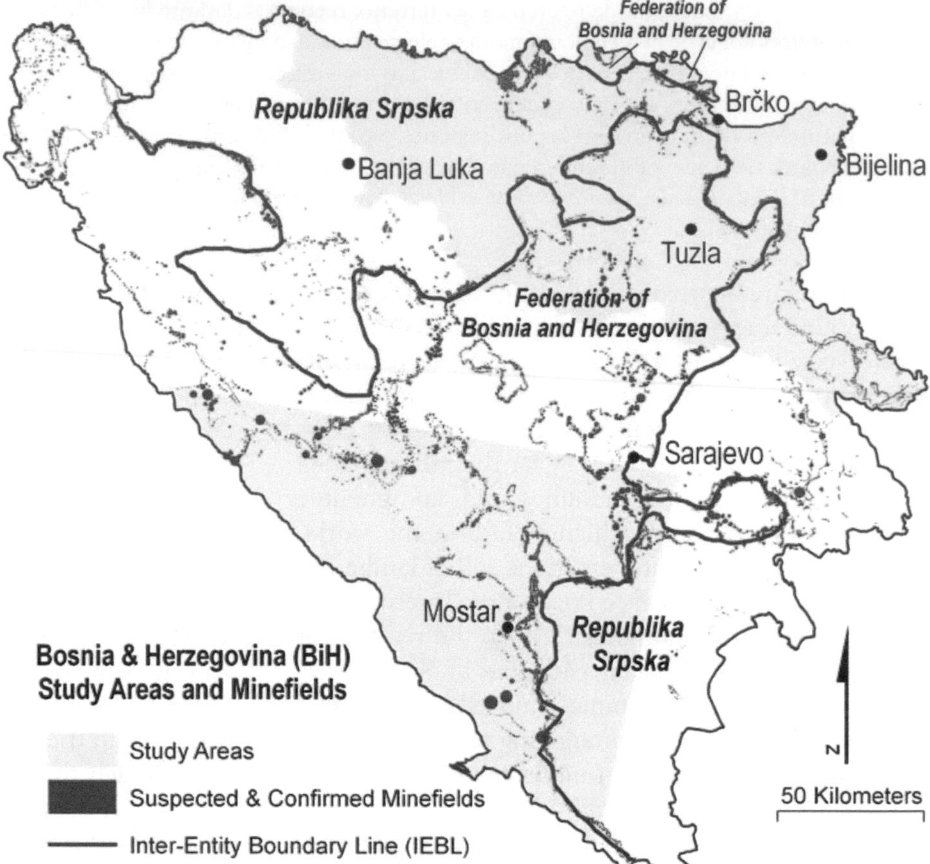

Figure 1. Bosnia-Herzegovina. Location of identified minefields, study regions, and the Inter-Entity Boundary Line established at the Dayton Peace Accords in November 2005.

arrangements can be devised to encourage cross-ethnic reconciliation and shared governance. In BiH, the complicated political arrangements agreed to at Dayton gave a large amount of autonomy to the two entities (Serb and Croat/Bosniak), designed a power-sharing arrangement for the whole country, and allowed local communities a wide range of governmental functions. The arrangement is supported widely by Serbs, moderately by Bosniaks, and weakly by Croats (who did not get their own separate political entity; Ó Tuathail, O'Loughlin, and Djipa 2006).

The war in BiH was marked by a strong international presence, both during the war and in its aftermath. Nearly fifteen years after the ceasefire, the Office of High Representative (OHR) of the United Nations (UN) still remains effectively in charge of the country, the Hague Tribunal on war crimes in the former Yugoslavia continues to prosecute major figures (the latest being that of Radovan Karadžič, former president of the Serb Republic), and thousands of UN peacekeepers remain on patrol. International nongovernmental and governmental agencies operate to identify the bodies of missing persons, to promote grassroots democracy, to settle refugees and displaced persons in their home communities, to rebuild economic infrastructures and promote development, and to raise funds for the identification and removal of the landmines that litter the country. In this article, we extend prior research (Witmer 2008) by comparing the results of a satellite data analysis and field checks from southern and northeast BiH regarding the effects of landmining on agricultural land abandonment in the aftermath of war.[1]

Remote Sensing and the Study of Conflicts

Academic research on war and conflict that examines remote sensing data is still surprisingly limited. Since 1960, satellite reconnaissance has played a significant role in providing information concerning enemy missiles, troop deployments, and military positionings, and the Cold War years saw major technological advancements by American and (later) other military interests (Corson and Palka 2004). Military uses of remote

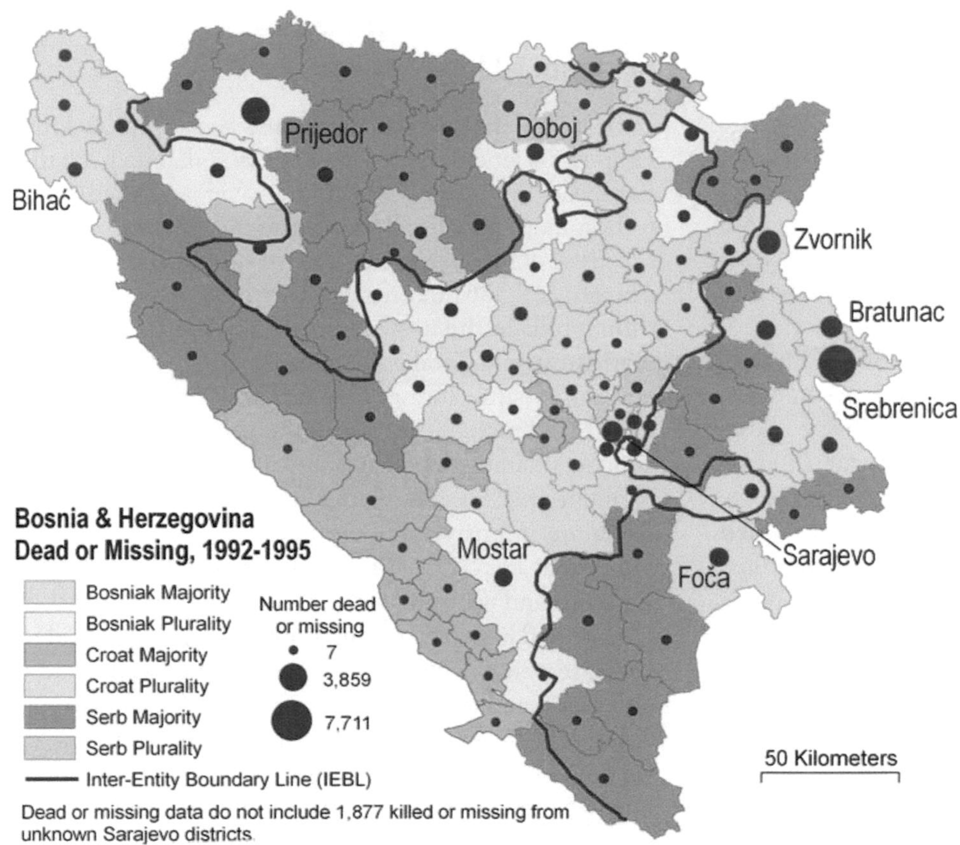

Figure 2. Distribution of deaths by *opština* during the 1992–1995 war and prewar ethnic composition in Bosnia-Herzegovina.

sensing technology are driven by strategic battlefield goals, with little attention given to broader war impacts. The uses of satellite imagery in conflict study tend to concentrate on the immediate impacts of military action, including identification of bomb damage, monitoring of fires and military movements, and mapping minefields.

Public attention was drawn to the use of nonmilitary satellite data during the 1991 Gulf War in Iraq and Kuwait due to extensive interest in the war's environmental consequences spurred by the massive impacts resulting from vehicle movements, hundreds of oil well fires, and numerous oil lakes (Stephens and Matson 1993; El-Baz and Makharita 1994; El-Gamily 2007). Further analyses identified the distribution of burning wells in different oilfields and produced estimates of flow rates and emissions of gaseous pollutants and particulates by incorporating the different spatial and temporal resolutions of AVHRR, Meteosat, Landsat Thematic Mapper (TM), and SPOT data (Husain 1994; Kwarteng 1998; Abuelgasim et al. 1999). The oil trench fires around Baghdad in March 2003 at the start of the second U.S.–Iraq conflict prompted further satellite monitoring using Landsat and IKONOS multispectral imagery (United Nations Environment Programme [UNEP] 2003).

Attention to other conflicts has focused on vegetation changes caused by military actions. The use of herbicides by the United States during the Vietnam War reduced the total mangrove area by one third in South Vietnam (Thu and Populus 2007). In Zimbabwe, a study of land cover changes identified the construction of service roads along minefields, vegetation removal from minefield perimeters, and vegetation regeneration of mined agricultural areas using remotely sensed data (Maathuis 2003). During the 1970s' Sandinista insurgency in Nicaragua, early Landsat MultiSpectral Scanner imagery captured the 45 to 55 percent reduction in agricultural productivity on the plantations (Howes 1979). Isolated areas of abandoned land were visible much later (1986–1996) using Landsat, SPOT, and AVHRR data as long-term results of the Contra war (Smith 1998).

Among the few applications of remote sensing data in conflict study, urban infrastructural and housing impacts of North Atlantic Treaty Organization (NATO) bombings of Yugoslav cities during the 1999 Kosovo war are evident in both Landsat TM (30-m resolution) and Indian Remote Sensing (IRS) 6-m imagery (United Nations Environment Programme and United Nations Centre for Human Settlements 1999). Very fine-resolution imagery is necessary for detecting impacts on the urban built environment. Structural damage from Israeli incursions into Jenin (West Bank) and NATO bombings in Macedonia were documented using IKONOS 1–2-m data (Al-Khudhairy, Caravaggi, and Glada 2005). Repressive actions by the Zimbabwean government resulting in destruction of dwellings were documented using Quickbird imagery (60-cm resolution) from 2002 and 2006 (Lempinen 2006). The U.S. Holocaust Museum's Web site allows visitors to see village destruction in Darfur (Sudan) by putting fine-resolution imagery online via Google Earth (http://www.ushmm.org/googleearth/); when coupled with pictures of destroyed homes identified on the images, viewers can easily see the efforts of Sudanese and paramilitary brutality. Recently, Human Rights Watch highlighted the ethnic cleansing of ethnic Georgian villages in South Ossetia, showing burning and destroyed homes after the August 2008 fighting in WorldView-1 & Formosat-2 satellite imagery (http://unosat.web.cern.ch/unosat/).

Displaced persons fleeing conflict zones often relocate to large refugee camps, where international assistance can more easily be administered and accessed. Remote sensing has been increasingly used to monitor the spatial extent of these camps for more efficient aid management, population estimates, and impacts to adjacent forests (Lodhi, Echavarria, and Keithley 1998; Bjorgo 2000; UN General Assembly Economic and Social Council 2000). Fine-resolution satellite imagery such as IKONOS panchromatic (1 m) and JERS (5–10 m) has been used to count refugee tents to estimate populations at risk (Giada et al. 2003) and locate hidden water sources to site refugee camps for Sudanese in eastern Chad (Bally et al. 2005). Although the UN is the leader in the humanitarian use of satellite data (Bjorgo 2002), little academic research has used the imagery to analyze the effects of war from such a synoptic view. Given the increased availability of mid- to fine-resolution (better than 36 m pixels) satellite imagery since the early 1990s (Stoney 2008), detailed analysis of war effects on the environment (both natural and anthropogenic) is long past due. In this vein, we examine the ability of satellite imagery to provide detailed information that can assist in war recovery by testing the accuracy of an abandoned agricultural land classification in Bosnia-Herzegovina.

The Environmental Context of Bosnia-Herzegovina

Bosnia-Herzegovina covers just over 51,000 km^2. Draping its land cover features (CORINE data from the European Environment Agency) over a shaded relief terrain map in Figure 3 shows the rugged terrain characteristic of BiH's southern and western regions rising to a height of 2,386 m, contrasting with the flat terrain and rolling hills in the northeast. Artificial surfaces are primarily urban and industrial areas but also include mine and dump sites; forests encompass coniferous, broadleafed, and mixed forests; and the agricultural category consists of arable land (irrigated and nonirrigated) and permanent crops. The flatter terrain in northeast BiH has more intensive agricultural land use, whereas the mountains of central BiH remain largely forested. The southern and western regions stand out both in terms of climate and land cover, consisting of poor-quality soil in karst areas supporting only shrubs and grasslands. These differences in land use reflect, in part, differences in soil depth (Figure 4), with southern BiH characterized by very shallow soil (less than 40 cm deep) and most of northeast BiH covered by a moderate soil depth (more than 60 cm deep).

The climate of our southern BiH study region is mediterranean, with temperatures averaging about 2°C in January and 23 to 26°C in July and rainfall between 1,500 and 2,000 mm. By contrast, the northern study region is characterized by a temperate continental climate with January temperatures about 1°C and July temperatures of 20 to 22°C and average rainfall of 800 mm. Despite the relatively abundant precipitation in the south, higher evapotranspiration rates and karst topography result in a higher water deficit or irrigation requirement there (Custovic 2005). The rural population density distribution (not shown but derived from 1991 census data by settlement or village with a buffer used to exclude all settlements within 1 km of the boundaries of urban areas) indicates a northern region from Bihać to Brčko, Tuzla, and Zvornik with higher population densities and corresponds with the higher agricultural productivity expected from the land use map.

Methodology and Ground-Referencing Procedures

Detecting changes to agricultural land use separately for both the northeast and southern study areas from

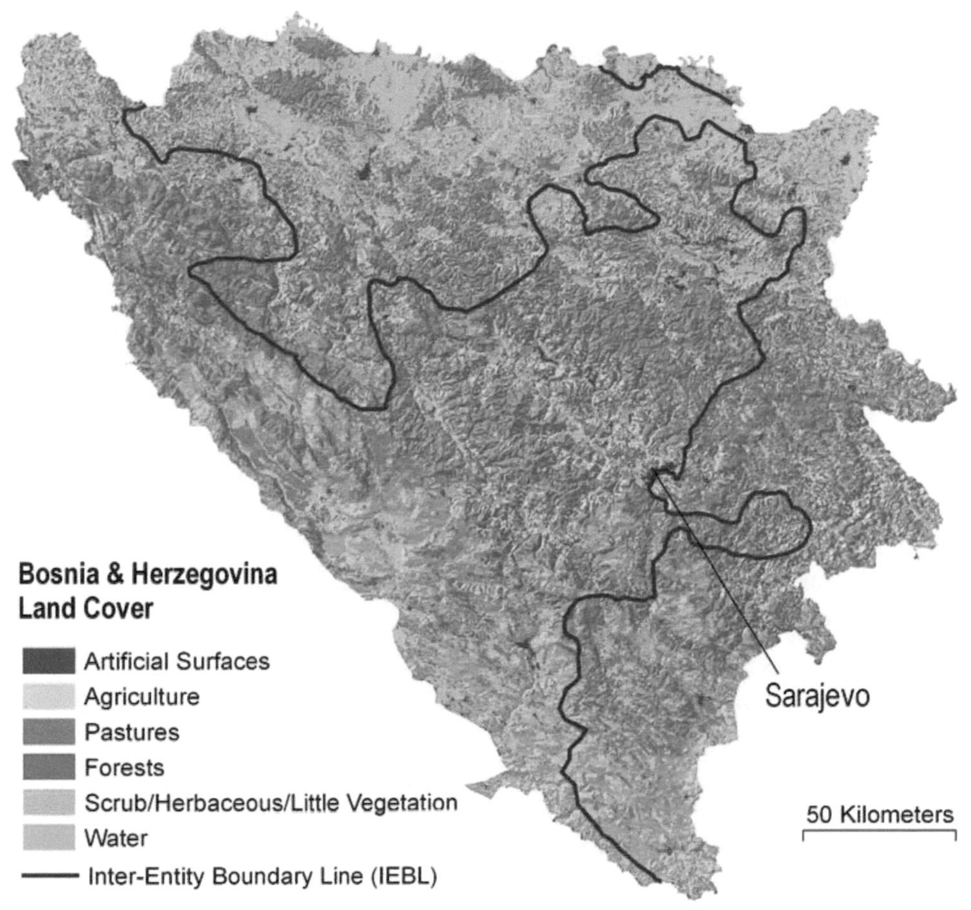

Figure 3. Land cover of Bosnia-Herzegovina draped over a shaded relief topographic map derived from digital elevation data. Land cover data are from the European Environment Agency (http://dataservice.eea.europa.eu/dataservice/) and elevation data from the Consultative Group on International Agricultural Research (http://srtm.csi.cgiar.org/). These primary land cover data were then overlaid onto topographic relief data collected from the Space Shuttle Radar Topography Mission in 2000. The digital elevation data were captured at a resolution of three arc seconds (nominal 90-m pixel resolution) and projected to UTM zone 33N for processing and display.

satellite imagery requires three steps: obtaining and preprocessing the satellite imagery from before and after the 1992–1995 war, selecting appropriate change detection methods, and conducting an accuracy assessment of the results (Lu et al. 2004). Complete details on the competing change detection methods used for the analysis of the northeast study area can be found in Witmer (2008).

Satellite Imagery and Preprocessing

Based on spatial, spectral, and temporal resolution requirements for this vegetation study, Landsat Thematic Mapper (TM) data were used. The TM sensor records information in the visible wavelengths (blue, green, and red) and near-infrared wavelengths, necessary for measuring vegetation health, which reflects strongly in the near-infrared portion of the spectrum. Because Landsat TM data consist of 30-m pixels with each scene covering 185 × 185 km of the Earth's surface, the northeast study area could be analyzed using two sets of overlapping scenes; one set of scenes was sufficient for the southern study area (see Figure 1). The 30-m pixel size enables us to identify medium to large agricultural fields and to reduce the number of mixed pixels that occurs with coarser imagery (see Figure 5b for an example of this scale; Müller and Munroe [2008] also use these satellite data for studying land abandonment in Albania).

The temporal frequency of the satellite data is dictated by the length and timing of the agricultural growing season and the availability of Landsat scenes. In agricultural land use studies, vegetation phenology is especially important due to the sudden changes in reflectance associated with crop planting and harvesting (Bauer 1985). Because we are detecting revegetation associated with abandoned agricultural land, the imagery should be separated by at least three years (Bauer 1985; Coppin et al. 2004). Landsat images over the previous fifteen years for spring (April, May, June) and summer (July, August, September) were acquired for the years 1990 to 1992 (just before the war) and recent postwar years (2002, 2004, and 2005), to ensure identification of plowed and harvested fields, whether sown with winter or summer

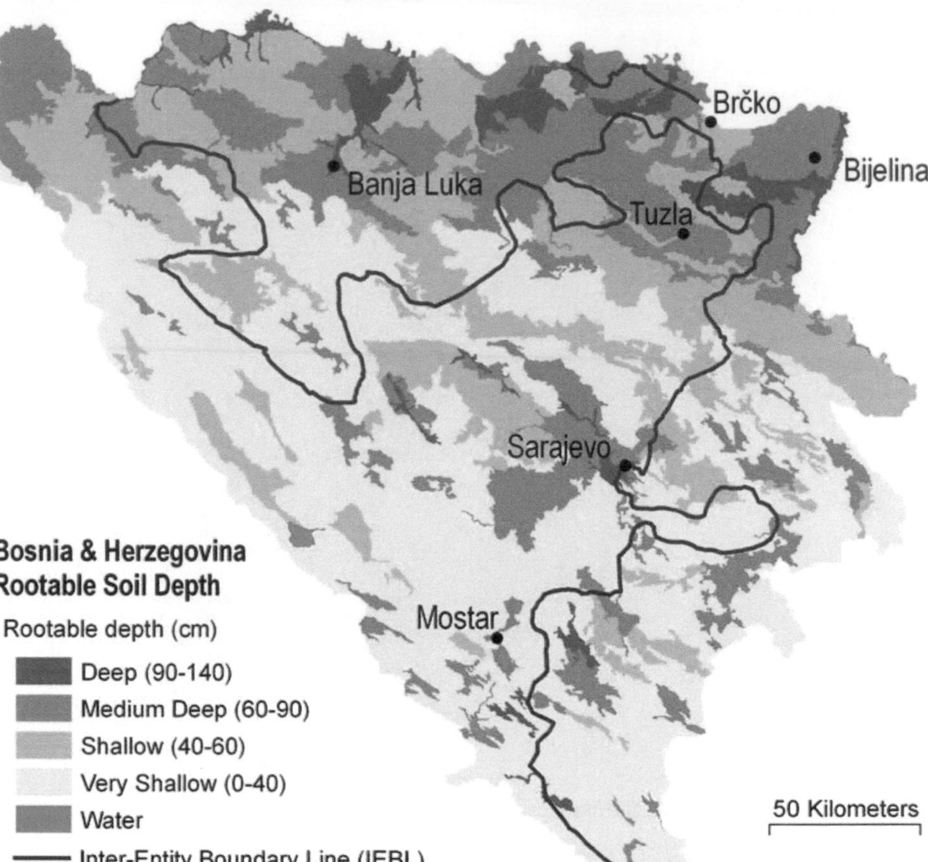

Figure 4. Rootable soil depth for Bosnia-Herzegovina. Data from the Participatory Land Use Development project of the UN Food and Agriculture Organization (http://www.plud.ba/).

crops. The Landsat imagery was acquired from the University of Maryland's Earth Science Data Interface at the Global Land-Cover Facility, Yale University's Center for Earth Observation, the U.S. Geological Survey, and a purchase from Eurimage. All scenes were registered to one of the University of Maryland's Landsat scenes, which have already been ortho-rectified and georegistered as part of their Landsat Geocover project (http://www.landcover.org/portal/geocover/). Co-registering each set of scenes ensures that pixels from different scenes align spatially and that changes detected are due to actual changes in vegetation and not registration errors. Additional processing identified the few pixels contaminated with clouds (mostly over the central BiH mountains) and excluded them from the analysis.

Change Detection Methods

Multiple change detection methods were tested for each study area with the minimum distance supervised classification found to perform best (for details, see Witmer 2008). This method requires the use of training data as input to the classifier. For each study area, thirty to sixty pixels associated with known areas of abandoned agricultural land were identified and used to train the classifier, which then identified all other agricultural pixels that exhibited similar spectral properties associated with abandoned agricultural land. For the northeast study area, the Landsat training pixels were identified using fine-resolution Quickbird imagery (60-cm pixels) available in Google Earth that is sufficient to distinguish the smooth texture of recently plowed, active agricultural land from the rough texture characteristic of shrubs and trees growing on abandoned agricultural land.

Because no fine-resolution imagery was available in southern BiH, fieldwork was conducted to identify areas of both abandoned and active agricultural land. To ensure an even distribution of field data, 200 random points were sampled from the agricultural land covered by the southern set of Landsat scenes. From the collected field data in the south, three areas of abandoned agricultural land (thirty to ninety pixels each)

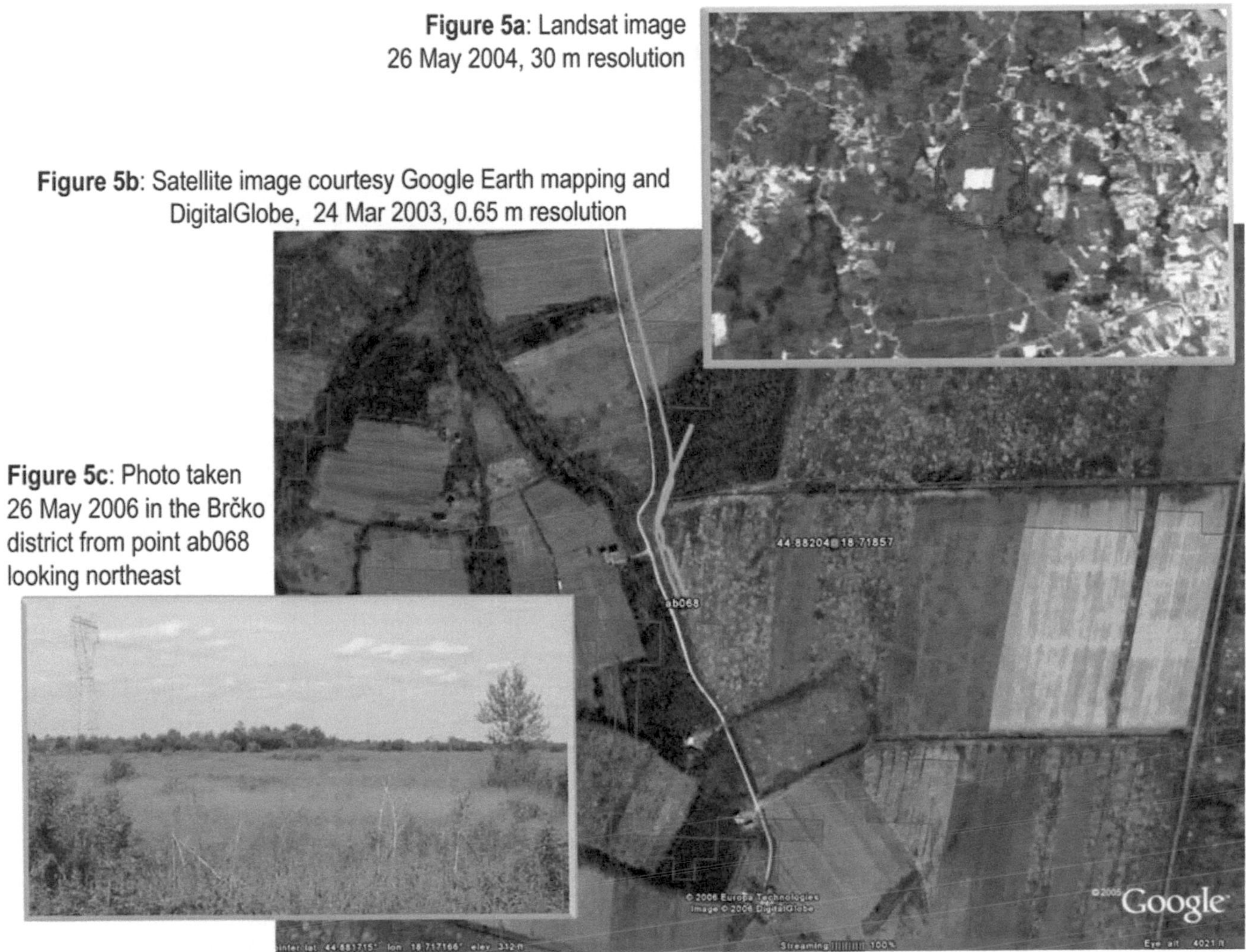

Figure 5. Images of the Brčko district show the general method for collecting ground reference data. Google Earth image, used with permission.

were used to train the classifier and the remainder withheld to conduct an accuracy assessment of the resulting classification.

Ground Reference Data and Accuracy Assessment

A persistent challenge in using satellite imagery is obtaining ground reference data for use in supervised classifications and accuracy assessments (Congalton and Green 1999; Cihlar 2000). Collecting ground reference data is especially problematic when using historical remote sensing data for which firsthand observations are unavailable. For the northeast study area, we chose a postchange detection approach similar to Serneels, Said, and Lambin (2001) and Nordberg and Evertson (2005) where ground reference data are collected after the classification analysis is complete. Following the general guideline of collecting at least fifty samples for each land-cover category (Congalton and Green 1999; Czaplewski 2003), we identified 150 "abandoned land" sites and 100 "active agriculture" sites from the classified imagery using a stratified random sampling approach. Because there is more cultivated land than abandoned land, abandoned land sites were disproportionately sampled to ensure that the smaller category (by area) is not misclassified (Khorram 1999). For the southern study area, observations from five sites (for three contiguous areas) were used to train the classifier and the remaining ninety-two ground observations were used to verify the classification accuracy.

Sample field site coordinates were transferred to a Global Positioning System (GPS) receiver that was then used to locate the points precisely. A detailed roadmap of BiH, a digital road network of paved and farm dirt roads from GISData (Zagreb), and printed Google Earth maps (both Terra Metrics 15 m and Quickbird 60 cm) helped in our navigation to the field sites. For each site visited, the state of the agricultural land was recorded on a field validation form. Because minefields, both confirmed and suspected, still effectively restrict access to much of BiH, only sites visible from paved roads or well-worn farm roads could be visited as part of the fieldwork. Mark Reed, head of the Roehll demining team in the Brčko municipality, indicated that even well-traveled dirt farm roads can contain mines that simply have not detonated due to chance (interview, 26 May 2006). In total, eighty-four field reference points were checked in northeast BiH and ninety-seven field sites were checked in the south. The identification and checking of the classification of sample points is illustrated in Figure 5 for a site in the Brčko region. Figure 5a is one of the Landsat scenes used to detect the abandoned agricultural land. Figure 5b shows Quickbird imagery overlaid with detected abandoned land (blue shaded/outlined), a sampled field site (blue dot with coordinates), and our GPS car track (green) imagery in Google Earth (the field circled in red corresponds to the active field in Figure 5b). The offset green car track and blue abandoned land reflect a georegistration error of the Quickbird imagery. Figure 5c is the photograph taken from point a "ab068" looking northeast toward the sampled field site.

Results

Although the overall land use classification accuracy for both regions is more than 80 percent, this figure hides significant differences between the two regions (Table 1). The producer's accuracy is calculated from the respective column totals in each matrix, the user's accuracy from the row totals, and the total accuracy along the diagonal. Because our aim is to assess the impact of the war (through landmines) on agricultural land, we are most interested in the results as an end user with the individual user's accuracies as the most appropriate measures. For the northeast region, the May 2006 field reference data reveal a user's classification accuracy of abandoned agricultural land of almost 82 percent. In contrast, the May 2007 field reference data for the southern region show a user's accuracy of less than 16 percent. Of the nineteen sites classified as abandoned agricultural land in the south, only three were actually confirmed as abandoned by the field reference data.

The results from Fisher's exact test of independence[2] further highlight the differences in effectiveness of identifying abandoned agricultural land between the study areas. For the northeast, the Fisher's exact test p value is highly significant (the classification is significantly better than random). In contrast, the p value of 0.058 for the southern region indicates that the classification is not significantly different from what could be expected from a random classification (at the 95 percent confidence level).[3]

Multiple reasons explain the differences in classification success. The significant differences in climate and soil conditions mean that vegetation grows much faster and denser in the northeast region than in the south, readily apparent from field observations in both regions. After nearly fifteen years, much of the abandoned land in the northeast was overgrown with dense brush dotted with trees 3 to 5 m tall, whereas abandoned agricultural land in the southern region mostly consisted of short grasses and small bushes with growth inhibited by the

Table 1. Accuracy assessment results for the northeast and southern study areas of Bosnia-Herzegovina

	2006 northeast field data				2007 southern field data		
	AB	NA	Total		AB	NA	Total
Classified data							
AB	49	11	60	AB	3	16	19
NA	2	22	24	NA	2	71	73
Total	51	33	84	Total	5	87	92
	Producer's accuracy		User's accuracy		Producer's accuracy		User's accuracy
	AB = 96.1%		AB = 81.7%		AB = 60.0%		AB = 15.8%
	NA = 66.7%		NA = 91.7%		NA = 81.6%		NA = 97.3%
	Total accuracy = 84.5%				Total accuracy = 80.4%		
	p value = 0.000				p value = 0.058		

Notes: Land cover categories: AB = abandoned; NA = nonabandoned. The reported p value is for Fisher's exact test.

poor soils and rainfall deficit. The shallow soils of southern BiH (Figure 4) were clearly visible along highway roadcuts at not only less than 40 cm deep but often less than 15 cm. This combination of shallow soil, poor soil quality (karst), and limited rainfall explain much of the error in detecting abandoned agricultural land from satellite data in the south.

In addition to these environmental reasons, the nature of the war in the two study areas also affected the ability of the satellite imagery to detect abandoned land. As is evident in Figures 1 and 2, the northeast study area has a much higher density of landmines along the war's frontlines (which shifted frequently) and around major military targets, especially major towns. Many more deaths are reported in the *opštini* in this region as well—and the number of returnees after ethnic cleansing is significantly greater than in the south. In the south, with the exception of the intense fighting for control of Mostar, the fighting was reduced by early demarcation of the zones of control for the respective militias.

Conclusions and Lessons for Other War Zones

Our study was designed to see to what extent a method using satellite imagery for identifying abandoned land due primarily to landmines in one region of Bosnia-Herzegovina could be applied in another region with different terrain, soils, climate, and agricultural profiles. The limitations of the procedure due to the difficulty of detecting slow regrowth in vegetation are evident in the unreliability of the classification. Our study suggests that the procedure that we propose in this article can be extended to other environments with a rich vegetation signal (like northeast Bosnia-Herzegovina) but its use is limited in vegetation-poor regions, like the Sahel. For study areas with relatively rapid vegetation regrowth, moderate-resolution (20–30 m) imagery is sufficient for detecting war-induced abandoned agricultural land. With good knowledge of the climate and soils of a study area, it is possible to study war zones that are still too dangerous for the conduct of research. When available, the use of Quickbird imagery can facilitate detection of land use changes, preferably supplemented with firsthand field data. This is especially important for long-term conflicts (more than five years) where conducting extensive fieldwork is still too risky but sufficient time has passed for land cover changes associated with human abandonment to become visible.

Acknowledgments

This research was supported by National Science Foundation grants from the Human and Social Dynamics Initiative (No. 0433927) and the Georaphy and Regional Science Program (No. 0623654). Thanks to Nancy Thorwardson of the Institute of Behavioral Science at the University of Colorado for preparing the maps for publication; to the *Annals* reviewers and editor, Audrey Kobayashi, for comments that improved the article; and to Digital Globe for permission to reproduce the Google Earth image (Figure 5).

Notes

1. There is no evidence of widespread abandoned agricultural land before the war began in 1992. Citing Food and Agriculture Organization experts, Bolton (2003) reports that in the Brčko municipality, landmines are the principal cause of agricultural land abandonment.
2. The chi-square test is similar to Fisher's exact test but was not used due to expected values in the error matrix less than five.
3. The error matrix significance was also tested using the kappa (K-hat) measure and associated Z statistic. Results from this measure mirrored those from Fisher's exact test but are not reported because the kappa statistic is designed for multinomial distributions.

References

Abuelgasim, A. A., W. D. Ross, S. Gopal, and C. E. Woodcock. 1999. Change detection using adaptive fuzzy neural networks: Environmental damage assessment after the Gulf War. *Remote Sensing of Environment* 70 (2): 208–23.

Al-Khudhairy, D. H. A., I. Caravaggi, and S. Glada. 2005. Structural damage assessments from IKONOS data using change detection, object-oriented segmentation, and classification techniques. *Photogrammetric Engineering and Remote Sensing* 71 (7): 825–37.

Bally, P., J. Bequignon, O. Arino, and S. Briggs. 2005. Remote sensing and humanitarian aid—A life-saving combination. *ESA Bulletin—European Space Agency* 122:36–41.

Bauer, M. E. 1985. Spectral inputs to crop identification and condition assessment. *Proceedings of IEEE* 73 (6): 1071–85.

Bjorgo, E. 2000. Using very high spatial resolution multispectral satellite sensor imagery to monitor refugee camps. *International Journal of Remote Sensing* 21 (3): 611–16.

———. 2002. *Space aid: Current and potential uses of satellite imagery in UN humanitarian organizations*. Geneva, Switzerland: United Nations. http://www.usip.org/virtualdiplomacy/publications/reports/12.html (last accessed 17 December 2008).

Bolton, M. 2003. Mine action in Bosnia's special district: A case study. *Journal of Mine Action* 7 (2). http://www.maic.jmu.edu/Journal/7.2/focus/bolton/bolton.htm (last accessed 15 August 2008)

Cihlar, J. 2000. Land cover mapping of large areas from satellites: Status and research priorities. *International Journal of Remote Sensing* 21 (6–7): 1093–114.

Congalton, R. G., and K. Green. 1999. *Assessing the accuracy of remotely sensed data: Principles and practices, mapping sciences series*. Boca Raton, FL: Lewis.

Coppin, P., I. Jonckheere, K. Nackaerts, B. Muys, and E. Lambin. 2004. Digital change detection methods in ecosystem monitoring: A review. *International Journal of Remote Sensing* 25 (9): 1565–96.

Corson, M. W., and E. J. Palka. 2004. Geotechnology, the U.S. military, and war. In *Geography and technology*, ed. S. D. Brunn, S. L. Cutter, and J. J.W. Harrington, 401–27. Dordrecht, The Netherlands: Kluwer Academic.

Custovic, H. 2005. An overview of general land and soil water conditions in Bosnia and Herzegovina. In *Soil resources of Europe*, 2nd ed., ed. R. J. A. Jones, B. Houšková, P. Bullock, and L. Montanarella, 73–82. Luxembourg: European Commission Joint Research Center, European Soil Bureau.

Czaplewski, R. L. 2003. Accuracy assessment of maps of forest condition: Statistical design and methodological considerations. In *Remote sensing of forest environments: Concepts and case studies*, ed. M. A. Wulder and S. E. Franklin, 114–40. Boston: Kluwer Academic.

El-Baz, F., and R. M. Makharita. 1994. *The Gulf War and the environment*. Amsterdam: Gordon and Breach.

El-Gamily, H. I. 2007. Utilization of multi-dates Landsat-TM data to detect and quantify the environmental damages in the southeastern region of Kuwait from 1990 to 1991. *International Journal of Remote Sensing* 28 (8): 1773–88.

Fitzgerald, K. 2007. Bosnia and Herzegovina. *Journal of Mine Action* (11) 1. http://maic.jmu.edu/journal/11.1/profiles/bih/bih.htm (last accessed 20 August 2008).

Giada, S., T. D. Groeve, D. Ehrlich, and P. Soille. 2003. Information extraction from very high resolution satellite imagery over Lukole refugee camp, Tanzania. *International Journal of Remote Sensing* 24 (22): 4251–66.

Howes, D. W. 1979. The mapping of an agricultural disaster with Landsat MSS data: The disruption of Nicaraguan agriculture. PhD dissertation, The University of Wisconsin, Milwaukee, WI.

Husain, T. 1994. Kuwaiti oil fires—Source estimates and the plume characterization. *Atmospheric Environment* 28 (13): 2149–58.

International Campaign to Ban Landmines (ICBL). 2002. Bosnia and Herzegovina: Landmine Monitor Report 2002. http://www.icbl.org/lm/2002/bosnia.html (last accessed 20 August 2008).

Khorram, S. 1999. *Accuracy assessment of remote sensing-derived change detection*. Bethesda, MD: American Society for Photogrammetry and Remote Sensing.

Kwarteng, A. Y. 1998. Multitemporal remote sensing data analysis of Kuwait's oil lakes. *Environment International* 24 (1–2): 121–37.

Lempinen, E. W. 2006. *Pioneering AAAS project finds strong evidence of Zimbabwe repression*. Washington, DC: American Association for the Advancement of Science.

Lodhi, M. A., F. R. Echavarria, and C. Keithley. 1998. Using remote sensing data to monitor land cover changes near Afghan refugee camps in northern Pakistan. *Geocarto International* 13 (1): 33–39.

Lu, D., P. Mausel, E. Brondizio, and E. Moran. 2004. Change detection techniques. *International Journal of Remote Sensing* 25 (12): 2365–407.

Maathuis, B. H. P. 2003. Remote sensing based detection of minefields. *Geocarto International* 18 (1): 51–60.

Mather, C. 2002. Maps, measurements, and landmines: The global landmines crisis and the politics of development. *Environment and Planning A* 34:239–50.

Mitchell, S. K. 2004. Death, disability, displaced persons and development: The case of landmines in Bosnia and Herzegovina. *World Development* 32 (12): 2105–20.

Müller, D., and D. K. Munroe. 2008. Changing rural landscapes in Albania: Cropland abandonment and forest clearing in the postsocialist transition. *Annals of the Association of American Geographers* 98 (4): 855–76.

Nordberg, M., and J. Evertson. 2005. Vegetation index differencing and linear regression for change detection in a Swedish mountain range using Landsat TM and ETM^+ imagery. *Land Degradation and Development* 16 (2): 129–49.

Ó Tuathail, G., and C. Dahlman. 2006. The "West Bank of the Drina": Land allocation and ethnic engineering in Republika Srpska. *Transactions of the Institute of British Geographers NS* 31 (3): 304–22.

Ó Tuathail, J. O'Loughlin, and D. Djipa. 2006. Bosnia-Herzegovina ten years after Dayton: Constitutional change and public opinion. *Eurasian Geography and Economics* 47 (1): 61–75.

Serneels, S., M. Y. Said, and E. F. Lambin. 2001. Land cover changes around a major east African wildlife reserve: The Mara Ecosystem (Kenya). *International Journal of Remote Sensing* 22 (17): 3397–420.

Smith, J. H. 1998. Land cover changes in the Bosawas region of Nicaragua: 1986–1995/1996. PhD dissertation, University of Georgia, Athens.

Stephens, G., and M. Matson. 1993. Monitoring the Persian Gulf War with NOAA AVHRR data. *International Journal of Remote Sensing* 14 (7): 1423–29.

Stoney, W. E. 2008. *ASPRS guide to land imaging satellites*. Falls Church, VA: Nobilus, Inc.

Thu, P. M., and J. Populus. 2007. Status and changes of mangrove forest in Mekong Delta: Case study in Tra Vinh, Vietnam. *Estuarine Coastal and Shelf Science* 71 (1–2): 98–109.

UN General Assembly Economic and Social Council. 2000. *Strengthening of the coordination of emergency humanitarian assistance of the United Nations*. Geneva, Switzerland: United Nations.

United Nations Environment Programme (UNEP). 2003. *Desk study on the environment in Iraq*. Geneva, Switzerland: United Nations Environment Programme.

United Nations Environment Programme and United Nations Centre for Human Settlements. 1999. *The Kosovo conflict—Consequences for the environment and human settlements*. Geneva, Switzerland: United Nations Environment Programme, United Nations Centre for Human Settlements (Habitat).

United Nations High Commissioner for Refugees. 2004. *Return statistics in 2004*. Sarajevo, Bosnia-Herzegovina: United Nations High Commissioner for Refugees. http://www.unhcr.ba/return/2004.htm (last accessed 16 December 2008).

Unruh, J. D., N. C. Heynen, and P. Hossler. 2003. The political ecology of recovery from armed conflict: The case of landmines in Mozambique. *Political Geography* 22 (8): 841–61.

Witmer, F. 2008. Detecting war-induced abandoned agricultural land in northeast Bosnia using multispectral, multitemporal Landsat TM imagery. *International Journal of Remote Sensing* 29 (13): 3805–31.

After Ethnic Cleansing: Return Outcomes in Bosnia-Herzegovina a Decade Beyond War

Gearóid Ó Tuathail and John O'Loughlin

Ethnic cleansing is a violent geopolitical practice designed to separate and segregate ethnic groups. This article describes both the war aims that justified ethnic cleansing in Bosnia-Herzegovina and the efforts by the international community to enable victims of ethnic cleansing to return to their homes. It considers the trends and geography of population returns ten years after the war, before presenting original survey research results on displacement and return experiences. An overwhelming majority of Bosnians reclaimed their prewar property and a majority of these actually returned to their homes. Those self-identifying as Bosnian Serbs were more likely to sell their prewar homes than were members of other ethnic groups; they also tend to be less interested in multiethnicity. The poor were more likely to reclaim and return to live in their houses than were richer groups. Those with strong attachment to their home villages were more likely to return. Despite more than a million returns, nearly half of whom are officially minority returns, Bosnia continues to grapple with the divisive legacy of ethnic cleansing. *Key Words: Bosnia-Herzegovina, ethnic cleansing, population returns, public opinion survey.*

种族清洗是一种暴力的地缘政治做法，用以隔离和切断民族群体。本文介绍了波斯尼亚－黑塞哥维那双方的战争目标，种族清洗的理由，以及国际社会的努力，使得种族清洗的受害者可以重返家园。本文研究了战后 10 年人口恢复的趋势和地理特征，才提交了关于流离失所和返回经验的原始调查成果。波斯尼亚的绝大多数人重新申报了他们的战前财产，这些人里的多数实际上返回了他们的家园。那些自我定位为波黑塞族的人们，比其他民族群体的成员更多地出售了他们战前的家园，他们也往往对多种族聚居缺乏兴趣。相对于比较富裕的群体，穷人更倾向于重新申报并且回到他们原来的房子。那些强烈依恋自己家乡的人更有可能回来居住。尽管有超过一百多万的人口回归，其中有近一半的回归人口是正式的少数民族，波黑将继续背负种族清洗的分裂遗产。*关键词：波斯尼亚和黑塞哥维那，种族清洗，人口恢复，民意调查。*

La limpieza étnica es una violenta práctica geopolítica diseñada para separar y segregar grupos étnicos. Este artículo describe tanto los designios de guerra con los que se pretendió justificar la limpieza étnica en Bosnia-Herzegovina, como los esfuerzos de la comunidad internacional por ayudar a las víctimas de tal práctica a regresar a sus hogares. El artículo toma en consideración las tendencias y geografía del retorno de población diez años después de la guerra, antes de presentar resultados originales de la investigación sobre las experiencias del desplazamiento y el regreso. La inmensa mayoría de los bosnios reclamaron sus propiedades de antes de la guerra y la mayoría de ellos realmente retornaron a sus hogares. Aquellos que se identificaron a sí mismos como serbo-bosnios estuvieron más inclinados a vender sus casas de antes de la guerra que los miembros de otros grupos étnicos; en aquéllos también se notaba menor interés en multietnicidad. Los pobres mostraban mayor inclinación a reclamar y a regresar a sus casas que los grupos más ricos. También exhibían mayores probabilidades de retorno quienes mostraban mayor apego a sus aldeas de origen. Pero a pesar de más de un millón de regresos, casi la mitad de los cuales son oficialmente considerados retornos de minorías, Bosnia sigue atada al legado divisivo de la limpieza étnica. *Palabras clave: Bosnia-Herzegovina, limpieza étnica, retornos de población, encuesta de opinión pública.*

Thirteen years after his indictment on charges of genocide, crimes against humanity, and crimes perpetrated against civilians and places of worship across the Republic of Bosnia and Herzegovina, the Bosnian Serb politician Radovan Karadžić was finally captured in Belgrade in July 2008 and extradited to face trial before the International Criminal Tribunal for the Former Yugoslavia. Karadžić's reappearance was a reminder of the cruelties of the Bosnian war of 1992 to 1995 and the term associated with the war: *ethnic cleansing*. The phrase has since transcended its origins and become a general description of "the removal by members of one group of another group from a locality they define as their own" (Mann 2004, 9). This definition problematically accedes to a groupist conception of conflict (Brubaker 2004)—politicized armed formations, not groups, are the actors that ethnically cleanse—and underappreciates the politico-geographic dimensions of the practice. Ethnic cleansing is ethnoterritorial geopolitics, an attack on an existing spatial order, and the imposition of one organized around ethnic division and segregation. The practice, as Mann and others note, is a dark feature of the age of nationalism: *Territory* and *terror* have common linguistic roots, with territory definitionally a place from which people have been frightened (Bhabha 1994). Drawing on results from a December 2005 survey, this article examines the dynamics of displacement and return after ethnic cleansing in Bosnia-Herzegovina. The survey provides insights beyond official statistics and competing claims about return outcomes and offers a profile of the returnees and their motivations.

Ethnic cleansing was given ostensible meaning by a series of socio-biological and politico-geographic attitudes that radical nationalists promoted amid the severe economic and constitutional crisis of the Socialist Federal Republic of Yugoslavia (SFRY) in the late 1980s and early 1990s (Dragović-Soso 2002). The first proposition was that group identity was the primordial axis of life in Yugoslavia. Politics was about groups, what the SFRY, following Soviet nationality policy, termed *constituent nations*, *national communities*, and *national minorities*. Asserted in the face of the historical contingencies of identity, groups were held to be transhistoric and quasi-biological entities. The second was the proposition that each group had a natural homeland, an ethnoterritorial place of origin. This belief was complicated by a third proposition, namely that throughout history groups tended to be in perpetual competition over land. A partial ethnoterritorial logic shaped the political geography of Tito's Yugoslavias, an internal republican structure that implicitly affirmed some polities as ethnoterritorial homelands. Yet, it also qualified that recognition with a commitment to a broader inclusive ideal, to the "brotherhood and unity" of "nations" to realize a common socialist community (Lampe 1996). The Republic of Bosnia-Herzegovina confounded the ethnoterritorial principle because it was the common home of two and then three "constituent peoples," Bosnian Muslims (1,902,956 people according to the 1991 census; only recognized as a "people" in 1968 and from 1993 self-described as Bosniaks), Bosnian Serbs (1,366,104), and Bosnian Croats (760,852). For the Partisans and the ruling Communist Party of Bosnia-Herzegovina, Bosnia-Herzegovina was neither a Muslim or a Serb or a Croat republic but also a Muslim, also a Serb, and also a Croat republic (Andjelic 2003). Bosnia-Herzegovina was different, a historic region known for a tradition of common life and neighborliness, although traumatic memories from the World War II bloodbath in the region were latent (Donia and Fine 1994; Hayden 1994).

The Context of Ethnic Cleansing in Bosnia-Herzegovina

Democratization amid economic and constitutional crisis enabled ethnonationalist parties representing the three dominant groups in Bosnia-Herzegovina to oust the ruling communists and gain power after republic-wide elections in December 1990. Despite preelection pledges to work together, the three parties were soon at loggerheads. Two of the parties—Karadžić's SDS and the Croat nationalist HDZ—were, respectively, under the direct influence and control of the neighboring Milošević regime in Serbia (which funded the SDS and later armed militias associated with it) and the Tudjman regime in Croatia (which established the Bosnian HDZ, chose its leaders, funded its activities, and armed its HVO militia). The third party, the Muslim-dominated SDA, had its own ethnonationalist aspirations (Hoare 2007). The outbreak of warfare between Serbia and Croatia in 1991 radicalized Bosnia-Herzegovina and, amid a drive for independence by Bosnia's Muslim/Bosniak and Croat leaders, the Milošević regime helped direct a military assault by the Yugoslav army and Serbian militias on Bosnian towns that plunged the republic into generalized warfare. With war unfolding across Bosnia-Herzegovina in May 1992, Radovan Karadžić addressed the secessionist Bosnian Serb parliament he presided over in Pale outside Sarajevo and articulated six "strategic goals" for what became the VRS (*Vojska Republika Srpska*, the Army of Republika Srpska), the new name for the Yugoslav People's Army in Bosnia. The goals reflected the propositions Karadžić and other radical ethnonationalists held. The first and principal goal was that Bosnia-Herzegovina's three constituent peoples should be "unmixed" and separated. The rest of the war aims concerned the borders necessary for that separation. There should be a corridor between Semberija and the Krajina (northwestern Bosnia and adjacent areas of Croatia). The Drina should be eliminated as a "border between worlds," the peculiarly articulated aim a conscious refutation of how the historic Croatian nationalist Ustaša conceptualized the significance of the river. A border should be established on the Una and

Neretva rivers. The city of Sarajevo should be divided into Serbian and Muslim parts. Karadžić's list ends with a classic geopolitical trope and old Serbian nationalist demand: The Serbian Republic of Bosnia and Herzegovina should have access to the sea (Donia 2003). These wartime aims were in keeping with a larger vision emanating from the Serbian Ministry of the Interior and other agencies controlled by the Milošević regime: the creation through force of arms of an ethnically homogeneous statelet in Bosnia, Republika Srpska, adjacent to a similar entity in Croatia, Republika Srpska Krajina, setting up the potential unification of both into a reconstituted Yugoslavia centered on Serbia and an exclusivist Serbian identity (Treanor, 2002).

This was the Milošević/Karadžić geopolitical vision that justified ethnic cleansing, the violent expulsion of ethnic others from their homes to create swaths of territory with ethnically homogeneous populations (Burg and Shoup 1999). But so-called ethnic cleansing was often about more than ethnicity. Opportunities for criminal activities, local grievances, revenge, and sheer nihilistic destructiveness fueled by alcohol and drugs were important aspects of the violent geopolitical engineering it represented (Cigar 1995). The majority of the ethnic cleansing in Bosnia-Herzegovina was perpetrated by armed formations affiliated with the wartime goals of the SDS and VRS. The HDZ and HVO also engaged in ethnic cleansing, especially in central Bosnia during the Croat-Muslim war of 1993–1994 (Shrader 2003). The Bosnian and Croatian armies were also guilty of acts of ethnic cleansing. By 1995, the former territory of Bosnia-Herzegovina was a patchwork of new ethnically cleansed spaces, with over half the prewar population of 4,365,574 displaced from their homes. More than a million became refugees and an estimated million people remained internally displaced within the country. The largest new territorial space was occupied by Republika Srpska, which consolidated its control over eastern Bosnia with a paroxysm of violence at Srebrenica in July 1995 that left more than 7,000 Muslim boys and men murdered. All told, an estimated 100,000 Bosnians lost their lives in the war, which ended with the Dayton Peace Accords of November 1995 that divided Bosnia-Herzegovina into two ethnoterritorial entities, the Republika Srpska (RS) and the Federation of Bosnia and Herzegovina, an entity internally divided into designated Croat, Bosniak, and mixed cantonal spaces. The control of the strategic municipality of Brčko was left undetermined and it later became a separate district.

Annex VII of the Dayton Peace Accords outlined principles for the potential reversal of the demographic consequences of ethnic cleansing. Paragraph one of article one declared, "All refugees and displaced persons have the right freely to return to their homes of origin. They shall have the right to have restored to them property of which they were deprived in the course of hostilities since 1991 and to be compensated for any property that cannot be restored to them." It took years for the international administrative structure established to supervise the implementation of the Dayton Peace Accords to create the political, legal, and administrative capacity to implement this provision and others outlined in Annex VII (Ó Tuathail and Dahlman 2010). Obstruction to returns, particularly so-called minority returns or returns of ethnic groups to settlement areas where they are no longer the majority, was frequently violent and hard-line from 1996 to 1999, an assertion of ethnic borders in the face of the right of the displaced to return to their prewar homes. Later obstructionism became more administrative and stealthy, with ethnocratic local administrations reluctant to implement the property laws designed to facilitate returns. Some administrations responded by granting land plots and building alternative accommodation for the co-ethnics they wanted to retain (to consolidate the wartime population displacements; Ó Tuathail and Dahlman 2006b). Only in recent years, as it became evident that the returns process was not going to substantially alter the demographic legacy of ethnic cleansing, have nationalist controlled municipalities demonstrated a politically expedient openness to returnees.

The Returns Process After the Dayton Accords

At the time of our December 2005 survey, official United Nations High Commissioner for Refugees (UNHCR) figures showed a total of 454,220 minority returns across Bosnia-Herzegovina and total majority and minority returns exceeding 1 million persons. The temporal distribution of all returns, summarized in Figure 1, is the result of an initial wave of returning refugees between 1996 and 1999 as they were repatriated by neighboring European countries. Return was not necessarily to home communities but to housing in segregated ethno-spaces, which often meant squatting on other people's property, thus hindering potential internal return axes. Minority returns were slow to take off and

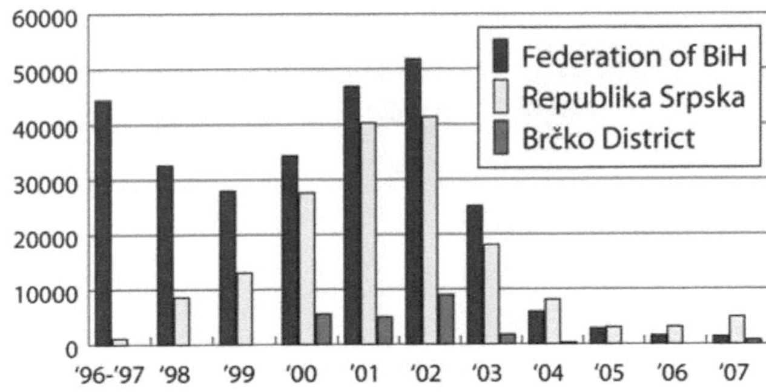

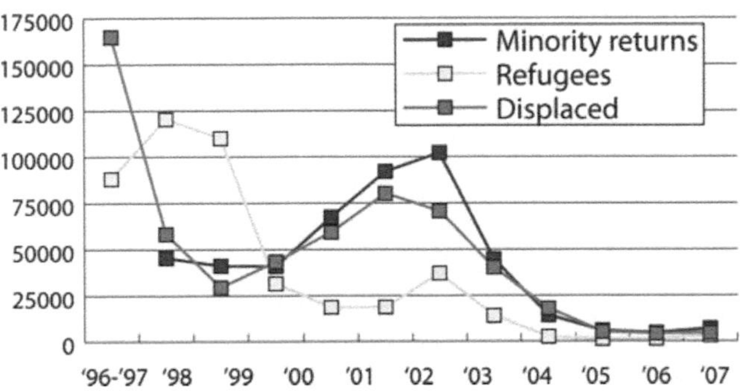

Figure 1. Yearly returns after the 1995 Dayton Accords by governmental entity and by ethnicity. BiH = Bosnia-Herzegovina.

initially dominated by intra-Federation displacements associated with the war in central Bosnia. Sometimes returns were very localized, with people moving only a few kilometers and "remixing" to live where they had before. In the Brčko district, the returns process did not begin in earnest until its political status was stabilized and a cross-ethnic power-sharing agreement under international supervision was put in place after 1997; it was 2000 before large numbers of returns occurred there (Dahlman and Ó Tuathail 2006). As the international community built capacity for minority returns and opposition to them was ameliorated, waves of returnees crossed entity boundary lines (evident in the graphs for the years 1999–2003). Since 2003, the pool of likely returnees has been reduced significantly, with less than 10,000 in total moving back in recent years. Nevertheless, just over 117,000 internally displaced persons were still registered in mid 2009 awaiting housing or assistance in reclaiming their properties (all data from the UNHCR, http://www.unhcr.ba).

The official number of minority returns to each of the municipalities (općine) mapped in Figure 2 is a function of several factors. The first is prewar settlement size. As one might expect, cities like Sarajevo (Serbs returning to a Bosniak-majority community), Banja Luka and Prijedor (Bosniaks returning to Serb-majority areas), and Mostar (generalized returns) have the largest absolute numbers of returnees. Smaller settlements, like Zvornik, Doboj, and Bijeljina (Bosniaks returning); Travnik and Bugojno (Croats returning); and those in the Una-Sana region (Serbs returning to Bosniak- and Croat-controlled areas), experienced severe ethnic cleansing yet many displaced chose to return to these areas in the hope of reconstituting vibrant prewar communities. (Zvornik, Doboj, and Bijeljina [town only] had Muslim majorities in 1991 but because they were Serb-majority after ethnic cleansing, Bosniak returns there were classified as minority returns.)

The second factor is the socioeconomic context. Minority returns required economic support. Without it, many families would not have had the means to return and reconstruct their former properties. Not all those wanting to return could acquire funding. Also, some families ended up in displacement locations that had better economic conditions than their home areas. Some did not gamble on return to areas where important sources of local employment were in the hands of wartime profiteers and political figures. Other "social service" explanations, such as the nonavailability of pensions, access to health care, and education in one's own language, culture, and traditions are important also (Ó Tuathail and Dahlman 2004). A third factor is local receptivity to returns. Although many municipalities represented themselves as open to return, local political environments differed greatly. The HDZ control of the mayor's office and municipal assembly of Jajce until 2004 left that city reluctant to accept Bosniak and Serb returns, so much so that it was deemed a "black sheep" municipality by the OHR (Dahlman and Ó Tuathail 2005).

There is considerable debate on the significance of return outcomes. For international officials like Paddy Ashdown (the head of the Office of the High Representative [OHR] from May 2002 to December 2005) Annex VII's implementation has "invented a new human right—the right to return home after a war."[1] More precisely, Annex VII permitted "domicile return," return not simply to one's country after war but the right to claim and return to one's prewar home. This provision was new in postconflict arrangements; the establishment of a comprehensive property law implementation process (PLIP) made it real for thousands of Bosnians who were able to file claims for their property (including "lease ownership" of apartments associated with state

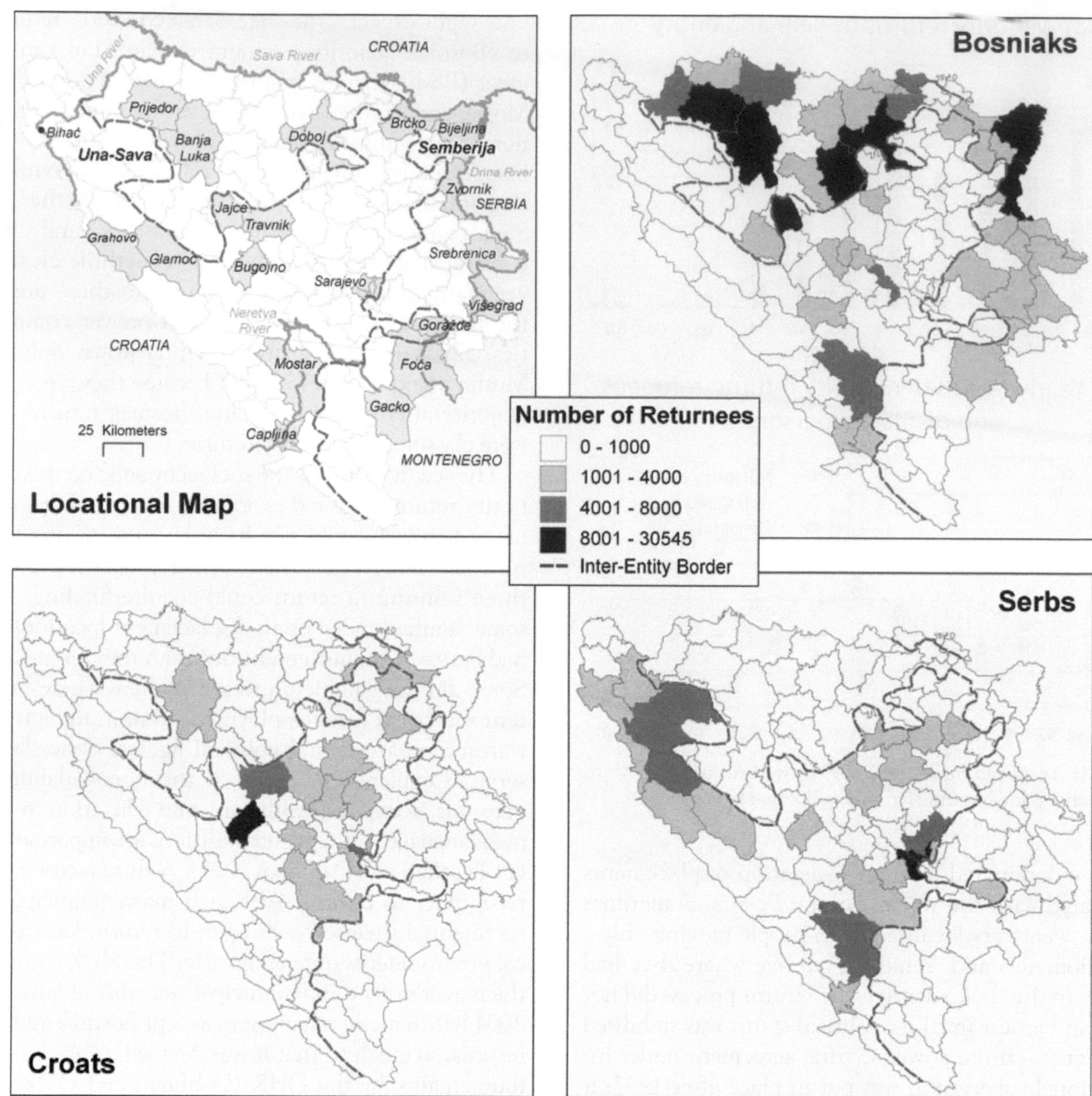

Figure 2. Distribution of returnees in Bosnia-Herzegovina 1996–2007, by *općina* and by ethnicity.

enterprises) and eventually have this claim adjudicated. In many instances, the property had been destroyed and owners required international humanitarian assistance to rebuild it. In other cases, their property was occupied, often by other displaced persons. They were required to vacate the property and move into alternative accommodations. Returnees were then able to reoccupy the property, although whether they did so or whether they rented or sold, was a matter on which there was considerable speculation but little reliable information.

The UNHCR and the Bosnian Ministry for Human Rights and Returns continue to produce return figures and to help coordinate sustainable return initiatives funded by international donors. Within the UNHCR, at least, however, there is an acknowledgment that official minority return figures are misleading. The UNHCR has conceded that many returnees have opted not to return permanently and have sold, exchanged, or rented their repossessed property. A briefing note from October 2007 claims that

> those who have returned permanently tend to be older and in rural areas where they depend upon agriculture. Many young IDPs have remained in their place of displacement

seeking education, social and economic opportunities that are scarcer in their communities of origin. The trend seems to be that people are remaining in (and moving to) areas where they can live amongst their own ethnic group. (UNHCR 2007)

The most pessimistic perspective on return outcomes is that of the Bosnia-Herzegovina Helsinki Committee, a nongovernmental group of lawyers in Sarajevo that issues annual reports on the state of human rights across the country. Their reports are suffused with claims that Bosnia-Herzegovina remains an ethnically cleansed country despite Annex VII. In 2008, for example, they claimed: "BiH is today divided into almost ethnically pure territories, while consequences of war migrations have only deepened through long standing obstructions and administrative barriers of authorities at all levels [*sic*]. Reliable data on true returns cannot be obtained in the field, or figures about restored or repaired houses that their owners sold or exchanged" (Helsinki Committee for Human Rights in Bosnia and Herzegovina 2008). This pessimistic picture is also articulated by the Bosniak (Bosnian Muslim) member of the tripartite Bosnian Presidency, the former wartime foreign and prime minister Haris Silajdžić. In an address in Washington, DC, in May 2008 he declared that "we have hundreds of thousands of our citizens around the world and the consequence of this is that we have now territories without its citizens [*sic*], without the inhabitants." In a follow-up interview, Silajdžić described Bosnia-Herzegovina as a patient that had its limbs amputated.[2] It was still alive but it was missing important parts of itself. In this imagining of state territory as a geo-body, Republika Srpska is an amputation of a previously whole and healthy Bosnia-Herzegovina. After ethnic cleansing, Bosnia-Herzegovina is a crippled country.

Who Are the Returnees and What Motivates Them?

As the Helsinki Committee notes, there is an absence of information on return outcomes beyond an official classified act of return. Bosnia-Herzegovina has not had a census since 1991 and its two entites do not release data on housing sales. In an effort to get at the motivations beyond official return figures, we directed Prism Research to conduct a doorstep survey of 2,000 BiH (Bosnia-Herzegovina) adults in early December 2005. The sample was based on a stratification of the *opštini* using the 1991 census data and proportionately sampled for the three ethnic communities. Of the more than 100 questions in the survey that focused on war experiences, interethnic reconciliation, attitudes toward political developments, migration intentions, and the usual sociodemographic measures, we report here on the responses to questions about displacement and the experiences of respondents in claiming their properties and returning to them. Our survey offers the depth of information that is absent from politically driven speculation about who is returning, who is not, and levels of fear about harassment and hostility.

Migration Intentions and Attitudes Toward Minority Returns

Despite the well-documented poverty and deprivation in Bosnia-Herzegovina (the most pressing problem according to our and other surveys), only 20.9 percent of the adults surveyed in late 2005 answered "yes" when asked if they would move if they had the opportunity. Significantly, in the context of the whole survey, only a few respondents (eleven people) would consider moving because of an unsafe or discriminatory environment. Parallel to the question about possible moves, we asked about experiences of discrimination and only 5.6 percent of the overall sample (112 people) felt a sense of discrimination during the previous year.

Evidence for optimism about the impact of Annex VII is the response to a question about possible minority returns. Three quarters of the sample agreed (24 percent) or strongly agreed (55 percent) that "people coming from the minority population who lived in this municipality before the war can return to their homes." Even allowing for possible bias to this sensitive question (a possible overreporting of the level of acceptance), there is a certain inevitability that the returns process will be completed. There remain, however, signs of opposition in some communities (Gacko reported only 15 percent support, Capljina 32 percent, and Mostar 54 percent). The locus of opposition in Serb-dominated communities in the southern and eastern parts of the country, although much diminished from the immediate postwar years, remains evident.

Movements During and After the War

Similar to the aggregate statistics collected by the UNHCR and other agencies, half of our sample (23 percent one move and 27 percent more than once) indicated they were forced to move during the war. Their reception into communities to which they fled, usually with a similar ethnic profile, was not always easy. Just over half of respondents (60 percent) said that they welcomed members of the same ethnicity who were displaced by the fighting. Many migrants went abroad for the duration of the conflict (5 percent of our total sample) but the majority of those forced to move (59 percent) lived with friends or relatives or arranged their own private accommodation for the duration of the war and its immediate aftermath. The properties of the displaced were occupied by others of a different ethnicity (26 percent), destroyed (33 percent), or damaged (13 percent). (Respondents could not provide an answer for the remaining cases.) The war was marked by widespread destruction or damage of houses as well as organized attacks on cultural and religious iconographic monuments (Ó Tuathail and Dahlman 2006a).

What Happened When Refugees and Displaced Persons Reclaimed Their Property?

Taking advantage of the PLIP, 85 percent of our sample who were displaced by the war laid a legal claim to their property. The remainder was still in the process of trying to begin reconstruction of the property or resolve the legal difficulties that pervade the returns process (Williams 2006). Our survey results show that the majority of people (62 percent) making property claims return to live in their houses. Only 13 percent sold their homes after reclaim, and 8 percent rent the homes out or use them on a part-time basis. In contrast to the pessimistic reports of the Helsinki Committee and the BiH (Bosnia-Herzegovina) authorities, our data indicates evidence for some successful return outcomes with a high rate of return of homeowners to their previous residences.

Who Has Returned to Their Homes?

Examination of the disposition of reclaimed properties according to their characteristics of their owners yielded just a few significant differences. The numbers in the subsamples are too small for many of the thirty-five communities sampled to yield reliable estimates, although differences among major towns and cities are noteworthy. In the predominantly Serb communities of Banja Luka (40 percent) and Bijeljina (11 percent), the lowest ratios of reclaim and return to the property are reported by the respondents. In Bijeljina, 72 percent of the returnees stated that they sold their properties after reclaim, by far the highest proportion in the overall sample. In contrast, high ratios of reclaim and returning to live are noted for the ethnically mixed cities of Brčko (87 percent), Bihac (83 percent), and Mostar (82 percent) in relatively prosperous locations, although they have their own microgeographies of settlement segregation.

The most dramatic difference in the return to ownership is the ethnic one between Serbs (33 percent) on the one hand and Croats (87 percent) and Bosniaks (76 percent) on the other. In fact, the ratio of Serbs who reclaimed and lived again in their homes is only slightly more than the ratio (32 percent) who sold the property after reclaim, a clear indication that Bosnian Serbs tend to be more invested in the idea of separate ethnoterritorial homelands. The ratio of Serbs waiting to reclaim their properties is 17 percent, much higher than the other groups, and suggests that many Bosnian Serbs came late to the returns process. This lag can be explained by the active campaign against return among Bosnian Serbs led by the ethnoterritorial minded group *Ostanak* (Stay). Our survey also indicated that Bosnian Serbs are less open to multiethnicity than are Bosniaks or Bosnian Croats. In response to the survey item "I would like to have more friends among people of different nationalities in this region," only 41 percent of Bosnian Serbs agreed or strongly agreed with the statement, compared to 62 percent of Bosniaks and 82 percent of Bosnian Croats. Similarly, Bosnian Serbs have a high ratio of agreeing or strongly agreeing (63 percent) with a proposed option of separate ethnic territories as a way to improve ethnic relations, that "ethnic relations in my locality will improve when all nationalities are separated into territories that belong only to them." Comparative ratios are 48 percent for Bosniaks and 67 percent for Bosnian Croats, but differences within each of the three groups according to material status, religiosity, and war experiences are greater than the interethnic variations in preferences for separatism (O'Loughlin and Ó Tuathail 2009).

Further examination of the disposition to reclaim one's home shows significant differences according to material status, experiences of discrimination, and conceptions of home. There are a clear relationship with self-reported purchasing power. The poorest segment of the population are more likely to reclaim and live in their houses than other richer groups (67 percent for those who answered "we do not have enough money to live" compared to 62 percent who have "only enough money for food" and 57 percent who answered "we can purchase all we need except durable items"). Although there have been claims (echoed by the UNHCR) that the oldest populations are more likely to return to live in their homes, we found no differences between the age groups. The only socioeconomic differentiation on the issue of return is according to wealth. This suggests that return might not be the "choice" it is often represented as being. With few options or alternatives, many of the poorest take what assistance programs there are available and return to locations where they knew better times and where they might have local self-sufficiency options (like farming and gardening).

Returnees who felt victimized by ethnic discrimination, presumably in their temporary war and postwar communities, are also more likely to return to live in their former homes. Although only a small ratio reports discrimination, this effect is important in determining the rate of minority return (80 percent compared to 62 percent). It is not evident from the responses why the respondents felt victimized—as newcomers to the community, as unemployed, as residents of temporary quarters, as recipients of governmental aid, as members of an ethnic group, or as rural migrants—but the effects on the returns preferences are significant.

Finally, the respondents' conceptions of "home" are a significant determinant of return (on the complexities of this notion see Jansen 2009). Those who answered that their home was "the village where I was born" showed a rate of reclaim and return to their prewar homes of 76 percent compared to 48 percent who gave a different response (where their children live, where there is a better life, or "the territory that belongs to my people"). Interpretation of this large percentage difference revolves around a sense of attachment to place. Those displaced by the war from villages to which they feel a strong attachment are more likely to return, regardless of ethnicity. From these results, we can draw a profile of the typical returnee: more attached to their prewar village and home, among the poorest in a poor country, victims of discrimination, and either Bosniak or Bosnian Croat.

Conclusion

As the returns process winds down with the declining pool of refugees and internally displaced persons, future numbers will be in the hundreds rather than the tens of thousands of each nationality that marked the peak years from 1999 to 2003. Although the reception of returnees across Bosnia-Herzegovina has been varied, the presence of the international organizations and officials—UNHCR, OHR, OSCE, and EUFOR officials—has nudged the process toward completion

Figure 3. Defaced road sign outside the Bosniak return village of Urković, in eastern Republika Srpska, Bosnia-Herzegovina. International officials responded by demanding local officials paint over the defacement with yellow paint to conceal it, which they eventually did.

and managed to diffuse potential flashpoints; however, although some communities are mixed again, this is no guarantee of interethnic harmony. In plenty of communities, returnees are reminded of their second-class status as minority returnees and not as ordinary prewar residents of their own hometowns. In a small village in eastern Bosnia inside Republika Srpska, we encountered a mundane example of this. The settlement is called Urkovici, but the road sign to it was consistently altered by the addition of a "T" by local Bosnian Serb nationalists so that it read TUrkovici[3] (Figure 3). The alteration repositioned the local returnees inside the dominant outgroup category of Serb nationalist discourse, that of "Turks." In this minor gesture is the whole geopolitical violence of ethnic cleansing: the imposition of othering on neighbors and the (re)placing of them as outsiders and foreigners in their own land. Bosnia is not yet beyond ethnic cleansing.

Acknowledgments

This research was supported by Human and Social Dynamics Initiative of the U.S. National Science Foundation, Grant 0433927. Thanks are due to Prism Research, Dr. Carl Dahlman, and many others for their assistance during fieldwork in Bosnia-Herzegovina and interviews in Washington, DC.

Notes

1. Interview, Jury's Hotel, Washington, DC, 3 April 2009.
2. Interview, Melrose Hotel, Washington, DC, 21 May 2008.

3. Interview with Charlie Powell, Head of the OHR regional office, Bratunac, 25 June 2005.

References

Andjelic, N. 2003. *Bosnia-Herzegovina: The end of a legacy*. Portland, OR: Frank Cass.

Bhabha, H. 1994. *The location of culture*. London and New York: Routledge.

Brubaker, R. 2004. *Ethnicity without groups*. Cambridge, MA: Harvard University Press.

Burg, S. L., and P. S. Shoup. 1999. *The war in Bosnia-Herzegovina: Ethnic conflict and international intervention*. Armonk, NY: M. E. Sharpe.

Cigar, N. 1995. *Genocide in Bosnia: The policy of "ethnic cleansing."* College Station: Texas A&M University Press.

Dahlman, C., and G. Ó Tuathail. 2005. Broken Bosnia: The localized geopolitics of displacement and return in two Bosnian places. *Annals of the Association of American Geographers* 95 (3): 644–62.

———. 2006. Bosnia's third geopolitical space: Nationalist separatism and international supervision in Bosnia's Brčko District. *Geopolitics* 11 (4): 651–75.

Donia, R. J. 2003. *The assembly of Republika Srpska, 1992–95: Highlights and excerpts, 2989167–2989282*. The Hague, The Netherlands: The International Criminal Tribunal for the Former Yugoslavia.

Donia, R. J., and J. V. A. Fine. 1994. *Bosnia and Hercegovina: A tradition betrayed*. London: Hurst.

Dragović-Soso, J. 2002. *Saviours of the nation: Serbia's intellectual opposition and the revival of nationalism*. Montreal: McGill-Queen's University Press.

Hayden, R. M. 1994. Recounting the dead: The rediscovery and redefinition of wartime massacres in late- and post-Communist Yugoslavia. In *Memory, history and opposition under state socialism*, ed. R. S. Watson, 167–84. Santa Fe, NM: School of American Research Press.

Helsinki Committee for Human Rights in Bosnia and Herzegovina. 2008. *Report on the status of human rights in Bosnia and Herzegovina (Analysis for the period January–December 2007)*. Sarajevo, Bosnia-Herzegovina: Helsinki Committee for Human Rights in Bosnia and Herzegovina. http://www.bh-hcr.org/Reports/reportHR2007.htm (last accessed 25 September 2009).

Hoare, M. A. 2007. *The history of Bosnia: From the Middle Ages to the present day*. London: Saqi.

Jansen, S. 2009. Troubled locations: Return, the life course and transformations of home in Bosnia-Herzegovina. In *Struggles for home: Violence, hope and the movement of people*, ed. S. Jansen and S. Löfving, 43–64. New York: Berghahn.

Lampe, J. 1996. *Yugoslavia as history: Twice there was a country*. Cambridge, UK: Cambridge University Press.

Mann, M. 2004. *The dark side of democracy: Explaining ethnic cleansing*. New York: Cambridge University Press.

O'Loughlin, J., and G. Ó Tuathail. 2009. Accounting for separatist sentiment: Bosnia-Herzegovina and the North Caucasus of Russia compared. *Ethnic and Racial Studies* 32: 591–615.

Ó Tuathail, G., and C. Dahlman. 2004. The effort to reverse ethnic cleansing in Bosnia-Herzegovina: The limits of returns. *Eurasian Geography and Economics* 45 (6): 439–64.

———. 2006a. Post-domicide Bosnia and Herzegovina: Homes, homelands and one million returns. *International Peacekeeping* 13 (2): 242–60.

———. 2006b. "The west bank of the Drina": Land allocation and ethnic engineering in Republika Srpska. *Transactions, Institute of British Geographers NS* 31 (3): 304–22.

———. 2010. *Bosnia remade: Ethnic cleansing and returns*. Oxford, UK: Oxford University Press.

Shrader, C. R. 2003. *The Muslim-Croat civil war in central Bosnia: A military history, 1992–1994*. College Station: Texas A&M University Press.

Treanor, P. 2002. The Bosnian Serb leadership 1990–1992. Research report prepared for the case of Krajišnik and Placšić (IT-00-39 and 40). International Criminal Tribunal on the Former Yugoslavia, The Hague, The Netherlands.

United Nations High Commissioner for Refugees (UNHCR). 2007. Briefing note on UNHCR and Annex 7 in Bosnia and Herzegovina. Sarajevo, Bosnia-Herzegovina: UNHCR.

Williams, R. C. 2006. The significance of property restitution to sustainable return in Bosnia and Herzegovina. *International Migration* 44 (3): 39–61.

Index

Page numbers in *Italics* represent tables.
Page numbers in **Bold** represent figures.

T - #0962 - 101024 - C0 - 276/216/11 - PB - 9781138853362 - Gloss Lamination